Inorganic Electrochemistry
Theory, Practice, and Application

Inorganic Electrochemistry
Theory, Practice, and Application

Contributors :
Christian Lopez,
Chiara Baroni, *et al.*

KOROS PRESS LIMITED
London, UK

Inorganic Electrochemistry : *Theory, Practice, and Application*
Contributors : Christian Lopez *and* Chiara Baroni, *et al.*

Published by Koros Press Limited

www.korospress.com

United Kingdom

Copyright 2016

Printed in 2017 for Sale in the Indian Subcontinent

Notice

Contributors, whose names have been given on the book cover, are not associated with the Publisher. The editors and the Publisher have attempted to trace the copyright holders of all material reproduced in this publication and apologise to copyright holders if permission has not been obtained. If any copyright holder has not been acknowledged, please write to us so we may rectify.

Reasonable efforts have been made to publish reliable data. The views articulated in the chapters are those of the individual contributors, and not necessarily those of the editors or the Publisher. Editors and/or the Publisher are not responsible for the accuracy of the information in the published chapters or consequences from their use. The Publisher accepts no responsibility for any damage or grievance to individual(s) or property arising out of the use of any material(s), instruction(s), methods or thoughts in the book.

Inorganic Electrochemistry : *Theory, Practice, and Application*

ISBN: 978-1-78163-553-7

British Library Cataloguing in Publication Data
A CIP record for this book is available from the British Library

Exclusively distributed by CBS Publishers & Distributors Pvt. Ltd.

Sales & Distribution Rights only for India, Pakistan, Bangladesh, Sri Lanka, Nepal and Bhutan.This book is not to be sold outside these territories.

PREFACE

Electrochemistry is the branch of chemistry that involves chemical reactions taking place in a solution at the interface of conductors. It also involves the study of transfer of electrons along with the study of movement and separation of charge in matter. Electrochemistry can be an elegant and essential support to synthetic inorganic chemistry. However, it is often perceived as a difficult technique.

This book aims to introduce inorganic chemists to electrochemical investigations in as straightforward a way as possible.

This page left intentionally blank.

CONTENTS

This page left intentionally blank.

LIST OF CONTRIBUTORS

Christian Lopez

Laboratory of Electrochemistry and Physical-chemistry of Materials and Interfaces, UMR 5279, CNRS-Grenoble INP-Université de Savoie-Université Joseph Fourier, BP75, 38402 Saint Martind d'Hères, France; E-Mail: christian.lopez@lepmi.grenoble-inp.fr

Chiara Baroni

Department of Applied Science and Technology, Politecnico di Torino, INSTM Reference Laboratory for Ceramics Engineering, Corso Duca degli Abruzzi 24, 10129 Torino, Italy; E-Mail: chiara.baroni@polito.it

Jean-Marc Tulliani

Department of Applied Science and Technology, Politecnico di Torino, INSTM Reference Laboratory for Ceramics Engineering, Corso Duca degli Abruzzi 24, 10129 Torino, Italy; E-Mail: chiara.baroni@polito.it

This page left intentionally blank.

Chapter 1

FUNDAMENTAL OF ELECTRODE REACTIONS

ELECTRON TRANSFER

Electron transfer (ET) occurs when an *electron* moves from an *atom* or a *chemical species* (*e.g.* a *molecule*) to another atom or chemical species. ET is a mechanistic description of the thermodynamic concept of *redox*, wherein the oxidation states of both reaction partners change.

Numerous biological processes involve ET reactions. These processes include oxygen binding, photo-synthesis, respiration, and detoxification. Additionally, the process of *energy transfer* can be formalized as a two-electron exchange (two concurrent ET events in opposite directions) in case of small distances between the transferring molecules. ET reactions commonly involve transition metal complexes, but there are now many examples of ET in *organic chemistry*.

Classes of Electron Transfer

There are several classes of electron transfer, defined by the state of the two redox centers and their connectivity

Inner Sphere Electron Transfer

Inner sphere or **bonded electron transfer** is a *redox* chemical reaction that proceeds *via* a *covalent* linkage — a strong electronic interaction — between the oxidant and the reductant reactants. In Inner Sphere (IS) electron transfer (ET), a *ligand* bridges the two metal redox centers during the electron transfer event. Inner sphere reactions are inhibited by large ligands, which prevent the formation of the crucial bridged intermediate. Thus, IS ET is rare in biological systems, where redox sites are often shielded by bulky proteins. However, redox centers can consist of organic groups rather than metal centers.

The bridging *ligand* could be virtually any entity that can convey electrons. Typically, such a ligand has more than one *lone electron pair*, such that it can serve as an electron donor to both the reductant and the oxidant. Common bridging ligands include the *halides* and the *pseudohalides* such as *hydroxide* and *thiocyanate*. More complex bridging ligands are also well known including *oxalate, malonate,* and *pyrazine*. Prior to ET, the bridged complex must form, and such processes are often highly reversible. Electron transfer occurs through the bridge once it is established. In some cases, the stable bridged structure may exist in the ground state; in other cases, the bridged structure may be a transiently-formed intermediate, or else as a transition state during the reaction.

The alternative to inner sphere electron transfer is *outer sphere electron transfer*. In any transition metal redox process, the mechanism can be assumed to be outer sphere unless the conditions of the inner sphere are met. Inner sphere electron transfer is generally *enthalpically* more favourable than outer sphere electron transfer due to a larger degree of interaction between the metal centers involved, however, inner sphere electron transfer is usually *entropically* less favourable since the two sites involved must become more ordered (come together *via* a bridge) than in outer sphere electron transfer.

Taube's Experiment

The discoverer of the inner sphere mechanism was *Henry Taube*, who was awarded the *Nobel Prize in Chemistry* in 1983 for his pioneering studies. A particularly historic finding is summarized in the abstract of the seminal publication. "When $Co(NH_3)_5Cl^{++}$ is reduced by Cr^{++} in M {meaning 1M} $HClO_4$, 1 Cl^- appears attached to Cr for each Cr(III) which is formed or Co(III) reduced. When the reaction is carried on in a medium containing radioactive Cl, the mixing of the Cl^- attached to Cr(III) with that in solution is less than 0.5%. This experiment shows that transfer of Cl to the reducing agent from the oxidizing agent is direct..." The paper and the excerpt above can be described with the following equation :

$$[CoCl(NH_3)_5]^{2+} + [Cr(H_2O)_6]^{2+} \rightarrow [Co(NH_3)_5(H_2O)]^{2+} + [CrCl(H_2O)_5]^{2+}$$

The point of interest is that the chloride that was originally bonded to the cobalt, the oxidant, becomes bonded to chromium, which in its +3 oxidation state, forms kinetically inert bonds to its *ligands*. This observation implies the intermediacy of the bimetallic complex $[Co(NH_3)_5(\mu\text{-}Cl)Cr(H_2O)_5]^{4+}$, wherein "$\mu$-Cl" indicates that the chloride bridges between the Cr and Co atoms, serving as a ligand for both. This chloride serves as a conduit for electron flow from Cr(II) to Co(III), forming Cr(III) and Co(II).

The Creutz-Taube Ion

In the preceding example, the occurrence of the chloride bridge is *inferred* from the product analysis, but it was *not* observed. One complex that serves as a model for the bridged intermediate is the "*Creutz Taube complex*," $[(NH_3)_5RuNC_4H_4NRu(NH_3)_5]^{5+}$. This species is named after Carol Creutz, who prepared the ion during her

Ph.D. studies with *Henry Taube*. The bridging ligand is the heterocycle *pyrazine*, 1,4-$C_4H_4N_2$. In the Creutz-Taube Ion, the average oxidation state of Ru is 2.5+. *Spectroscopic* studies, however, show that the two Ru centers are equivalent, which indicates the ease with which the electron hole communicates between the two metals. The significance of the Creutz-Taube ion is its simplicity, which facilitates theoretical analysis, and its high symmetry, which ensures a high degree of delocalization. Many more complex mixed valence species are known both as molecules and polymeric materials.

Mixed Valence Compounds

Mixed valence compounds contain an *element* which is present in more than one *oxidation state*. Well-known mixed valence compounds include the *Creutz-Taube complex*, *Prussian blue* and *Molybdenum blue*. Many solids are mixed-valency including *indium chalcogenides*. Mixed valency is required for *organic metals* to exhibit electrical conductivity.

As the extinction coefficient decreases, the coupling constant decreases, influencing the angle to increase.

Mixed-valence compounds are sub-divided into three groups, according to the **Robin-Day Classification** :

- Class I, where the valences are "trapped," or localized on a single site, such as Pb_3O_4 and *antimony tetroxide*. There are distinct sites with different specific valences in the complex that cannot easily interconvert.
- Class II, which are intermediate in character. There is some localization of distinct valences, but there is a low *activation energy* for their interconversion. Some thermal activation is required to induce electron transfer from one site to another *via* the bridge. These species exhibit an intense *Intervalence charge transfer* (IT or IVCT) band, a broad intense absorption in the IR-or visible part of the spectrum, and also exhibit magnetic exchange coupling at low temperatures. The degree of interaction between the metal sites can be estimated from the absorption profile of the IVCT band and the spacing between the sites. This type of complex is common when metals are in different ligand fields. For example, *Prussian blue* is an iron (II,III)–*cyanide* complex in which there is an iron (II) atom surrounded by six carbon atoms of six *cyanide* ligands bridged to an iron (III) atom by their nitrogen ends. In the *Turnbull's blue* preparation,

an iron (II) solution is mixed with an iron (III) cyanide (c-linked) complex. An electron-transfer reaction occurs *via* the cyanide ligands to give iron (III) associated with an iron (II)-cyanide complex.

- Class III, wherein mixed valence is not distinguishable by spectroscopic methods as the valence is completely delocalized. The Creutz-Taube Ion is an example of this class of complexes. These species also exhibit an IT band. Each site exhibits an intermediate oxidation state, which can be half-integer in value. This class is possible when the ligand environment is similar or identical for each of the two metal sites in the complex. The bridging ligand needs to be very good at electron transfer, be highly conjugated, and be easily reduced.

Organic mixed valence compounds are also known. Examples are the oxidized form of *tetrathiafulvalene* and the radical cation of *N,N,N',N'-tetramethyl-p-phenylenediamine*.

Outer Sphere Electron Transfer

Outer sphere refers to an *electron transfer* (ET) event that occurs between chemical species that remain separate intact before, during, and after the ET event. In contrast, for *inner sphere electron transfer* the participating redox sites undergoing ET become connected by a chemical bridge. Because the ET in outer sphere electron transfer occurs between two non-connected species, the electron is forced to move through space from one *redox* center to the other.

Marcus Theory

The main theory that describes the rates of outer-sphere electron transfer was developed by *Rudolph A. Marcus* in the 1950s. A major aspect of *Marcus theory* is the dependence of the electron transfer rate on the thermodynamic driving force (difference in the redox potentials of the electron-exchanging sites). For most reactions, the rates increase with increased driving force. A second aspect is that the rate of outer-sphere electron-transfer depends inversely on the "re-organizational energy." Re-organization energy describes the changes in bond lengths and angles that are required for the oxidant and reductant to switch their oxidation states. This energy is assessed by measurements of the self-exchange rates. Outer sphere electron transfer is the most common type of electron transfer, especially in *biochemistry*, where redox centers are separated by several (up to about 11) angstroms by intervening protein. In bio-chemistry, there are two main types of outer sphere ET : ET between two biological molecules or fixed distance electron transfer, in which the electron transfers within a *single* biomolecule (*e.g.*, intraprotein).

Examples

Self-exchange

Outer sphere electron transfer can occur between chemical species that are identical save for their oxidation state. This process is termed self-exchange. An example is the *degenerate* reaction between the tetrahedral ions *permanganate* and *manganate* :

$$[MnO_4]^- + [Mn^*O_4]^{2-} \rightarrow [MnO_4]^{2-} + [Mn^*O_4]^-$$

For octahedral metal complexes, the rate constant for self-exchange reactions correlates with changes the population of the e_g orbitals, the population of which most strongly affects the length of metal-ligand bonds :

- For the $[Co(bipy)_3]^+ / [Co(bipy)_3]^{2+}$ pair, self-exchange proceeds at 10^9 $M^{-1}s^{-1}$. In this case, the electron configuration changes from Co(I) : $(t_{2g})^6(e_g)^2$ to Co(II) : $(t_{2g})^5(e_g)^2$.
- For the $[Co[bipy)_3]^{2+} / [Co(bipy)_3]^{3+}$ pair, self-exchange proceeds at 18 $M^{-1}s^{-1}$. In this case, the electron configuration changes from Co(II) : $(t_{2g})^5(e_g)^2$ to Co(III) : $(t_{2g})^6(e_g)^0$.

Iron-sulfur Proteins

Outer sphere ET is the basis of the biological function of the *iron-sulfur proteins*. The Fe centers are typically further co-ordinated by cysteinyl ligands. The $[Fe_4S_4]$ electron-transfer proteins ($[Fe_4S_4]$ *ferredoxins*) may be further sub-divided into low-potential (bacterial-type) and *high-potential ferredoxins*. Low-and high-potential ferredoxins are related by the following redox scheme :

Because of the small structural differences between the individual redox states, ET is rapid between these clusters.

Heterogeneous Electron Transfer

In heterogeneous electron transfer, an electron moves between a chemical species and a solid-state *electrode*. Theories addressing heterogeneous electron transfer have applications in *electro-chemistry* and the design of *solar cells*.

THEORY

The first generally accepted theory of ET was developed by *Rudolph A. Marcus* to address *outer-sphere electron transfer* and was based on a *transition-state theory* approach. The Marcus theory of electron transfer was then extended to include *inner-sphere electron transfer* by *Noel Hush* and Marcus. The resultant theory, called *Marcus-Hush theory*, has guided most discussions of electron transfer ever since. Both theories are, however, semi-classical in nature, although they have been extended to fully *quantum mechanical* treatments by *Joshua Jortner*, *Alexender M. Kuznetsov*, and others proceeding from *Fermi's Golden Rule* and following earlier work in *non-radiative transitions*. Furthermore, theories have been put forward to take into account the effects of *vibronic coupling* on electron transfer; in particular, the *PKS theory of electron transfer*.

Before 1991, ET in *metalloproteins* was thought to affect primarily the diffuse, averaged properties of the non-metal atoms forming an insulated barrier between the metals, but Beratan, Betts and Onuchic subsequently showed that the ET rates are governed by the bond structures of the proteins--that the electrons, in effect, tunnel through the bonds comprising the chain structure of the proteins.

ELECTRON EQUIVALENT

Electron Equivalent is a concept commonly used in *redox chemistry*, reactions involving *electron transfer*, to define a quantity (*e.g.* energy or moles) relative to one electron. *Energies of formation* are often given as kilojoules per electron equivalent to enable calculation of specific reaction energies on a "per electron" basis. Reactions containing movement of electrons are often balanced such that reaction quantities are given in relation to the transfer of a single electron, allowing quantification of reactants and products in relation to a single electron transfer.

ELECTRODE

An **electrode** is an *electrical conductor* used to make contact with a non-metallic part of a *circuit* (*e.g.* a *semi-conductor*, an *electrolyte* or a *vacuum*). The word was coined by the scientist *Michael Faraday* from the *Greek* words *elektron* (meaning *amber*, from which the word *electricity* is derived) and *hodos*, a way.

Anode and Cathode in Electro-chemical Cells

An electrode in an *electro-chemical cell* is referred to as either an *anode* or a *cathode* (words that were also coined by Faraday). The anode is now defined as the electrode at which *electrons* leave the cell and *oxidation* occurs, and the cathode as the electrode at which electrons enter the cell and *reduction* occurs. Each electrode may become either the anode or the cathode depending on the direction of current through the cell. A bipolar electrode is an electrode that functions as the anode of one cell and the cathode of another cell.

Primary Cell

A *primary cell* is a special type of electro-chemical cell in which the reaction cannot be reversed, and the identities of the anode and cathode are therefore fixed. The anode is always the negative electrode. The cell can be discharged but not recharged.

Secondary Cell

A *secondary cell*, for example a *rechargeable battery*, is one in which the chemical reactions are reversible. When the cell is being charged, the anode becomes the positive (+) and the cathode the negative (−) electrode. This is also the case in an *electrolytic cell*. When the cell is being discharged, it behaves like a primary cell, with the anode as the negative and the cathode as the positive electrode.

Other Anodes and Cathodes

In a *vacuum tube* or a *semi-conductor* having polarity (*diodes, electrolytic capacitors*) the anode is the positive (+) electrode and the cathode the negative (−). The electrons enter the device through the cathode and exit the device through the anode. Many devices have other electrodes to control operation, *e.g.*, base, gate, control grid.

In a three-electrode cell, a counter electrode, also called an *auxiliary electrode*, is used only to make a connection to the electrolyte so that a current can be applied to the working electrode. The counter electrode is usually made of an inert material, such as a *noble metal* or *graphite*, to keep it from dissolving.

Welding Electrodes

In *arc welding* an electrode is used to conduct current through a workpiece to fuse two pieces together. Depending upon the process, the electrode is either consumable, in the case of *gas metal arc welding* or *shielded metal arc welding*, or non-consumable, such as in *gas tungsten arc welding*. For a direct current system the weld rod or stick may be a cathode for a filling type weld or an anode for other welding processes. For an alternating current arc welder the welding electrode would not be considered an anode or cathode.

Alternating Current Electrodes

For electrical systems which use *alternating current* the electrodes are the connections from the circuitry to the object to be acted upon by the electric current but are not designated anode or cathode because the direction of flow of the electrons changes *periodically*, usually many *times per second*.

Uses

Electrodes are used to provide current through non-metal objects to alter them in numerous ways and to measure conductivity for numerous purposes. Examples include :

* Electrodes for *fuel cells*
* Electrodes for medical purposes, such as *EEG, ECG, ECT, defibrillator*
* Electrodes for *electrophysiology* techniques in bio-medical research
* Electrodes for execution by the *electric chair*
* Electrodes for *electroplating*
* Electrodes for *arc welding*
* Electrodes for *cathodic protection*
* Electrodes for *grounding*

- Electrodes for *chemical analysis* using *electro-chemical* methods
- Inert electrodes for *electrolysis* (made of *platinum*)
- *Membrane electrode assembly.*

Chemically Modified Electrodes

Chemically modified electrodes are electrodes that have their surfaces chemically modified to change the electrode's *physical, chemical, electro-chemical, optical, electrical,* and *transport* properties. These electrodes are used for advanced purposes in research and investigation.

Auxiliary Electrode

The **Auxiliary electrode**, often also called the **counter electrode**, is an *electrode* used in a three electrode electro-chemical cell for *voltammetric analysis* or other reactions in which an electrical *current* is expected to flow. The auxiliary electrode is distinct from the *reference electrode*, which establishes the electrical potential against which other potentials may be measured, and the *working electrode*, at which the cell reaction takes place.

In a two-electrode system, either a known current or potential is applied between the working and auxiliary electrodes and the other variable may be measured. The auxiliary electrode functions as a cathode whenever the working electrode is operating as an anode and *vice versa*. The auxiliary electrode often has a surface area much larger than that of the working electrode to ensure that the half-reaction occurring at the auxiliary electrode can occur fast enough so as not to limit the process at the working electrode.

When a *three electrode cell* is used to perform electro-analytical chemistry, the auxiliary electrode, along with the working electrode, provides circuit over which current is either applied or measured. Here, the potential of the auxiliary electrode is usually not measured and is adjusted so as to balance the reaction occurring at the working electrode. This configuration allows the potential of the working electrode to be measured against a known reference electrode without compromising the stability of that reference electrode by passing current over it.

The auxiliary electrode may be isolated from the working electrode using a glass frit. Such isolation prevents any by-products generated at the auxiliary electrode from contaminating the main test solution : for example, if a reduction is being performed at the working electrode in aqueous solution, oxygen may be evolved from the auxiliary electrode. Such isolation is crucial during the bulk electrolysis of a species which exhibits reversible redox behaviour.

Auxiliary electrodes are often fabricated from electro-chemically inert materials such as gold, platinum, or carbon.

Working Electrode

The working electrode is the electrode in an electro-chemical system on which the reaction of interest is occurring. The working electrode is often used in conjunction with an auxiliary electrode, and a reference electrode in a three electrode system. Depending on whether the reaction on the electrode is a reduction or an oxidation, the working electrode can be referred to as either cathodic or anodic. Common working electrodes can consist of inert metals such as gold, silver or platinum, to inert carbon such as glassy carbon or pyrolytic carbon, and mercury drop and film electrodes. Chemically modified electrodes are employed for the analysis of both organic molecules as well as metal ions. Special types of working electrodes :

- *Ultra-micro-electrode* (UME)
- *Rotating disk electrode* (RDE)
- *Rotating ring-disk electrode* (RRDE)
- *Hanging mercury drop electrode* (HMDE)
- *Dropping mercury electrode* (DME)

Ultra-micro-electrode

An **ultra-micro-electrode** (UME) is a *working electrode* used in a *volta-ammetry*. The small size of UME give them relatively large *diffusion layers* and small overall currents. These features allow UME to achieve useful steady-state conditions and very high scan rates (V/s) with limited distortion. UME were developed independently by *Wightman* and *Fleischmann* around 1980. Small current at UME enables electro-chemical measurements in low conductive media (organic solvents), where voltage drop associated with high solution resistance makes these experiments difficult for convention electrodes. Furthermore, small voltage drop at UME leads to a very small voltage distortion at the electrode-solution interface which allows using two-electrode setup in *voltammetric* experiment instead of conventional three-electrode setup.

Design

Ultra-micro-electrodes are often defined as electrodes which are smaller than the diffusion layer achieved in a readily accessed experiment. A working definition is an electrode that has at least one dimension (the critical dimension) smaller than 25 μm. *Platinum* electrodes with a radius of 5 μm are commercially available and electrodes with critical dimension of 0.1 μm have been made. Electrodes with even smaller critical dimension have been reported in the literature, but exist mostly as proofs of concept. The most common UME is a disk shaped electrode created by embedding a thin wire in glass, resin, or plastic. The resin is cut and polished to expose a cross-section of the wire. Other shapes, such as wires and rectangles, have also been reported. *Carbon-fiber micro-electrodes* are fabricated with conductive carbon carbon fibers sealed in glass capillary with exposed tips. These electrodes are frequently used with *in vivo voltammetry*.

Theory

Linear Region

Every electrode has a range of scan rates called the linear region. The response to a reversible redox couple in the linear region is a "diffusion controlled peak" which can be modelled with the *Cottrell equation*. The upper limit of the useful linear region is bound by an excess of changing current combined with distortions created from large peak currents and associated resistance. The charging current scales linearly with scan rate while the peak current, which contains the useful information, scales with the square root of scan rate. As scan rates increase, the relative peak response diminishes. Some of the charge current can be mitigated with RC compensation and/or mathematically removed after the experiment. However, the distortions resulting from increased current and the associated resistance cannot be subtracted. These distortions ultimately limit the scan rate for which an electrode is useful. For example, a working electrode with a radius of 1.0 mm is not useful for experiments much greater than 500 mV/s.

Moving to an UME drops the currents being passed and thus greatly increases the useful sweep rate up to 10 V/s. These faster scan rates allow the investigation of *electro-chemical reaction mechanisms* with much higher rates than can be explored with regular working electrodes. By adjusting the size of the working electrode an enormous *kinetic* range can be studied. For UME only the very fast reactions can be studied through peak current since the linear region only exists for UME at very high scan rates.

Steady-state Region

At scan rates slower than those of the linear region is a region which is mathematically complex to model and rarely investigated. At even slower scan rates there is the steady-state region. In the steady-state region linear sweeps traces display reversible redox couple as steps rather than peaks. These steps can readily be modelled for meaningful data.

To access the steady-state region the scan rate must be dropped. As scan rates are slowed, the relative currents also drop at a given point reducing the reliability of the measurement. The low ratio of diffusion layer volume to electrode surface area means regular stationary electrodes can not be dropped low enough before their current measurements become unreliable. In contrast, the ratio of diffusion layer volume to electrode surface area is much higher for UME. When the scan rate of UME is dropped it quickly enters the steady-state regime at useful scan rates. Even though UME supply small total currents their steady-state currents are high compared to regular electrodes.

Rotating Disk Electrode

A **rotating disk electrode** (RDE) is a *hydrodynamic working electrode* used in a *three electrode system*. The electrode rotates during experiments inducing a *flux* of *analyte* to the electrode. These working electrodes are used in *electro-chemical* studies when investigating *reaction mechanisms* related to *redox* chemistry, among other *chemical* phenomena. The more complex *rotating ring-disk electrode* can be used as a *rotating disk electrode* if the ring is left inactive during the experiment.

Structure

The electrode includes a conductive disk embedded in an inert non-conductive polymer or resin that can be attached to an electric motor that has very fine control of the electrode's rotation rate. The disk, like any working electrode, is generally made of a *noble metal* or *glassy carbon*, however any conductive material can be used based on specific needs.

Function

The disk's *rotation* is usually described in terms of *angular velocity*. As the disk turns, some of the solution described as the *hydrodynamic boundary layer* is dragged by the spinning disk and the resulting *centrifugal force* flings the solution away from the center of the electrode. Solution flows up, perpendicular to the electrode, from the bulk to replace the *boundary layer*. The sum result is a *laminar* flow of solution towards and across the electrode. The rate of the solution flow can be controlled by the electrode's angular velocity and modelled mathematically. This flow can quickly achieve conditions in which the steady-state current is controlled by the solution flow rather than diffusion. This is a contrast to still and unstirred experiments such as *cyclic voltammetry* where the steady-state current is limited by the diffusion of substrate.

By running *linear sweep voltammetry* and other experiments at various rotation rates, different electro-chemical phenomena can be investigated, including multi-electron transfer, the kinetics of a slow electron transfer, adsorption/desorption steps, and *electro-chemical reaction mechanisms*.

Limitation

Potential sweep reversals as used in *cyclic voltammetry* are not possible for a RDE system since the products of the potential sweep are continually swept away from the electrode. A reversal would simply produce an *i-E* curve which exactly matched the initial scan direction. Further complications arise in investigating the reactivity of these products, since they are swept away from the electrode. In contrast, the *rotating ring-disk electrode* is well suited to investigate this further reactivity.

Rotating Ring-disk Electrode

A **rotating ring-disk electrode** (RRDE) is double *working electrode* used in *hydrodynamic voltammetry*, very similar to a *rotating disk electrode* (RDE). The electrode actually rotates during experiments inducing a *flux* of *analyte* to the electrode. These working electrode are used in *electro-chemical* studies when investigating *reaction mechanisms* related to *redox* chemistry among other *chemical* phenomena.

Hanging Mercury Drop Electrode

The **hanging mercury drop electrode** (HMDE) is a *working electrode* variation on the *dropping mercury electrode* (DME). Experiments run with dropping mercury electrodes are referred to as forms of *polarography*. If the experiments are performed at an electrode with a constant surface (like the HMDE) it is referred as *voltammetry*.

Like other working electrodes these electrodes are used in *electro-chemical* studies using *three electrode systems* when investigating *reaction mechanisms* related to *redox* chemistry among other *chemical* phenomenon.

Distinction

The hanging mercury drop electrode (HDME) produces a partial *mercury* drop of controlled geometry and surface area at the end of a *capillary* in contrast to the *dropping mercury electrode* (DME) which steadily releases drops of mercury during an experiment. The disadvantages a DME experiences due to a constantly changing surface are not experienced by the HMDE since it has static surface area during an experiment. The static surface of the HDME means it is more likely to suffer from the surface *adsorption* phenomenon than a DME. Unlike solid electrodes which need to be cleaned and polished between most experiments, the self-renewing HMDE can simply release the contaminated drop and grow a clean drop between each experiment.

Dropping Mercury Electrode

The dropping mercury electrode (DME) is a working electrode made of mercury and used in polarography. Experiments run with mercury electrodes are referred to as forms of polarography even if the experiments are identical or very similar to a corresponding voltammetry experiment which uses solid working electrodes. Like other working electrodes these electrodes are used in electro-chemical studies using three electrode systems when investigating reaction mechanisms related to redox chemistry among other chemical phenomena.

Structure

A flow of *mercury* passes through an insulating *capillary* producing a droplet which grows from the end of the capillary in a reproducible way. Each droplet grows until it reaches a diameter of about a millimeter and releases. The released droplet is no longer in contact with the *working electrode* whose contact is above the capillary. As the electrode is used mercury collects in the bottom of the cell. In some cell designs this mercury pool is connected to a lead and used as the cell's *auxiliary electrode*. Each released drop is immediately followed by the formation of another drop. The drops are generally produced at a rate of about 0.2 Hz.

Considerations

A major advantage of the DME is that each drop has a smooth and uncontaminated surface free from any *adsorbed* analyte or impurity. The self-renewing electrode does not need to be cleaned or polished like a solid electrode. This advantage comes at the cost of a working electrode with a constantly changing surface area. Since the drops are produced predictably the changing surface area can be accounted for or even used advantageously. In addition the drops' growth causes capacitive current and *faradaic current*. These changing current effects combined with experiments where the potential is continuously changed can result in noisy traces. In some experiments the traces are continually sampled, showing all the current deviation resulting from the drop growth. Other sampling methods smooth the data by sampling the current at the electrode only once per drop at a specific size. The DME's periodic expansion into the solution and hemispherical shape also affects the way the analyte diffuses to the electrode surface. The DME consists of a fine capillary with a bore size of 20-50 μm.

Reference Electrode

A **reference electrode** is an *electrode* which has a stable and well-known *electrode potential*. The high stability of the electrode potential is usually reached by employing a *redox* system with constant (buffered or saturated) *concentrations* of each participants of the redox reaction.

There are many ways reference electrodes are used. The simplest is when the reference electrode is used as a *half cell* to build an *electro-chemical cell*. This allows the *potential* of the other half cell to be determined. An accurate and practical method to measure an electrode's potential in isolation (*absolute electrode potential*) has yet to be developed.

Non-aqueous Reference Electrodes

While it is convenient to compare between solvents to qualitatively compare systems it is not quantitatively meaningful. Much as pK_a are related between solvents, but not the same, so is the case with E°. While the SHE might seem to be a reasonable reference for non-aqueous work as it turns out the platinum is rapidly poisoned by many solvents including acetonitrile causing uncontrolled drifts in potential. Both the SCE and saturated Ag/AgCl are aqueous electrodes based around saturated aqueous solution. While for short periods it may be possible to use such aqueous electrodes as references with non-aqueous solutions the long-term results are not trustworthy. Using aqueous electrodes introduces undefined, variable, and unmeasurable junction potentials to the cell in the form of a liquid-liquid junction as well as different ionic composition between the reference compartment and the rest of the cell. The best argument against using aqueous reference electrodes with non-aqueous systems, as mentioned earlier, is that potentials measured in different solvents are not directly comparable.

The potential for the Fc0/+ couple is sensitive to solvent.

	$Fc^{0/+}$ couple, NBu_4PF_6 at 298°C
Solvent	E
MeCN	0.40
CH$_2$Cl$_2$	0.46

(Contd...)

(Contd...)

	$Fc^{0/+}$ couple, NBu_4PF_6 at 298°C
THF	0.56
DMF	0.45
acetone	0.48

A **Quasi-Reference Electrode** (QRE) avoids the issues mentioned above. A QRE with *ferrocene* or similar *internal standard* (*cobaltocene*) referenced back to ferrocene is ideal for non-aqueous work. Since the early 1960s ferrocene has been gaining acceptance as the standard reference for non-aqueous work for a number of reasons, and in 1984, IUPAC recommended ferrocene (II/III) as a standard redox couple. The preparation of the QRE electrode is simple, allowing for a fresh reference to be prepared with each set of experiments. Since QREs are made fresh, there is also no concern with improper storage or maintenance of the electrode. QREs are also more affordable than other reference electrodes.

To make a quasi-reference electrode (QRE) :

1. Insert a piece of silver wire into concentrated HCl then allow the wire to dry on a *KimWipe*. This forms an insoluble layer of AgCl on the surface of the electrode and gives you a Ag/AgCl wire. Repeat dipping every few months or if the QRE starts to drift.

2. Obtain a *Vycor* glass *frit* (4 mm diameter) and glass tubing of similar diameter. Attach Vycor glass frit to the glass tubing with heat shrink Teflon tubing.

3. Rinse then fill the clean glass tube with supporting electrolyte solution and insert Ag/AgCl wire.

4. The *ferrocene* (II/III) couple should lie around 400 mV versus this Ag/AgCl QRE in an acetonitrile solution. This potential will varying up to 200 mV with the specific undefined conditions. Thus adding an internal standard such as ferrocene at some point during the experiment is always necessary.

Pseudo-reference Electrodes

A pseudo-reference electrode is a term that is not well defined and borders on having multiple meanings since *pseudo* and *quasi* are often used interchangeably. They are a class of electrodes named pseudo-reference electrodes because they do not maintain a constant potential but vary predictably with conditions. If the conditions are known, the potential can be calculated and the electrode can be used as a reference. Most electrodes work over a limited range of conditions, such as pH or temperature, outside of this range the electrodes behaviour becomes unpredictable. The advantage of a pseudo-reference electrode is that the resulting variation is factored into the system allowing researchers to accurately study systems over a wide range of conditions.

Yttria-stabilized zirconia (*YSZ*) membrane electrodes were developed with a variety of redox couples, *e.g.*, Ni/NiO. Their potential depends on pH. When the pH value is known, these electrodes can be employed as a reference with notable applications at elevated temperatures.

Aqueous Reference Electrodes

Common reference electrodes and potential with respect to the standard hydrogen electrode :

- *Standard hydrogen electrode* (SHE) (E=0.000 V) activity of H^+=1
- *Normal hydrogen electrode* (NHE) (E ≈ 0.000 V) concentration H^+=1
- *Reversible hydrogen electrode* (RHE) (E=0.000 V-0.0591*pH)
- *Saturated calomel electrode* (SCE) (E=+0.241 V saturated)
- *Copper-copper(II) sulfate electrode* (CSE) (E=+0.314 V)
- *Silver chloride electrode* (E=+0.197 V saturated)
- *pH-electrode*
- *Palladium-hydrogen electrode*
- *Dynamic hydrogen electrode* (DHE).

Standard Hydrogen Electrode

The **standard hydrogen electrode** (abbreviated **SHE**), is a *redox electrode* which forms the basis of the *thermodynamic scale of oxidation-reduction potentials*. Its *absolute electrode potential* is estimated to be 4.44 ± 0.02 V at 25°C, but to form a basis for comparison with all other electrode reactions, hydrogen's *standard electrode potential* (E^0) is declared to be zero at all temperatures. Potentials of any other electrodes are compared with that of the standard hydrogen electrode at the same temperature.

Hydrogen electrode is based on the redox *half cell* :

$$2H^+(aq) + 2e^- \rightarrow H_2(g)$$

This redox reaction occurs at a platinized *platinum* electrode. The electrode is dipped in an acidic solution and pure hydrogen gas is bubbled through it. The concentration of both the reduced form and oxidised form is maintained at unity. That implies that the pressure of hydrogen gas is 1 bar and the activity of hydrogen ions in the solution is unity. The activity of hydrogen ions is their effective concentration, which is equal to the formal concentration times the activity coefficient. These unit-less activity coefficients are close to 1.00 for very dilute water solutions, but are usually lower for more concentrated solutions. The *Nernst equation* should be written as :

$$E = \frac{RT}{F} \ln \frac{a_{H^+}}{(p_{H_2}/p^0)^{1/2}}$$

or

$$E = -\frac{2.303RT}{F} pH - \frac{RT}{2F} \ln pH_2 / p^0$$

where :

- a_{H^+} is the *activity* of the hydrogen ions, $a_H^+ = f_H^+ \, C_H^+ / C^0$
- p_{H2} is the partial pressure of the *hydrogen* gas, in *pascals*, Pa

- R is the *universal gas constant*
- T is the temperature, in *kelvins*
- F is the *Faraday constant* (the charge per a mole of electrons), equal to $9.6485309*10^4$ C mol^{-1}
- p^0 is the standard pressure 10^5 in Pa.

Relationship between the Normal Hydrogen Electrode (NHE) and the Standard Hydrogen Electrode (SHE)

During the early development of electro-chemistry, researchers used the normal hydrogen electrode as their standard for zero potential. This was convenient because it could *actually be constructed* by having "[immersing] a platinum electrode into a solution of 1N strong acid and [bubbling] hydrogen gas through the solution at about 1 atm pressure". However, this electrode/solution interface was not entirely reproducible, so the standard for zero potential was later changed. What replaced it was a theoretical electrode/solution interface, where the concentration of H$^+$ was $1m$, but the H$^+$ ions were assumed to have no interaction with other ions (a condition not physically attainable at those concentrations). To differentiate this new standard from the previous one it was given the name 'Standard Hydrogen Electrode'.

In summary,

NHE : Potential of a platinum electrode in 1N acid solution (historical standard, no longer in use)

SHE : Potential of a platinum electrode in a theoretical solution (the current **standard** for zero potential)

Choice of platinum

The choice of platinum for the hydrogen electrode is due to several factors :

- Inertness of platinum (it does not corrode)
- The capability of platinum to catalyze the reaction of proton reduction
- A high intrinsic *exchange current density* for proton reduction on platinum
- Excellent reproducibility of the potential (bias of less than 10 µV when two well-made hydrogen electrodes are compared with one another)

The surface of platinum is platinized (*i.e.*, covered with *platinum black*) to :

- Increase total surface area. This improves reaction kinetics and maximum possible current
- Use a surface material that absorbs hydrogen well at its interface. This also improves reaction kinetics

Other metals can be used for building electrodes with a similar function such as the *palladium-hydrogen electrode*.

Interference

Because of the high adsorption activity of the platinized platinum electrode, it's very important to protect electrode surface and solution from the presence of organic substances as well as from atmospheric oxygen. Inorganic ions that can reduce to a lower valency state at the electrode also have to be avoided (*e.g.*, Fe^{3+}, CrO_4^{2-}). A number of organic substances are also reduced by hydrogen at a platinum surface, and these also have to be avoided.

Cations that can reduce and deposit on the platinum can be source of interference : silver, mercury, copper, lead, cadmium and thallium.

Substances that can inactivate ("poison") the catalytic sites include arsenic, sulfides and other sulfur compounds, colloidal substances, alkaloids, and material found in living systems.

Isotopic Effect

The standard redox potential of the deuterium couple is slightly different from that of proton couple (ca.-0.0044 V *vs* SHE). Various values in this range have been obtained-0.0061 V,-0.00431 V,-0.0074.

$$2D^+(aq) + 2e^- \rightarrow D_2(g)$$

Construction

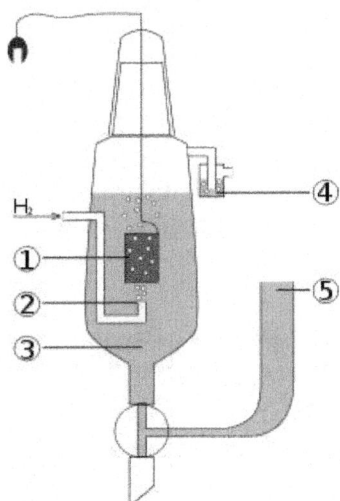

The scheme of the standard hydrogen electrode :
1. *Platinized* platinum electrode
2. Hydrogen blow
3. Solution of the acid with activity of $H^+ = 1$ mol dm^{-3}
4. Hydroseal for prevention of the oxygen interference
5. Reservoir through which the second half-element of the galvanic cell should be attached. The connection can be direct, through a narrow tube to reduce mixing, or through a *salt bridge*, depending on the other electrode and solution. This creates an ionically conductive path to the working electrode of interest.

Reversible Hydrogen Electrode

A **reversible hydrogen electrode** (RHE) is a *reference electrode,* more specific a sub-type of the *standard hydrogen electrodes* for *electro-chemical* processes and differs from the standard hydrogen electrode by the fact that the measured *potential* does not change with the *pH* so that they can be directly used in the electrolyte.

The name refers to the fact that the electrode is in the actual electrolyte solution and not separated by a salt bridge. The hydrogen ion concentration is therefore not 1, but corresponds to that of the electrolyte solution; in this way we can achieve a stable potential with a changing pH value. The potential of the RHE correlates to the pH value :

$$E_0 = 0.000 - 0.059 * pH$$

In general, for hydrogen electrodes in which the reaction :

$$2\,H_3O^+ + 2\,e^- \rightleftharpoons H_2 + 2\,H_2O$$

expires, the following dependence of the *equilibrium* potential E, hydrogen pressure $p[H_2]$ and the activity of a $a[H_3O^+]$ of the *hydronium* ions :

$$E = E_{00} + \frac{RT}{F}\left(\ln(a[H_3O^+]) - \frac{1}{2}\ln(p[H_2]) \right)$$

Here E_{00} is the standard reduction potential (this is by definition equal to zero), R is the *universal gas constant,* T the *absolute temperature* and F is the *Faraday constant.*

Surges occur in the *electrolysis of water* which means that the required *cell voltage* due to kinetic inhibition is higher than the equilibrium potential. The voltage increases with increasing current density at the electrodes. The measurement of equilibrium potentials is therefore possible without power.

Principle

The reversible hydrogen electrode is a fairly practical and reproducible electrode "standard." The term refers to a hydrogen electrode immersed in the electrolyte solution actually used.

The benefit of that electrode is that no *salt bridge* is needed :

- No contamination of the electrolyte by Cl^- or SO_4^{2-}
- No diffusion potentials at the electrolyte bridge (*liquid junction potential*). This is important at temperature different to 25°C.
- Long time measurements possible (no electrolyte bridge means no maintenance of the bridge).

Saturated Calomel Electrode

The **Saturated calomel electrode** (SCE) is a *reference electrode* based on the reaction between elemental *mercury* and *mercury (I) chloride.* The aqueous phase in contact with the mercury and the mercury (I) chloride (Hg_2Cl_2, *"calomel"*) is a saturated solution of *potassium* chloride in water. The electrode is normally linked *via* a porous frit to the solution in which the other electrode is immersed. This porous frit is a *salt bridge*.

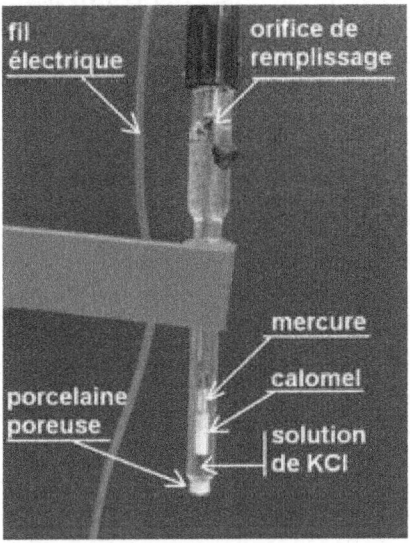

Fig. : Saturated calomel electrode.

In *cell notation* the electrode is written as :

$Cl^-(4M) \,|\, Hg_2Cl_2(s) \,|\, Hg(l) \,|\, Pt$

Theory of Operation

The electrode is based on the redox reaction :

$$Hg_2^{2+} + 2e^- \rightleftarrows 2Hg(l)$$

The *Nernst equation* for this reaction is :

$$E = E^0_{Hg_2^{2+}/Hg} - \frac{RT}{2F}\ln\frac{1}{a_{Hg_2^{2+}}}$$

where E^0 is the *standard electrode potential* for the reaction and a_{Hg} is the *activity* for the mercury cation (the activity for a liquid of 1 Molar is 1). This activity can be found from the *solubility product* of the reaction :

$$Hg_2^{2+} + 2Cl^- \rightleftarrows Hg_2Cl_2(s), \quad K_{sp} = a_{Hg_2^{2+}} + a_{Cl^-}^2$$

By replacing the activity in the Nernst equation with the value in the solubility equation, we get :

$$E = E^0_{Hg_2^{2+}/Hg} + \frac{RT}{2F}\ln K_{sp} - \frac{RT}{2F}\ln a_{Cl^-}^2$$

The only variable in this equation is the activity (or concentration) of the chloride anion. But since the inner solution is saturated with potassium chloride, this activity is fixed by the solubility of potassium chloride. When saturated the *redox potential* of the calomel electrode is +0.2444 V *vs. SHE* at 25°C, but slightly higher when the chloride solution is less than saturated. For example, a 3.5M KCl electrolyte solution increases the reference potential to +0.250 V *vs.* SHE at 25°C, and a 0.1 M solution to +0.3356 V at the same temperature.

Application

The SCE is used in *pH* measurement, *cyclic voltammetry* and general aqueous *electro-chemistry.*

This electrode and the *silver/silver chloride reference electrode* work in the same way. In both electrodes, the activity of the metal ion is fixed by the solubility of the metal salt.

The calomel electrode contains mercury, which poses much greater health hazards than the silver metal used in the Ag/AgCl electrode.

Theory of Operation

The electrode is based on the redox reaction :

$$Hg_2^{2+} + 2e^- \rightleftharpoons 2Hg(l)$$

The *Nernst equation* for this reaction is :

$$E = E^0_{Hg_2^{2+}/Hg} - \frac{RT}{2F} \ln \frac{1}{a_{Hg_2^{2+}}}$$

where E^0 is the *standard electrode potential* for the reaction and a_{Hg} is the *activity* for the mercury cation (the activity for a liquid of 1 Molar is 1). This activity can be found from the *solubility product* of the reaction :

$$Hg_2^{2+} + 2Cl^- \rightleftharpoons Hg_2Cl_2(s), \qquad K_{sp} = a_{Hg_2^{2+}} a_{Cl^-}^2$$

By replacing the activity in the Nernst equation with the value in the solubility equation, we get :

$$E = E^0_{Hg_2^{2+}/Hg} + \frac{RT}{2F} \ln K_{sp} - \frac{RT}{2F} \ln a_{Cl^-}^2$$

The only variable in this equation is the activity (or concentration) of the chloride anion. But since the inner solution is saturated with potassium chloride, this activity is fixed by the solubility of potassium chloride. When saturated the *redox potential* of the calomel electrode is +0.2444 V *vs. SHE* at 25°C, but slightly higher when the chloride solution is less than saturated. For example, a 3.5M KCl electrolyte solution increases the reference potential to +0.250 V *vs.* SHE at 25°C, and a 0.1 M solution to +0.3356 V at the same temperature.[1]

Application

The SCE is used in *pH* measurement, *cyclic voltammetry* and general aqueous *electro-chemistry.*

This electrode and the *silver/silver chloride reference electrode* work in the same way. In both electrodes, the activity of the metal ion is fixed by the solubility of the metal salt.

The calomel electrode contains mercury, which poses much greater health hazards than the silver metal used in the Ag/AgCl electrode.

Silver Chloride Electrode

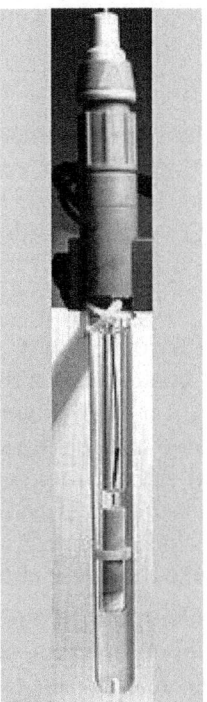

Fig. : Ag-AgCl reference electrode.

A **silver chloride electrode** is a type of *reference electrode*, commonly used in *electro-chemical* measurements. For example, it is usually the internal reference electrode in *pH meters*. As another example, the silver chloride electrode is the most commonly used reference electrode for testing *cathodic protection corrosion* control systems in *sea water* environments.

The electrode functions as a *redox electrode* and the reaction is between the *silver* metal (Ag) and its salt — *silver chloride* (AgCl, also called silver(I) chloride).

The corresponding equations can be presented as follows :

$$Ag^+ + e^- \rightleftharpoons Ag(s)$$
$$AgCl(s) \rightleftharpoons Ag^+ + Cl^-$$

or an overall reaction can be written :

$$AgCl(s) + e^- = Ag(s) + Cl^-$$

This reaction is characterized by fast electrode kinetics, meaning that a sufficiently high *current* can be passed through the electrode with the 100% efficiency of the redox reaction (*dissolution* of the metal or cathodic *deposition* of the silver-ions). The reaction has been proved to obey these equations in solutions of pH values between 0 and 13.5.

The *Nernst equation* below shows the dependence of the potential of the silver-silver(I) chloride electrode on the *activity* or effective *concentration* of chloride-ions :

$$E = E^0 - \frac{RT}{F} \ln a_{Cl^-}$$

The standard electrode potential E^0 against *standard hydrogen electrode* (SHE) is 0.230V ± 10mV. The potential is however very sensitive to traces of bromide ions which make it more negative. (The more exact standard potential given by an IUPAC review paper is 0.22249 V, with a standard deviation of 0.13 mV at 25°C.)

Applications

Commercial reference electrodes consist of a plastic tube electrode body. The electrode is a silver wire that is coated with a thin layer of silver chloride, either physically by dipping the wire in molten silver chloride, or chemically by electroplating the wire in concentrated hydrochloric acid. A porous plug on one end allows contact between the field environment with the silver chloride *electrolyte*. An insulated lead wire connects the silver rod with measuring instruments. A *voltmeter* negative lead is connected to the test wire. The reference electrode contains *potassium chloride* to stabilize the silver chloride concentration.

The potential of a silver : silver chloride reference electrode with respect to the *standard hydrogen electrode* depends on the electrolyte composition.

Table : Reference Electrode Potentials.

Electrode	Potential $E^0 + E_{lj}$	Temperature Coef.
	(V) at 25°C	at around 25°C
SHE	0.000	0.000
Ag/AgCl/Sat. KCl	+0.197	-1.01
Ag/AgCl/3.5 *mol/kg* KCl	+0.205	-0.73
Ag/AgCl/3.0 *mol/kg* KCl	+0.210	?
Ag/AgCl/1.0 *mol/kg* KCl	+0.235	+0.25
Ag/AgCl/0.6 *mol/kg* KCl	+0.25	
Ag/AgCl (Seawater)	+0.266	

Notes to the Table : (1) The table data source is, except where a separate reference is given. (2) E_{lj} is the potential of the liquid junction between the given electrolyte and the electrolyte with the activity of chloride of 1 mol/kg.

The electrode has many features making is suitable for use in the field :

- Simple construction
- Inexpensive to manufacture
- Stable potential
- Non-toxic components.

They are usually manufactured with saturated potassium chloride electrolyte, but can be used with lower concentrations such as 1 *mol/kg* potassium chloride. As noted above, changing the electrolyte concentration changes the electrode potential. Silver chloride is slightly soluble in strong potassium chloride solutions, so it is sometimes recommended the potassium chloride be saturated with silver chloride to avoid stripping the silver chloride off the silver wire.

Elevated Temperature Application

When appropriately constructed, the silver chloride electrode can be used up to 300°C. The standard potential (*i.e.*, the potential when the chloride activity is 1 mol/kg) of the silver chloride electrode is a function of temperature as follows :

Table : Temperature Dependence of the Standard Potential of the
Silver/Silver Chloride Electrode.

Temperature	Potential E^0
°C	V *versus SHE* at the same temperature
25	0.22233
60	0.1968
125	0.1330
150	0.1032
175	0.0708
200	0.0348
225	−0.0051
250	−0.054
275	−0.090

Bard *et. al.* give the following correlations for the standard potential of the silver chloride electrode as a function of temperature (where t is temperature in°C) :

$E^0(V) = 0.23695 - 4.8564 \times 10^{-4}t - 3.4205 \times 10^{-6}t^2 - 5.869 \times 10^{-9}t^3$ for 0 < t < 95°C.

The same source also gives the fit to the high-temperature potential, which reproduces the data in the table above :

$E^0(V) = 0.23735 - 5.3783 \times 10^{-4}t - 2.3728 \times 10^{-6}t^2$ for 25 < t < 275°C.

The extrapolation to 300°C gives E^0 of -0.138 V.

Farmer gives the following correlation for the potential of the silver chloride electrode with 0.1 mol/kg KCl solution, accounting for the activity of Cl⁻ at the elevated temperature :

$$E^{0.1\,mol/kg\,KCl}(V) = 0.23735\text{-}5.3783 \times 10^{-4}t\text{-}2.3728 \times 10^{-6}t^2 + 2.2671 \times 10^{-4}\text{ for } 25 < t < 275°C.$$

Glass Electrode

A **glass electrode** is a type of *ion-selective electrode* made of a doped glass membrane that is sensitive to a specific ion. It is an important part of the instrumentation for chemical analysis and physico-chemical studies. In modern practice, widely used membranous ion-selective electrodes (ISE, including glasses) are part of a galvanic cell. The electric potential of the electrode system in solution is sensitive to changes in the content of a certain type of ions, which is reflected in the dependence of the *electromotive force* (EMF) of galvanic element concentrations of these ions.

History

The first studies of glass electrodes (GE) found different sensitivities of different glasses to change of the medium's acidity (pH), due to effects of the alkali metal ions.

* 1906 — M. Cremer determined that the electric potential that arises between parts of the fluid, located on opposite sides of the glass membrane is proportional to the concentration of acid (hydrogen ion concentration).
* 1909 — S. P. L. Sørensen introduced the concept of *pH*.
* 1909 — F. Haber and Z. Klemensiewicz publicized on 28 January, 1909 results of their research on the glass electrode in The Society of Chemistry in *Karlsruhe* (first publication — The Journal of Physical Chemistry by W. Ostwald and J. H. van 't Hoff) — 1909).
* 1922 — W. S. Hughes showed that the alkali-silicate GE are similar to hydrogen electrode, reversible with respect to H⁺.

Applications

Glass electrodes are commonly used for pH measurements. There are also specialized ion sensitive glass electrodes used for determination of concentration of lithium, sodium, ammonium, and other ions. Glass electrodes have been utilized in a wide range of applications — from pure research, control of industrial processes, to analyze foods, cosmetics and comparison of indicators of the environment and environmental regulations : a micro-electrode measurements of membrane electrical potential of a biological cell, analysis of soil acidity, etc.

Types

Almost all commercial electrodes respond to single charged *ions*, like H⁺, Na⁺, Ag⁺. The most common glass electrode is the *pH*-electrode. Only a few *chalcogenide glass* electrodes are sensitive to double-charged ions, like Pb^{2+}, Cd^{2+} and some others.

There are two main glass-forming systems :

* *Silicate* matrix based on molecular network of *silicon dioxide* (SiO_2) with additions of other metal oxides, such as Na, K, Li, Al, B, Ca, etc.

* *Chalcogenide* matrix based on molecular network of AsS, AsSe, AsTe.

Interfering Ions

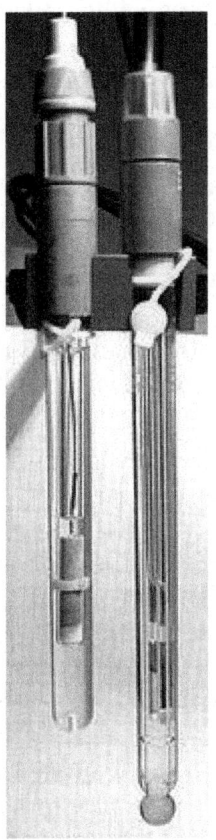

Fig. : A *silver chloride* reference electrode (left) and glass pH electrode.

Because of the *ion-exchange* nature of the glass membrane, it is possible for some other ions to concurrently interact with ion-exchange centers of the glass and to distort the linear dependence of the measured electrode potential on pH or other electrode function. In some cases it is possible to change the electrode function from one ion to another. For example, some silicate pNa electrodes can be changed to pAg function by soaking in a silver salt solution.

Interference effects are commonly described the semi-empirical *Nicolsky*-Eisenman equation (also known as *Nikolsky*-Eisenman equation),[4] an extension to the *Nernst equation*. It is given by :

$$E = E^0 + \frac{RT}{z_iF}\ln\left[a_i + \sum_j (k_{ij}a_j^{z_i/z_j})\right]$$

where E is the emf, E^0 the *standard electrode potential*, z the ionic valency including the sign, a the *activity*, i the ion of interest, j the interfering ions and k_{ij} is the selectivity coefficient. The smaller the selectivity coefficient, the less is the interference by j.

To see the interfering effect of Na^+ to a pH-electrode :

$$E = E^0 + \frac{RT}{F}\ln\left(a_{H^+} + k_{H^+,Na^+}a_{Na^+}\right)$$

Metallic Function of the Glass Electrode

Before the 1950s, there was no explanation for some important aspects of the behaviour of glass electrodes (GE) and the factual reversibility of this behaviour. Some authors have refuted the existence of a particular function at GE in such solutions where they do not behave fully as hydrogen electrode, denying the phenomena of these functions, which were attributed by researchers as an incorrect interpretation of the structural changes in the surface layers of glass; it was mistakenly attributing the changes of EMF element to change in the capacity of GE, and therefore received too large values of pH.

George Eisenman wrote in his retrospective review :

The pioneering studies of Lengiel and Blum were extended by others, who were primarily interested in the existence of Na^+ sensitivity *per se* (*i.e.*, the Na^+ selectivity relative to H^+ only) and in establishing whether or not the electrodes were reversible in the thermodynamic sense. A review of this work is given by *Shultz*, whose studies and those of Nicolskii and Tolmacheva are noteworthy. In fact, Shultz was the first to demonstrate, by direct comparison with a sodium amalgam electrode, that glasses behave as reversible electrodes for Na^+ at neutral and alkaline pH.

In 1951 *Mikhail Schultz*, first proved rigorously the thermodynamical reversibility of the Na-function of different glasses in different pH ranges (later the functions for other metal ions) that confirmed the validity of one of the key hypotheses of ion-exchange theory, now the Nikolsky-*Shultz*-Eisenman thermodynamic ion-exchange theory of GE.

This fact is important because ion-exchange theory was confirmed after a thermodynamically rigorous experimental confirmation of metallic function only. Before, it could be called only as a hypothesis (an *epistemological*). This opened the way for industrial technology GE, forming *ionometry* with them, later with membrane electrodes. In the context of "generalized" theory of glass electrodes, Shultz has created a framework for interpreting a mechanism of the influence of diffusion processes in glasses and *resin* in their electrode properties, giving new *quantitative relationship*, which take into account the dynamic and energetic characteristics of ion exchangers. Schulz introduced a thermodynamic consideration of processes in the membranes. Considering the different abilities of the dissociation of ionogenic groups of the glasses, his theory allows a rigorous analytical way to connect the electrode properties of glasses and ion-exchange resins with their chemical characteristics.

Range of a pH Glass Electrode

The pH range at constant *concentration* can be divided into 3 parts :

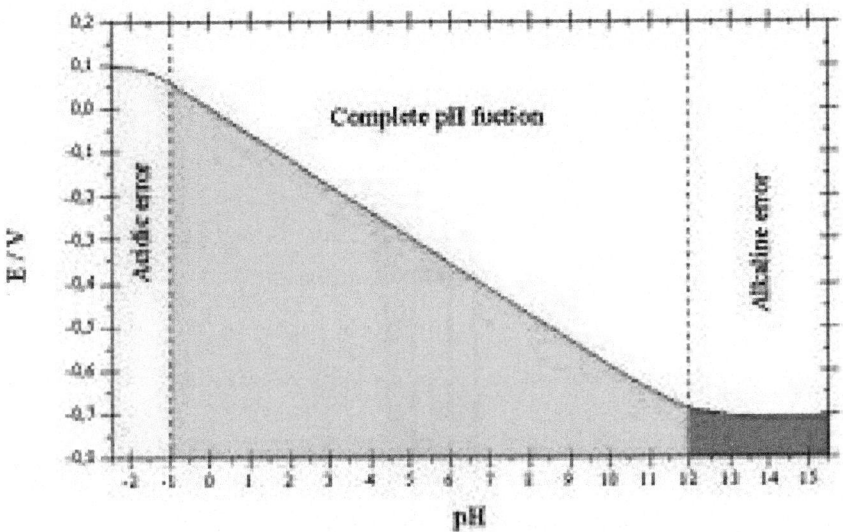

Fig. : Scheme of the typical dependence E (Volt)–pH for *ion-selective electrode.*

- Complete realization of general electrode function, where potential depends linearly on pH, realizing an *ion-selective electrode* for *hydronium.*

$$E = E^0 - \frac{2.303RT}{F}\text{pH}$$

where F is Faraday's constant.

- Alkali error range-at low concentration of *hydrogen ions* (high values of pH) contributions of interfering *alkali metals* (like Li, Na, K) are comparable with the one of hydrogen ions. In this situation dependence of the potential on pH become non-linear.

  The effect is usually noticeable at pH > 12, and concentrations of lithium or sodium ions of 0.1 moles per litre or more. Potassium ions usually cause less error than sodium ions.

- *Acidic* error range–at very high concentration of hydrogen ions (low values of pH) the dependence of the electrode on pH becomes non-linear and the influence of the *anions* in the solution also becomes noticeable. These effects usually become noticeable at pH <-1.

There are different types of pH glass electrode, some of them have improved characteristics for working in alkaline or acidic media. But almost all electrodes have sufficient properties for working in the most popular pH range from pH = 2 to pH = 12. Special electrodes should be used only for working in aggressive conditions.

Most of text written above is also correct for any ion-exchange electrodes.

Construction

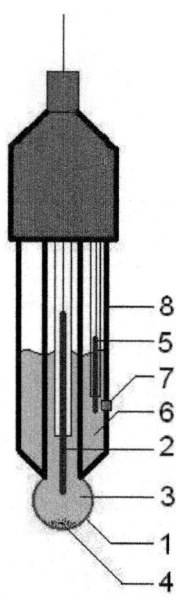

Fig. : Scheme of typical pH glass electrode.

A typical modern pH probe is a combination electrode, which combines both the glass and reference electrodes into one body. The combination electrode consists of the following parts :

1. A sensing part of electrode, a bulb made from a specific glass
2. Internal electrode, usually *silver chloride electrode* or *calomel electrode*
3. Internal solution, usually a pH=7 *buffered* solution of 0.1 *mol/L* KCl for pH electrodes or 0.1 mol/L MeCl for pMe electrodes
4. When using the *silver chloride electrode*, a small amount of AgCl can precipitate inside the glass electrode
5. Reference electrode, usually the same type as 2
6. Reference internal solution, usually 0.1 mol/L KCl
7. Junction with studied solution, usually made from *ceramics* or capillary with *asbestos* or quartz fiber.
8. Body of electrode, made from non-conductive glass or plastics.

The bottom of a pH electrode balloons out into a round thin glass bulb. The pH electrode is best thought of as a tube within a tube. The innermost tube (the inner tube) contains an unchanging 1×10^{-7} mol/L HCl solution. Also inside the inner tube is the cathode terminus of the reference probe. The anodic terminus wraps itself around the outside of the inner tube and ends with the same sort of reference probe as was on the inside of the inner tube. It is filled with a reference solution of 0.1 mol/L KCl and has contact with the solution on the outside of the pH probe by way of a porous plug that serves as a salt bridge.

Galvanic Cell Schematic Representation

This device is essentially a *galvanic cell* that can be schematically represented as :

Glass electrode | | Reference Solution | | **Test Solution** | | Glass electrode

Ag(s) | AgCl(s) | KCl(*aq*) | | 1×10^{-7}M H$^+$ solution | | glass membrane | | **Test Solution** | | ceramic junction | | KCl(*aq*) | AgCl(s) | Ag(s)

In this schematic representation of the galvanic cell, one will note the rigorous symmetry between the left and the right members as seen from the center of the row occupied by the "Test Solution" (the solution whose the pH must be measured). In other words, the glass membrane and the ceramic junction occupies both the same relative place in each respective electrode (indicative (sensing) electrode or reference electrode). The double "pipe symbol" (| |) brackets a diffusive barrier avoiding (glass membrane), or slowing down (ceramic junction), the mixing of the different solutions. By using the same electrodes on the left and right, any potential generated

The measuring part of the electrode, the glass bulb on the bottom, is coated both inside and out with a ~10 nm layer of a hydrated *gel*. These two layers are separated by a layer of dry glass. The silica glass structure (that is, the conformation of its atomic structure) is shaped in such a way that it allows Na^+ ions some mobility. The metal cations (Na^+) in the hydrated gel diffuse out of the glass and into solution while H$^+$ from solution can diffuse into the hydrated gel. It is the hydrated gel, which makes the pH electrode an ion-selective electrode.

H$^+$ does not cross through the glass membrane of the pH electrode, it is the Na$^+$ which crosses and allows for a change in *free energy*. When an ion diffuses from a region of activity to another region of activity, there is a free energy change and this is what the pH meter actually measures. The hydrated gel membrane is connected by Na$^+$ transport and thus the concentration of H$^+$ on the outside of the membrane is 'relayed' to the inside of the membrane by Na$^+$.

All glass pH electrodes have extremely high *electric resistance* from 50 to 500 MΩ. Therefore, the glass electrode can be used only with a high input-impedance measuring device like a *pH meter*, or, more generically, a high input-impedance voltmeter which is called an *electrometer*.

Storage

Between measurements any glass and membrane electrodes should be kept in the solution of its own ion (Ex. pH glass electrode should be kept in 0.1 mol/L HCl or 0.1 mol/L H_2SO_4). It is necessary to prevent the glass membrane from drying out.

Palladium-hydrogen Electrode

The **palladium-hydrogen electrode** (abbreviation : Pd/H_2) is one of the common *reference electrodes* used in *electro-chemical study*. Most of its characteristics are similar to the *standard hydrogen electrode* (with *platinum*). But *palladium* has one significant feature — the capability to absorb (dissolve into itself) molecular *hydrogen*.

Electrode Operation

Two phases can coexist in palladium when hydrogen is absorbed :

- Alpha-phase at hydrogen concentration less than 0.025 atoms per atom of palladium
- Beta-phase at hydrogen concentration corresponding to the *non-stoichiometric* formula $PdH_{0.6}$

The electro-chemical behaviour of a palladium electrode in equilibrium with H_3O^+ ions in solution parallels the behaviour of palladium with molecular hydrogen :

$$\frac{1}{2}H_2 = H_{ads} = H_{abs}$$

Thus the equilibrium is controlled in one case by the partial pressure or *fugacity* of molecular hydrogen and in other case — by *activity* of H^+-ions in solution.

$$E = E^0 + \frac{RT}{F}\ln\frac{a_{H^+}}{(pH_2)^{1/2}}$$

When palladium is electro-chemically charged by hydrogen, the existence of two phases is manifested by a constant *potential* of approximately +50 mV compared to the *reversible hydrogen electrode*. This potential is independent of the amount of hydrogen absorbed over a wide range. This property has been utilized in the construction of a palladium/hydrogen *reference electrode*. The main feature of such electrode is an absence of non-stop bubbling of molecular hydrogen through the solution as it is absolutely necessary for the *standard hydrogen electrode*.

Dynamic Hydrogen Electrode

A **dynamic hydrogen electrode** (DHE) is a *reference electrode*, more specific a sub-type of the *standard hydrogen electrodes* for *electro-chemical* processes by simulating a *reversible hydrogen electrode* with an approximately 20 to 40 mV more negative potential.

Principle

A separator in a glass tube connects two electrolytes and a small current is enforced between the *cathode* and *anode*.

Applications

* *In-situ* reference electrode for *direct methanol fuel cells*
* *Proton exchange membrane fuel cells.*

Electrode Potential

Electrode potential, E, in *electro-chemistry*, according to an *IUPAC* definition, is the *electromotive force* of a *cell* built of two *electrodes* :

* on the left-hand side is the *standard hydrogen electrode*, and
* on the right-hand side is the electrode the potential of which is being defined.

  By *convention* :

  $$E_{Cell} = E_{Cathode} - E_{Anode}$$

  From the above, for the cell with the *standard hydrogen electrode* (potential of 0 by convention), one obtains :

  $$E_{Cell} = E_{Right} - 0 = E_{Electrode}$$

  The left-right convention is consistent with the international agreement that redox potentials be given for reactions written in the form of reduction half-reactions.

  Electrode potential is measured in *volts* (V).

Origin and Interpretation

Electrode potential appears at the *interface* between an electrode and *electrolyte* due to the transfer of charged species across the interface, specific adsorption of ions at the interface, and *specific adsorption*/orientation of polar molecules, including those of the solvent.

Electrode potential is the electric potential on an electrode component. In a cell, there will be an electrode potential for the cathode, and an electrode potential for the anode. The difference between the two electrode potentials equals the cell potential.

$$E_{Cell} = E_{Cathode} - E_{Anode}$$

The measured electrode potential may be either that at *equilibrium* on the working electrode ("reversible potential"), or a potential with a non-zero net reaction on the working electrode but zero net current ("corrosion potential", "*mixed potential*"), or a potential with a non-zero net current on the working electrode (like in *galvanic corrosion* or *voltammetry*). Reversible potentials can be sometimes converted to the *standard electrode potential* for a given electro-active species by extrapolation of the measured values to the *standard state*.

The value of the electrode potential under non-equilibrium depends on the nature and composition of the contacting phases, and on the *kinetics of electrode reactions* at the interface.

Measurement

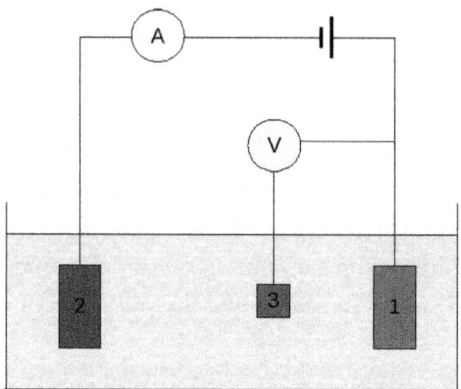

Fig. : Three-electrode setup for measurement of electrode potential.

The measurement is generally conducted using a three-electrode setup :

1. Working electrode
2. Counter electrode
3. *Reference electrode* (standard hydrogen electrode or an equivalent).

In case of non-zero net current on the electrode, it is essential to minimize the *ohmic IR-drop* in the electrolyte, *e.g.*, by positioning the reference electrode near the surface of the working electrode, or by using a *supporting electrolyte* of sufficiently high *conductivity*. The potential measurements are performed with the positive terminal of the *electrometer* connected to working electrode and the negative terminal to the reference electrode.

Potential Difference of a Cell Assembled of Two Electrodes

Potential of a cell assembled of two electrodes can be determined from the two individual electrode potentials using :

$$\Delta V_{cell} = E_{red,cathode} - E_{red,anode}$$

or, equivalently,

$$\Delta V_{cell} = E_{red,cathode} + E_{oxy,anode}$$

This follows from the IUPAC definition of the electric potential difference of a galvanic cell, according to which the electric potential difference of a cell is the difference of the potentials of the electrodes on the right and the left of the galvanic cell. When ΔV_{cell} is positive, then positive electrical charge flows through the cell from the left electrode (*anode*) to the right electrode (*cathode*).

Absolute Electrode Potential

Absolute electrode potential, in *electro-chemistry*, according to an *IUPAC* definition, is the *electrode potential* of a *metal* measured with respect to a universal reference system (without any additional metal–solution interface).

Definition

According to a more specific definition presented by Trasatti, the absolute electrode potential is the difference in electronic energy between a point inside the metal (*Fermi level*) of an *electrode* and a point outside the *electrolyte* in which the electrode is submerged (an electron at rest in vacuum).

This potential is difficult to determine accurately. For this reason, *standard hydrogen electrode* is typically used for reference potential. The absolute potential of the SHE is 4.44 ± 0.02 *V* at 25°C. Therefore, for any electrode at 25°C :

$$E_{(abs)}^{M} = E_{(SHE)}^{M} + (4.44 \pm 0.02) \text{ V}$$

where :

E is electrode potential

V is *volt*

M denotes the electrode made of metal M

(abs) denotes the absolute potential

(SHE) denotes the electrode potential relative to the standard hydrogen electrode.

A different definition for the absolute electrode potential (also known as absolute half-cell potential and single electrode potential) has also been discussed in the literature. In this approach, one first defines an isothermal absolute single-electrode process (or absolute half-cell process.) For example, in the case of a generic metal being oxidized to form a solution-phase ion, the process would be :

$$M \rightarrow M_{(solution)}^{+} + e_{(gas)}^{-}$$

For the *hydrogen* electrode, the absolute half-cell process would be :

$$1/2H_{2 \, (gas)} \rightarrow H_{(solution)}^{+} + e_{(gas)}^{-}$$

Other types of absolute electrode reactions would be defined analogously.

In this approach, all three species taking part in the reaction, including the electron, must be placed in thermodynamically well-defined states. All species, including the electron, are at the same temperature, and appropriate standard states for all species, including the electron, must be fully defined. The absolute electrode potential is then defined as the Gibbs free energy for the absolute electrode process. To express this in volts one divides the Gibb's free energy by the negative of Faraday's constant.

Rockwood's approach to absolute-electrode thermodynamics is easily expendable to other thermodynamic functions. For example, the absolute half-cell entropy has been defined as the entropy of the absolute half-cell process defined above. An alternative definition of the absolute half-cell entropy has recently been published by Fang *et. al.* who define it as the entropy of the following reaction (using the hydrogen electrode as an example) :

$$1/2H_{2 \text{ (gas)}} \rightarrow H^{+}_{\text{(solution)}} + e-$$

This approach differs from the approach described by Rockwood in the treatment of the electron, *i.e.* whether it is placed in the gas phase or in the metal.

Determination

The basis for determination of the absolute electrode potential under the Trasatti definition is given by the equation :

$$E^{M}(\text{abs}) = \phi^{M} + \Delta_{S}^{M}\psi$$

where :

$E^{M}(\text{abs})$ is the absolute potential of the electrode made of metal M

ϕ^{M} is the electron *work function* of metal M

$\Delta_{S}^{M}\psi$ is the *contact potential* difference at the metal(M)–solution(S) interface.

For practical purposes, the value of the absolute electrode potential of the standard hydrogen electrode is best determined with the utility of data for an *ideally-polarizable mercury* (Hg) electrode :

$$E^{\ominus}(H^{+}/H_{2})(\text{abs}) = \phi^{Hg} + \Delta_{S}^{Hg}\psi^{\ominus}_{\sigma=0} - E^{Hg}_{\sigma=0}(\text{SHE})$$

where :

$E^{\ominus}(H^{+}/H_{2})(\text{abs})$ is the absolute standard potential of the hydrogen electrode

$\sigma = 0$ denotes the condition of the *point of zero charge* at the interface.

The types of physical measurements required under the Rockwood definition are similar to those required under the Trasatti definition, but they are used in a different way, *e.g.* in Rockwood's approach they are used to calculate the equilibrium vapour pressure of the electron gas. The numerical value for the absolute potential of the standard hydrogen electrode one would calculate under the Rockwood definition is sometimes fortuitously close to the value one would obtain under the Trasatti definition. This near-agreement in the numerical value depends on the choice of ambient temperature and standard states, and is the result of the near-cancellation of certain terms in the expressions. For example, if a standard state of one atmosphere ideal gas is chosen for the electron gas then the cancellation of terms occurs at a temperature of 296 K, and the two definitions give an equal numerical result. At 298.15 K a near-cancellation of terms would apply and the two approaches would produce nearly the same numerical values. However, there is no fundamental significance to this near agreement because it depends on arbitrary choices, such as temperature and definitions of standard states.

Standard Electrode Potential

The values of standard electrode potentials are given in the table below in *volts* relative to the *standard hydrogen electrode* and are for the following conditions :

- A temperature of 298.15 K;
- An *effective concentration* of 1 mol/L for each aqueous species or a species in a mercury *amalgam*;
- A *partial pressure* of 101.325 *kPa* (absolute) (1 *atm*, 1.01325 *bar*) for each gaseous reagent. This pressure is used because most literature data are still given for this value rather than for the current standard of 100 kPa.
- An *activity* of unity for each pure solid, pure liquid, or for water (solvent).

    Legend : (*s*)–solid; (*l*)–liquid; (*g*)–gas; (*aq*)–aqueous (default for all charged species); (*Hg*)–amalgam.

Half-reaction			
Oxidant	$\rightleftharpoons$	*Reductant*	*E° (V)*
$Sr^+ + e^-$	$\rightleftharpoons$	Sr	−4.10
$Ca^+ + e^-$	$\rightleftharpoons$	Ca	−3.8
$Pr^{3+} + e^-$	$\rightleftharpoons$	Pr^{2+}	−3.1
$\frac{3}{2}N_2(g) + H^+ + e^-$	$\rightleftharpoons$	$HN_3(aq)$	−3.09
$Li^+ + e^-$	$\rightleftharpoons$	$Li(s)$	−3.0401
$N_2(g) + 4H_2O + 2e^-$	$\rightleftharpoons$	$2NH_2OH(aq) + 2OH^-$	−3.04
$Cs^+ + e^-$	$\rightleftharpoons$	$Cs(s)$	−3.026
$Ca(OH)_2 + 2e^-$	$\rightleftharpoons$	$Ca + 2\,OH^-$	−3.02
$Er^{3+} + e^-$	$\rightleftharpoons$	Er^{2+}	−3.0
$Ba(OH)_2 + 2e^-$	$\rightleftharpoons$	$Ba + 2\,OH^-$	−2.99
$Rb^+ + e^-$	$\rightleftharpoons$	$Rb(s)$	−2.98
$K^+ + e^-$	$\rightleftharpoons$	$K(s)$	−2.931
$Ba^{2+} + 2e^-$	$\rightleftharpoons$	$Ba(s)$	−2.912
$La(OH)_3(s) + 3e^-$	$\rightleftharpoons$	$La(s) + 3OH^-$	−2.90
$Fr^+ + e^-$	$\rightleftharpoons$	Fr	−2.9
$Sr^{2+} + 2e^-$	$\rightleftharpoons$	$Sr(s)$	−2.899
$Sr(OH)_2 + 2e^-$	$\rightleftharpoons$	$Sr + 2\,OH^-$	−2.88
$Ca^{2+} + 2e^-$	$\rightleftharpoons$	$Ca(s)$	−2.868
$Eu^{2+} + 2e^-$	$\rightleftharpoons$	$Eu(s)$	−2.812
$Ra^{2+} + 2e^-$	$\rightleftharpoons$	$Ra(s)$	−2.8
$Ho^{3+} + e^-$	$\rightleftharpoons$	Ho^{2+}	−2.8
$Bk^{3+} + e^-$	$\rightleftharpoons$	Bk^{2+}	−2.8
$Yb^{2+} + 2e^-$	$\rightleftharpoons$	Yb	−2.76

(Contd...)

(Contd...)

Half–reaction			
Oxidant	$\rightleftharpoons$	Reductant	$E°$ (V)
$Na^+ + e^-$	$\rightleftharpoons$	Na(s)	−2.71
$Mg^+ + e^-$	$\rightleftharpoons$	Mg	−2.70
$Nd^{3+} + e^-$	$\rightleftharpoons$	Nd^{2+}	−2.7
$Mg(OH)_2 + 2e^-$	$\rightleftharpoons$	Mg + 2 OH⁻	−2.690
$Sm^{2+} + 2e^-$	$\rightleftharpoons$	Sm	−2.68
$Be_2O_3^{2-} + 3\,H_2O + 4e^-$	$\rightleftharpoons$	2 Be + 6 OH⁻	−2.63
$Pm^{3+} + e^-$	$\rightleftharpoons$	Pm^{2+}	−2.6
$Dy^{3+} + e^-$	$\rightleftharpoons$	Dy^{2+}	−2.6
$No^{2+} + 2e^-$	$\rightleftharpoons$	No	−2.50
$HfO(OH)_2 + H_2O + 4e^-$	$\rightleftharpoons$	Hf + 4 OH⁻	−2.50
$Th(OH)_4 + 4e^-$	$\rightleftharpoons$	Th + 4 OH⁻	−2.48
$Md^{2+} + 2e^-$	$\rightleftharpoons$	Md	−2.40
$Tm^{2+} + 2e^-$	$\rightleftharpoons$	Tm	−2.4
$La^{3+} + 3e^-$	$\rightleftharpoons$	La(s)	−2.379
$Y^{3+} + 3e^-$	$\rightleftharpoons$	Y(s)	−2.372
$Mg^{2+} + 2e^-$	$\rightleftharpoons$	Mg(s)	−2.372
$ZrO(OH)_2(s) + H_2O + 4e^-$	$\rightleftharpoons$	Zr(s) + 4OH⁻	−2.36
$Pr^{3+} + 3e^-$	$\rightleftharpoons$	Pr	−2.353
$Ce^{3+} + 3e^-$	$\rightleftharpoons$	Ce	−2.336
$Er^{3+} + 3e^-$	$\rightleftharpoons$	Er	−2.331
$Ho^{3+} + 3e^-$	$\rightleftharpoons$	Ho	−2.33
$H_2AlO_3^- + H_2O + 3e^-$	$\rightleftharpoons$	Al + 4 OH⁻	−2.33
$Nd^{3+} + 3e^-$	$\rightleftharpoons$	Nd	−2.323
$Tm^{3+} + 3e^-$	$\rightleftharpoons$	Tm	−2.319
$Al(OH)_3(s) + 3e^-$	$\rightleftharpoons$	Al(s) + 3OH⁻	−2.31
$Sm^{3+} + 3e^-$	$\rightleftharpoons$	Sm	−2.304
$Fm^{2+} + 2e^-$	$\rightleftharpoons$	Fm	−2.30
$Am^{3+} + e^-$	$\rightleftharpoons$	Am^{2+}	−2.3
$Dy^{3+} + 3e^-$	$\rightleftharpoons$	Dy	−2.295
$Lu^{3+} + 3e^-$	$\rightleftharpoons$	Lu	−2.28
$Tb^{3+} + 3e^-$	$\rightleftharpoons$	Tb	−2.28
$Gd^{3+} + 3e^-$	$\rightleftharpoons$	Gd	−2.279
$H_2 + 2e^-$	$\rightleftharpoons$	2H⁻	−2.23

(Contd...)

(*Contd...*)

Half-reaction			
Oxidant	$\rightleftharpoons$	Reductant	$E°$ (V)
$Es^{2+} + 2e^-$	$\rightleftharpoons$	Es	−2.23
$Pm^{2+} + 2e^-$	$\rightleftharpoons$	Pm	−2.2
$Tm^{3+} + e^-$	$\rightleftharpoons$	Tm^{2+}	−2.2
$Dy^{2+} + 2e^-$	$\rightleftharpoons$	Dy	−2.2
$Ac^{3+} + 3e^-$	$\rightleftharpoons$	Ac	−2.20
$Yb^{3+} + 3e^-$	$\rightleftharpoons$	Yb	−2.19
$Cf^{2+} + 2e^-$	$\rightleftharpoons$	Cf	−2.12
$Nd^{2+} + 2e^-$	$\rightleftharpoons$	Nd	−2.1
$Ho^{2+} + 2e^-$	$\rightleftharpoons$	Ho	−2.1
$Sc^{3+} + 3e^-$	$\rightleftharpoons$	Sc(s)	−2.077
$AlF_6^{3-} + 3e^-$	$\rightleftharpoons$	Al + 6 F⁻	−2.069
$Am^{3+} + 3e^-$	$\rightleftharpoons$	Am	2.048
$Cm^{3+} + 3e^-$	$\rightleftharpoons$	Cm	−2.04
$Pu^{3+} + 3e^-$	$\rightleftharpoons$	Pu	−2.031
$Pr^{2+} + 2e^-$	$\rightleftharpoons$	Pr	−2.0
$Er^{2+} + 2e^-$	$\rightleftharpoons$	Er	−2.0
$Eu^{3+} + 3e^-$	$\rightleftharpoons$	Eu	−1.991
$Lr^{3+} + 3e^-$	$\rightleftharpoons$	Lr	−1.96
$Cf^{3+} + 3e^-$	$\rightleftharpoons$	Cf	−1.94
$Es^{3+} + 3e^-$	$\rightleftharpoons$	Es	−1.91
$Pa^{4+} + e^-$	$\rightleftharpoons$	Pa^{3+}	−1.9
$Am^{2+} + 2e^-$	$\rightleftharpoons$	Am	−1.9
$Th^{4+} + 4e^-$	$\rightleftharpoons$	Th	−1.899
$Fm^{3+} + 3e^-$	$\rightleftharpoons$	Fm	−1.89
$Np^{3+} + 3e^-$	$\rightleftharpoons$	Np	−1.856
$Be^{2+} + 2e^-$	$\rightleftharpoons$	Be	−1.847
$H_2PO_2^- + e^-$	$\rightleftharpoons$	P + 2 OH⁻	−1.82
$U^{3+} + 3e^-$	$\rightleftharpoons$	U	−1.798
$Sr^{2+} + 2e^-$	$\rightleftharpoons$	Sr/Hg	−1.793
$H_2BO_3^- + H_2O + 3e^-$	$\rightleftharpoons$	B + 4 OH⁻	−1.79
$ThO_2 + 4H^+ + 4e^-$	$\rightleftharpoons$	Th + 2 H₂O	−1.789
$HfO^{2+} + 2 H^+ + 4e^-$	$\rightleftharpoons$	Hf + H₂O	−1.724
$HPO_3^{2-} + 2 H_2O + 3e^-$	$\rightleftharpoons$	P + 5 OH⁻	−1.71

(*Contd...*)

(Contd...)

Half-reaction			
Oxidant	$\rightleftharpoons$	*Reductant*	*E° (V)*
$SiO_3^{2-} + H_2O + 4e^-$	$\rightleftharpoons$	Si + 6 OH$^-$	-1.697
$Al^{3+} + 3e^-$	$\rightleftharpoons$	Al(*s*)	-1.662
$Ti^{2+} + 2e^-$	$\rightleftharpoons$	Ti(*s*)	-1.63
$ZrO_2(s) + 4H^+ + 4e^-$	$\rightleftharpoons$	Zr(*s*) + $2H_2O$	-1.553
$Zr^{4+} + 4e^-$	$\rightleftharpoons$	Zr(*s*)	-1.45
$Ti^{3+} + 3e^-$	$\rightleftharpoons$	Ti(*s*)	-1.37
$TiO(s) + 2H^+ + 2e^-$	$\rightleftharpoons$	Ti(*s*) + H_2O	-1.31
$Ti_2O_3(s) + 2H^+ + 2e^-$	$\rightleftharpoons$	2TiO(*s*) + H_2O	-1.23
$Zn(OH)_4^{2-} + 2e^-$	$\rightleftharpoons$	Zn(*s*) + 4OH$^-$	-1.199
$Mn^{2+} + 2e^-$	$\rightleftharpoons$	Mn(*s*)	-1.185
$Fe(CN)_6^{4-} + 6H^+ + 2e^-$	$\rightleftharpoons$	Fe(*s*) + 6HCN(*aq*)	-1.16
$Te(s) + 2e^-$	$\rightleftharpoons$	Te^{2-}	-1.143
$V^{2+} + 2e^-$	$\rightleftharpoons$	V(*s*)	-1.13
$Nb^{3+} + 3e^-$	$\rightleftharpoons$	Nb(*s*)	-1.099
$Sn(s) + 4H^+ + 4e^-$	$\rightleftharpoons$	$SnH_4(g)$	-1.07
$SiO_2(s) + 4H^+ + 4e^-$	$\rightleftharpoons$	Si(*s*) + $2H_2O$	-0.91
$B(OH)_3(aq) + 3H^+ + 3e^-$	$\rightleftharpoons$	B(*s*) + $3H_2O$	-0.89
$Fe(OH)_2(s) + 2e^-$	$\rightleftharpoons$	Fe(*s*) + 2OH$^-$	-0.89
$Fe_2O_3(s) + 3H_2O + 2e^-$	$\rightleftharpoons$	$2Fe(OH)_2(s) + 2OH^-$	-0.86
$TiO^{2+} + 2H^+ + 4e^-$	$\rightleftharpoons$	Ti(*s*) + H_2O	-0.86
$2H_2O + 2e^-$	$\rightleftharpoons$	$H_2(g) + 2OH^-$	-0.8277
$Bi(s) + 3H^+ + 3e^-$	$\rightleftharpoons$	BiH_3	-0.8
$Zn^{2+} + 2e^-$	$\rightleftharpoons$	Zn(*Hg*)	-0.7628
$Zn^{2+} + 2e^-$	$\rightleftharpoons$	Zn(*s*)	-0.7618
$Ta_2O_5(s) + 10H^+ + 10e^-$	$\rightleftharpoons$	2Ta(*s*) + $5H_2O$	-0.75
$Cr^{3+} + 3e^-$	$\rightleftharpoons$	Cr(*s*)	-0.74
$[Au(CN)_2]^- + e^-$	$\rightleftharpoons$	Au(*s*) + 2CN$^-$	-0.60
$Ta^{3+} + 3e^-$	$\rightleftharpoons$	Ta(*s*)	-0.6
$PbO(s) + H_2O + 2e^-$	$\rightleftharpoons$	Pb(*s*) + 2OH$^-$	-0.58
$2TiO_2(s) + 2H^+ + 2e^-$	$\rightleftharpoons$	$Ti_2O_3(s) + H_2O$	-0.56
$Ga^{3+} + 3e^-$	$\rightleftharpoons$	Ga(*s*)	-0.53
$U^{4+} + e^-$	$\rightleftharpoons$	U^{3+}	-0.52
$H_3PO_2(aq) + H^+ + e^-$	$\rightleftharpoons$	P + $2H_2O$	-0.508

(Contd...)

(*Contd...*)

Half-reaction			
Oxidant	$\rightleftharpoons$	*Reductant*	*E° (V)*
$H_3PO_3(aq) + 2H^+ + 2e^-$	$\rightleftharpoons$	$H_3PO_2(aq) + H_2O$	−0.499
$H_3PO_3(aq) + 3H^+ + 3e^-$	$\rightleftharpoons$	$P(red) + 3H_2O$	−0.454
$Fe^{2+} + 2e^-$	$\rightleftharpoons$	$Fe(s)$	−0.44
$2CO_2(g) + 2H^+ + 2e^-$	$\rightleftharpoons$	$HOOCCOOH(aq)$	−0.43
$Cr^{3+} + e^-$	$\rightleftharpoons$	Cr^{2+}	−0.42
$Cd^{2+} + 2e^-$	$\rightleftharpoons$	$Cd(s)$	−0.40
$GeO_2(s) + 2H^+ + 2e^-$	$\rightleftharpoons$	$GeO(s) + H_2O$	−0.37
$Cu_2O(s) + H_2O + 2e^-$	$\rightleftharpoons$	$2Cu(s) + 2OH^-$	−0.360
$PbSO_4(s) + 2e^-$	$\rightleftharpoons$	$Pb(s) + SO_4^{2-}$	−0.3588
$PbSO_4(s) + 2e^-$	$\rightleftharpoons$	$Pb(Hg) + SO_4^{2-}$	−0.3505
$Eu^{3+} + e^-$	$\rightleftharpoons$	Eu^{2+}	−0.35
$In^{3+} + 3e^-$	$\rightleftharpoons$	$In(s)$	0.34
$Tl^+ + e^-$	$\rightleftharpoons$	$Tl(s)$	−0.34
$Ge(s) + 4H^+ + 4e^-$	$\rightleftharpoons$	$GeH_4(g)$	−0.29
$Co^{2+} + 2e^-$	$\rightleftharpoons$	$Co(s)$	−0.28
$H_3PO_4(aq) + 2H^+ + 2e^-$	$\rightleftharpoons$	$H_3PO_3(aq) + H_2O$	−0.276
$V^{3+} + e^-$	$\rightleftharpoons$	V^{2+}	−0.26
$Ni^{2+} + 2e^-$	$\rightleftharpoons$	$Ni(s)$	−0.25
$As(s) + 3H^+ + 3e^-$	$\rightleftharpoons$	$AsH_3(g)$	−0.23
$AgI(s) + e^-$	$\rightleftharpoons$	$Ag(s) + I^-$	−0.15224
$MoO_2(s) + 4H^+ + 4e^-$	$\rightleftharpoons$	$Mo(s) + 2H_2O$	−0.15
$Si(s) + 4H^+ + 4e^-$	$\rightleftharpoons$	$SiH_4(g)$	−0.14
$Sn^{2+} + 2e^-$	$\rightleftharpoons$	$Sn(s)$	−0.13
$O_2(g) + H^+ + e^-$	$\rightleftharpoons$	$HO_2 \bullet (aq)$	−0.13
$Pb^{2+} + 2e^-$	$\rightleftharpoons$	$Pb(s)$	−0.13
$WO_2(s) + 4H^+ + 4e^-$	$\rightleftharpoons$	$W(s) + 2H_2O$	−0.12
$P(red) + 3H^+ + 3e^-$	$\rightleftharpoons$	$PH_3(g)$	−0.111
$CO_2(g) + 2H^+ + 2e^-$	$\rightleftharpoons$	$HCOOH(aq)$	−0.11
$Se(s) + 2H^+ + 2e^-$	$\rightleftharpoons$	$H_2Se(g)$	−0.11
$CO_2(g) + 2H^+ + 2e^-$	$\rightleftharpoons$	$CO(g) + H_2O$	−0.11
$SnO(s) + 2H^+ + 2e^-$	$\rightleftharpoons$	$Sn(s) + H_2O$	−0.10
$SnO_2(s) + 4H^+ + 4e^-$	$\rightleftharpoons$	$SnO(s) + 2H_2O$	−0.09
$WO_3(aq) + 6H^+ + 6e^-$	$\rightleftharpoons$	$W(s) + 3H_2O$	−0.09

(*Contd...*)

(*Contd...*)

Half–reaction			
Oxidant	$\rightleftharpoons$	*Reductant*	*E° (V)*
$P + 3H^+ + 3e^-$	$\rightleftharpoons$	$PH_3(g)$	−0.063
$Fe^{3+} + 3e^-$	$\rightleftharpoons$	$Fe(s)$	−0.04
$HCOOH(aq) + 2H^+ + 2e^-$	$\rightleftharpoons$	$HCHO(aq) + H_2O$	−0.03
$2H^+ + 2e^-$	$\rightleftharpoons$	$\mathbf{H_2(g)}$	**0.0000**
$AgBr(s) + e^-$	$\rightleftharpoons$	$Ag(s) + Br^-$	+0.07133
$S_4O_6^{2-} + 2e^-$	$\rightleftharpoons$	$2S_2O_3^{2-}$	+0.08
$Fe_3O_4(s) + 8H^+ + 8e^-$	$\rightleftharpoons$	$3Fe(s) + 4H_2O$	+0.085
$N_2(g) + 2H_2O + 6H^+ + 6e^-$	$\rightleftharpoons$	$2NH_4OH(aq)$	+0.092
$HgO(s) + H_2O + 2e^-$	$\rightleftharpoons$	$Hg(l) + 2OH^-$	+0.0977
$Cu(NH_3)_4^{2+} + e^-$	$\rightleftharpoons$	$Cu(NH_3)_2^+ + 2NH_3$	+0.10
$Ru(NH_3)_6^{3+} + e^-$	$\rightleftharpoons$	$Ru(NH_3)_6^{2+}$	+0.10
$N_2H_4(aq) + 4H_2O + 2e^-$	$\rightleftharpoons$	$2NH_4^+ + 4OH^-$	+0.11
$H_2MoO_4(aq) + 6H^+ + 6e^-$	$\rightleftharpoons$	$Mo(s) + 4H_2O$	+0.11
$Ge^{4+} + 4e^-$	$\rightleftharpoons$	$Ge(s)$	+0.12
$C(s) + 4H^+ + 4e^-$	$\rightleftharpoons$	$CH_4(g)$	+0.13
$HCHO(aq) + 2H^+ + 2e^-$	$\rightleftharpoons$	$CH_3OH(aq)$	+0.13
$S(s) + 2H^+ + 2e^-$	$\rightleftharpoons$	$H_2S(g)$	+0.14
$Sn^{4+} + 2e^-$	$\rightleftharpoons$	Sn^{2+}	+0.15
$Cu^{2+} + e^-$	$\rightleftharpoons$	Cu^+	+0.159
$HSO_4^- + 3H^+ + 2e^-$	$\rightleftharpoons$	$SO_2(aq) + 2H_2O$	+0.16
$UO_2^{2+} + e^-$	$\rightleftharpoons$	UO_2^+	+0.163
$SO_4^{2-} + 4H^+ + 2e^-$	$\rightleftharpoons$	$SO_2(aq) + 2H_2O$	+0.17
$TiO^{2+} + 2H^+ + e^-$	$\rightleftharpoons$	$Ti^{3+} + H_2O$	+0.19
$SbO^+ + 2H^+ + 3e^-$	$\rightleftharpoons$	$Sb(s) + H_2O$	+0.20
$AgCl(s) + e^-$	$\rightleftharpoons$	$Ag(s) + Cl^-$	+0.22233
$H_3AsO_3(aq) + 3H^+ + 3e^-$	$\rightleftharpoons$	$As(s) + 3H_2O$	+0.24
$GeO(s) + 2H^+ + 2e^-$	$\rightleftharpoons$	$Ge(s) + H_2O$	+0.26
$UO_2^+ + 4H^+ + e^-$	$\rightleftharpoons$	$U^{4+} + 2H_2O$	+0.273
$Re^{3+} + 3e^-$	$\rightleftharpoons$	$Re(s)$	+0.300
$Bi^{3+} + 3e^-$	$\rightleftharpoons$	$Bi(s)$	+0.308
$VO^{2+} + 2H^+ + e^-$	$\rightleftharpoons$	$V^{3+} + H_2O$	+0.34
$Cu^{2+} + 2e^-$	$\rightleftharpoons$	$Cu(s)$	+0.340
$[Fe(CN)_6]^{3-} + e^-$	$\rightleftharpoons$	$[Fe(CN)_6]^{4-}$	+0.36

(*Contd...*)

(Contd...)

Half-reaction			
Oxidant	$\rightleftharpoons$	*Reductant*	*E° (V)*
$Fc^+ + e^-$	$\rightleftharpoons$	$Fc(s)$	+0.4
$O_2(g) + 2H_2O + 4e^-$	$\rightleftharpoons$	$4OH^-(aq)$	+0.401
$H_2MoO_4 + 6H^+ + 3e^-$	$\rightleftharpoons$	$Mo^{3+} + 2H_2O$	+0.43
$CH_3OH(aq) + 2H^+ + 2e^-$	$\rightleftharpoons$	$CH_4(g) + H_2O$	+0.50
$SO_2(aq) + 4H^+ + 4e^-$	$\rightleftharpoons$	$S(s) + 2H_2O$	+0.50
$Cu^+ + e^-$	$\rightleftharpoons$	$Cu(s)$	+0.520
$CO(g) + 2H^+ + 2e^-$	$\rightleftharpoons$	$C(s) + H_2O$	+0.52
$I_3^- + 2e^-$	$\rightleftharpoons$	$3I^-$	+0.53
$I_2(s) + 2e^-$	$\rightleftharpoons$	$2I^-$	+0.54
$[AuI_4]^- + 3e^-$	$\rightleftharpoons$	$Au(s) + 4I^-$	+0.56
$H_3AsO_4(aq) + 2H^+ + 2e^-$	$\rightleftharpoons$	$H_3AsO_3(aq) + H_2O$	+0.56
$[AuI_2]^- + e^-$	$\rightleftharpoons$	$Au(s) + 2I^-$	+0.58
$MnO_4^- + 2H_2O + 3e^-$	$\rightleftharpoons$	$MnO_2(s) + 4OH^-$	+0.59
$S_2O_3^{2-} + 6H^+ + 4e^-$	$\rightleftharpoons$	$2S(s) + 3H_2O$	+0.60
$H_2MoO_4(aq) + 2H^+ + 2e^-$	$\rightleftharpoons$	$MoO_2(s) + 2H_2O$	+0.65
$+ 2H^+ + 2e^-$	$\rightleftharpoons$		+0.6992
$O_2(g) + 2H^+ + 2e^-$	$\rightleftharpoons$	$H_2O_2(aq)$	+0.70
$Tl^{3+} + 3e^-$	$\rightleftharpoons$	$Tl(s)$	+0.72
$PtCl_6^{2-} + 2e^-$	$\rightleftharpoons$	$PtCl_4^{2-} + 2Cl^-$	+0.726
$H_2SeO_3(aq) + 4H^+ + 4e^-$	$\rightleftharpoons$	$Se(s) + 3H_2O$	+0.74
$PtCl_4^{2-} + 2e^-$	$\rightleftharpoons$	$Pt(s) + 4Cl^-$	+0.758
$Fe^{3+} + e^-$	$\rightleftharpoons$	Fe^{2+}	+0.77
$Ag^+ + e^-$	$\rightleftharpoons$	$Ag(s)$	+0.7996
$Hg_2^{2+} + 2e^-$	$\rightleftharpoons$	$2Hg(l)$	+0.80
$NO_3^-(aq) + 2H^+ + e^-$	$\rightleftharpoons$	$NO_2(g) + H_2O$	+0.80
$2FeO_4^{2-} + 5H_2O + 6e^-$	$\rightleftharpoons$	$Fe_2O_3(s) + 10\ OH^-$	+0.81
$[AuBr_4]^- + 3e^-$	$\rightleftharpoons$	$Au(s) + 4Br^-$	+0.85
$Hg^{2+} + 2e^-$	$\rightleftharpoons$	$Hg(l)$	+0.85
$[IrCl_6]^{2-} + e^-$	$\rightleftharpoons$	$[IrCl_6]^{3-}$	+0.87[4]
$MnO_4^- + H^+ + e^-$	$\rightleftharpoons$	$HMnO_4^-$	+0.90
$2Hg^{2+} + 2e^-$	$\rightleftharpoons$	Hg_2^{2+}	+0.91
$Pd^{2+} + 2e^-$	$\rightleftharpoons$	$Pd(s)$	+0.915
$[AuCl_4]^- + 3e^-$	$\rightleftharpoons$	$Au(s) + 4Cl^-$	+0.93

(Contd...)

(Contd...)

Half–reaction			
Oxidant	$\rightleftharpoons$	Reductant	$E°$ (V)
$MnO_2(s) + 4H^+ + e^-$	$\rightleftharpoons$	$Mn^{3+} + 2H_2O$	+0.95
$[AuBr_2]^- + e^-$	$\rightleftharpoons$	$Au(s) + 2Br^-$	+0.96
$[HXeO_6]^{3-} + 2H_2O + 2e^- +$	$\rightleftharpoons$	$[HXeO_4]^- + 4OH^-$	+0.99
$[VO_2]^+(aq) + 2H^+ + e^-$	$\rightleftharpoons$	$[VO]^{2+}(aq) + H_2O$	+1.0
$H_6TeO_6(aq) + 2H^+ + 2e^-$	$\rightleftharpoons$	$TeO_2(s) + 4H_2O$	+1.02
$Br_2(l) + 2e^-$	$\rightleftharpoons$	$2Br^-$	+1.066
$Br_2(aq) + 2e^-$	$\rightleftharpoons$	$2Br^-$	+1.0873
$IO_3^- + 5H^+ + 4e^-$	$\rightleftharpoons$	$HIO(aq) + 2H_2O$	+1.13
$[AuCl_2]^- + e^-$	$\rightleftharpoons$	$Au(s) + 2Cl^-$	+1.15
$HSeO_4^- + 3H^+ + 2e^-$	$\rightleftharpoons$	$H_2SeO_3(aq) + H_2O$	+1.15
$Ag_2O(s) + 2H^+ + 2e^-$	$\rightleftharpoons$	$2Ag(s) + H_2O$	+1.17
$ClO_3^- + 2H^+ + e^-$	$\rightleftharpoons$	$ClO_2(g) + H_2O$	+1.18
$[HXeO_6]^{3-} + 5H_2O + 8\,e^-$	$\rightleftharpoons$	$Xe(g) + 11OH^-$	+1.18
$Pt^{2+} + 2e^-$	$\rightleftharpoons$	$Pt(s)$	+1.188
$ClO_2(g) + H^+ + e^-$	$\rightleftharpoons$	$HClO_2(aq)$	+1.19
$2IO_3^- + 12H^+ + 10e^-$	$\rightleftharpoons$	$I_2(s) + 6H_2O$	+1.20
$ClO_4^- + 2H^+ + 2e^-$	$\rightleftharpoons$	$ClO_3^- + H_2O$	+1.20
$\mathbf{O_2(g) + 4H^+ + 4e^-}$	$\rightleftharpoons$	$2H_2O$	+1.229
$MnO_2(s) + 4H^+ + 2e^-$	$\rightleftharpoons$	$Mn^{2+} + 2H_2O$	+1.23
$[HXeO_4]^- + 3H_2O + 6\,e^-$	$\rightleftharpoons$	$Xe(g) + 7OH^-$	+1.24
$Tl^{3+} + 2e^-$	$\rightleftharpoons$	Tl^+	+1.25
$Cr_2O_7^{2-} + 14H^+ + 6e^-$	$\rightleftharpoons$	$2Cr^{3+} + 7H_2O$	+1.33
$Cl_2(g) + 2e^-$	$\rightleftharpoons$	$2Cl^-$	+1.36
$CoO_2(s) + 4H^+ + e^-$	$\rightleftharpoons$	$Co^{3+} + 2H_2O$	+1.42
$2[[Hydroxylamine \mid NH_3OH^+]] + H^+ + 2e^-$	$\rightleftharpoons$	$N_2H_5^+ + 2H_2O$	+1.42
$2HIO(aq) + 2H^+ + 2e^-$	$\rightleftharpoons$	$I_2(s) + 2H_2O$	+1.44
$BrO_3^- + 5H^+ + 4e^-$	$\rightleftharpoons$	$HBrO(aq) + 2H_2O$	+1.45
$ß-PbO_2(s) + 4H^+ + 2e^-$	$\rightleftharpoons$	$Pb^{2+} + 2H_2O$	+1.460
$a-PbO_2(s) + 4H^+ + 2e^-$	$\rightleftharpoons$	$Pb^{2+} + 2H_2O$	+1.468
$2BrO_3^- + 12H^+ + 10e^-$	$\rightleftharpoons$	$Br_2(l) + 6H_2O$	+1.48
$2ClO_3^- + 12H^+ + 10e^-$	$\rightleftharpoons$	$Cl_2(g) + 6H_2O$	+1.49
$HClO(aq) + H^+ + 2e^-$	$\rightleftharpoons$	$Cl^-(aq) + H_2O$	+1.49
$MnO_4^- + 8H^+ + 5e^-$	$\rightleftharpoons$	$Mn^{2+} + 4H_2O$	+1.51

(Contd...)

(*Contd...*)

Half-reaction			
Oxidant	⇌	Reductant	E° (V)
$HO_2^{\cdot} + H^+ + e^-$	⇌	$H_2O_2(aq)$	+1.51
$Au^{3+} + 3e^-$	⇌	$Au(s)$	+1.52
$NiO_2(s) + 4H^+ + 2e^-$	⇌	$Ni^{2+} + 2OH^-$	+1.59
$Ce^{4+} + e^-$	⇌	Ce^{3+}	+1.61
$2HClO(aq) + 2H^+ + 2e^-$	⇌	$Cl_2(g) + 2H_2O$	+1.63
$Ag_2O_3(s) + 6H^+ + 4e^-$	⇌	$2Ag^+ + 3H_2O$	+1.67
$HClO_2(aq) + 2H^+ + 2e^-$	⇌	$HClO(aq) + H_2O$	+1.67
$Pb^{4+} + 2e^-$	⇌	Pb^{2+}	+1.69
$MnO_4^- + 4H^+ + 3e^-$	⇌	$MnO_2(s) + 2H_2O$	+1.70
$AgO(s) + 2H^+ + e^-$	⇌	$Ag^+ + H_2O$	+1.77
$H_2O_2(aq) + 2H^+ + 2e^-$	⇌	$2H_2O$	+1.78
$Co^{3+} + e^-$	⇌	Co^{2+}	+1.82
$Au^+ + e^-$	⇌	$Au(s)$	+1.83
$BrO_4^- + 2H^+ + 2e^-$	⇌	$BrO_3^- + H_2O$	+1.85
$Ag^{2+} + e^-$	⇌	Ag^+	+1.98
$S_2O_8^{2-} + 2e^-$	⇌	$2SO_4^{2-}$	+2.010
$O_3(g) + 2H^+ + 2e^-$	⇌	$O_2(g) + H_2O$	+2.075
$HMnO_4^- + 3H^+ + 2e^-$	⇌	$MnO_2(s) + 2H_2O$	+2.09
$XeO_3(aq) + 6H^+ + 6\,e^-$	⇌	$Xe(g) + 3H_2O$	+2.12
$H_4XeO_6(aq) + 8H^+ + 8\,e^-$	⇌	$Xe(g) + 6\,H_2O$	+2.18
$FeO_4^{2-} + 3e^- + 8H^+$	⇌	$Fe^{3+} + 4H_2O$	+2.20
$XeF_2(aq) + 2H^+ + 2e^-$	⇌	$Xe(g) + 2HF(aq)$	+2.32
$H_4XeO_6(aq) + 2H^+ + 2e^-$	⇌	$XeO_3(aq) + H_2O$	+2.42
$F_2(g) + 2e^-$	⇌	$2F^-$	+2.87
$F_2(g) + 2H^+ + 2e^-$	⇌	$2HF(aq)$	+3.05

Clicking on this column to re-sort by potential didn't work in the *Safari web browser* in v. 4.0.3 or earlier (but works in v. 4.0.5). In this case just reload the page to restore the original order.

1. Not specified in the indicated reference, but assumed due to the difference between the value-0.454 and that computed by (2 × -0.499 +-0.508)/3 =-0.502 exactly matching the difference between the values for white (-0.063) and red (-0.111) phosphorus in equilibrium with PH_3.

Chapter 2

OHM'S LAW

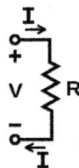

Fig. : V, I, and R, the parameters of Ohm's law.

Ohm's law states that the *current* through a conductor between two points is directly *proportional* to the *potential difference* across the two points. Introducing the constant of proportionality, the *resistance*, one arrives at the usual mathematical equation that describes this relationship :

$$I = \frac{V}{R},$$

where I is the current through the conductor in units of *amperes*, V is the potential difference measured *across* the conductor in units of *volts*, and R is the *resistance* of the conductor in units of *ohms*. More specifically, Ohm's law states that the R in this relation is constant, independent of the current.

The law was named after the German physicist *Georg Ohm*, who, in a treatise published in 1827, described measurements of applied voltage and current through simple electrical circuits containing various lengths of wire. He presented a slightly more complex equation than the one above to explain his experimental results. The above equation is the modern form of Ohm's law.

In physics, the term *Ohm's law* is also used to refer to various generalizations of the law originally formulated by Ohm. The simplest example of this is :

$$J = \sigma E,$$

where J is the *current density* at a given location in a resistive material, E is the electric field at that location, and σ is a material dependent parameter called the *conductivity*. This reformulation of Ohm's law is due to *Gustav Kirchhoff*.

HISTORY

In January 1781, before *Georg Ohm*'s work, *Henry Cavendish* experimented with *Leyden jars* and glass tubes of varying diameter and length filled with salt solution. He measured the current by noting how strong a shock he felt as he completed the circuit with his body. Cavendish wrote that the "velocity" (current) varied directly as the "degree of electrification" (voltage). He did not communicate his results to other scientists at the time, and his results were unknown until *Maxwell* published them in 1879.

Ohm did his work on resistance in the years 1825 and 1826, and published his results in 1827 as the book *Die galvanische Kette, mathematisch bearbeitet* ("The galvanic circuit investigated mathematically"). He drew considerable inspiration from *Fourier*'s work on heat conduction in the theoretical explanation of his work. For experiments, he initially used *voltaic piles*, but later used a *thermocouple* as this provided a more stable voltage source in terms of internal resistance and constant potential difference. He used a galvanometer to measure current, and knew that the voltage between the thermocouple terminals was proportional to the junction temperature. He then added test wires of varying length, diameter, and material to complete the circuit. He found that his data could be modelled through the equation.

$$x = \frac{a}{b+l},$$

where x was the reading from the *galvanometer*, l was the length of the test conductor, a depended only on the thermocouple junction temperature, and b was a constant of the entire setup. From this, Ohm determined his law of proportionality and published his results.

Ohm's law was probably the most important of the early quantitative descriptions of the physics of electricity. We consider it almost obvious today. When Ohm first published his work, this was not the case; critics reacted to his treatment of the subject with hostility. They called his work a "web of naked fancies" and the German Minister of Education proclaimed that "a professor who preached such heresies was unworthy to teach science." The prevailing scientific philosophy in Germany at the time asserted that experiments need not be performed to develop an understanding of nature because nature is so well ordered, and that scientific truths may be deduced through reasoning alone. Also, Ohm's brother Martin, a mathematician, was battling the German educational system. These factors hindered the acceptance of Ohm's work, and his work did not become widely accepted until the 1840s. Fortunately, Ohm received recognition for his contributions to science well before he died.

In the 1850s, Ohm's law was known as such and was widely considered proved, and alternatives, such as *"Barlow's law"*, were discredited, in terms of real applications to telegraph system design, as discussed by *Samuel F. B. Morse* in 1855.

While the old term for electrical conductance, the *mho* (the inverse of the resistance unit ohm), is still used, a new name, the *siemens*, was adopted in 1971, honouring Ernst Werner von Siemens. The siemens is preferred in formal papers.

In the 1920s, it was discovered that the current through a practical resistor actually has statistical fluctuations, which depend on temperature, even when voltage and resistance are exactly constant; this fluctuation, now known as *Johnson–Nyquist noise*, is due to the discrete nature of charge. This thermal effect implies that measurements of current and voltage that are taken over sufficiently short periods of time will yield ratios of V/I that fluctuate from the value of R implied by the time average or *ensemble average* of the measured current; Ohm's law remains correct for the average current, in the case of ordinary resistive materials.

Ohm's work long preceded *Maxwell's equations* and any understanding of frequency-dependent effects in AC circuits. Modern developments in electromagnetic theory and circuit theory do not contradict Ohm's law when they are evaluated within the appropriate limits.

SCOPE

Ohm's law is an *empirical law*, a generalization from many experiments that have shown that current is approximately proportional to electric field for most materials. It is less fundamental than *Maxwell's equations* and is not always obeyed. Any given material will *break down* under a strong-enough electric field, and some materials of interest in electrical engineering are "non-ohmic" under weak fields.

Ohm's law has been observed on a wide range of length scales. In the early 20th century, it was thought that Ohm's law would fail at the *atomic* scale, but experiments have not borne out this expectation. As of 2012, researchers have demonstrated that Ohm's law works for *silicon* wires as small as four atoms wide and one atom high.

MICROSCOPIC ORIGINS

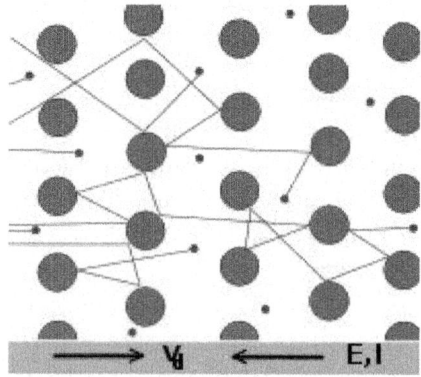

Fig. : Drude Model electrons (shown here in blue) constantly bounce among heavier, stationary crystal ions.

The dependence of the current density on the applied electric field is essentially *quantum mechanical* in nature; A qualitative description leading to Ohm's law can be based upon *classical mechanics* using the *Drude model* developed by *Paul Drude* in 1900.

The Drude model treats *electrons* (or other charge carriers) like pinballs bouncing among the *ions* that make up the structure of the material. Electrons will be accelerated in the opposite direction to the electric field by the average electric field at their location. With each collision, though, the electron is deflected in a random direction with a velocity that is much larger than the velocity gained by the electric field. The net result is that electrons take a zigzag path due to the collisions, but generally drift in a direction opposing the electric field.

The *drift velocity* then determines the electric *current density* and its relationship to *E* and is independent of the collisions. Drude calculated the average drift velocity from $p = -eE\tau$ where *p* is the average *momentum*, $-e$ is the charge of the electron and τ is the average time between the collisions. Since both the momentum and the current density are proportional to the drift velocity, the current density becomes proportional to the applied electric field; this leads to Ohm's law.

HYDRAULIC ANALOGY

A *hydraulic analogy* is sometimes used to describe Ohm's law. Water pressure, measured by *pascals* (or *PSI*), is the analog of voltage because establishing a water pressure difference between two points along a (horizontal) pipe causes water to flow. Water flow rate, as in *liters* per second, is the analog of current, as in *coulombs* per second. Finally, flow restrictors — such as apertures placed in pipes between points where the water pressure is measured — are the analog of resistors. We say that the rate of water flow through an aperture restrictor is proportional to the difference in water pressure across the restrictor. Similarly, the rate of flow of electrical charge, that is, the electric current, through an electrical resistor is proportional to the difference in voltage measured across the resistor.

Flow and pressure variables can be calculated in fluid flow network with the use of the hydraulic ohm analogy. The method can be applied to both steady and transient flow situations. In the linear *laminar flow* region, *Poiseuille's law* describes the hydraulic resistance of a pipe, but in the *turbulent flow* region the pressure–flow relations become non-linear.

The hydraulic analogy to Ohm's law has been used, for example, to approximate blood flow through the circulatory system.

CIRCUIT ANALYSIS

In *circuit analysis*, three equivalent expressions of Ohm's law are used interchangeably :

$$I = \frac{V}{R} \text{ or } V = IR \text{ or } R = \frac{V}{I}.$$

Each equation is quoted by some sources as the defining relationship of Ohm's law, or all three are quoted, or derived from a proportional form, or even just the two that do not correspond to Ohm's original statement may sometimes be given.

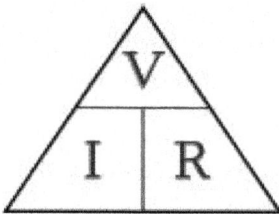

Fig. : Ohm's law triangle.

The interchangeability of the equation may be represented by a triangle, where V (*voltage*) is placed on the top section, the I (*current*) is placed to the left section, and the R (*resistance*) is placed to the right. The line that divides the left and right sections indicate multiplication, and the divider between the top and bottom sections indicates division (hence the division bar).

RESISTIVE CIRCUITS

Resistors are circuit elements that impede the passage of electric charge in agreement with Ohm's law, and are designed to have a specific resistance value R. In a schematic diagram the resistor is shown as a zig-zag symbol. An element (resistor or conductor) that behaves according to Ohm's law over some operating range is referred to as an *ohmic device* (or an *ohmic resistor*) because Ohm's law and a single value for the resistance suffice to describe the behaviour of the device over that range.

Ohm's law holds for circuits containing only resistive elements (no capacitances or inductances) for all forms of driving voltage or current, regardless of whether the driving voltage or current is constant (DC) or time-varying such as AC. At any instant of time Ohm's law is valid for such circuits.

Resistors which are in *series* or in *parallel* may be grouped together into a single "equivalent resistance" in order to apply Ohm's law in analyzing the circuit. This application of Ohm's law is illustrated with examples in "*How To Analyze Resistive Circuits Using Ohm's Law*" on *wikiHow*.

REACTIVE CIRCUITS WITH TIME-VARYING SIGNALS

When reactive elements such as capacitors, inductors, or transmission lines are involved in a circuit to which AC or time-varying voltage or current is applied, the relationship between voltage and current becomes the solution to a *differential equation*, so Ohm's law does not directly apply since that form contains only resistances having value R, not complex impedances which may contain capacitance ("C") or inductance ("L").

Equations for *time-invariant* AC circuits take the same form as Ohm's law, however, the variables are generalized to *complex numbers* and the current and voltage waveforms are *complex exponentials*.

In this approach, a voltage or current waveform takes the form Ae^{st}, where t is time, s is a complex parameter, and A is a complex scalar. In any *linear time-invariant system*, all of the currents and voltages can be expressed with the same s parameter as the input to the system, allowing the time-varying complex exponential term to be canceled out and the system described algebraically in terms of the complex scalars in the current and voltage waveforms.

The complex generalization of resistance is *impedance*, usually denoted Z; it can be shown that for an inductor,

$Z = sL$

and for a capacitor,

$$Z = \frac{1}{sC}.$$

We can now write,

$V = I \cdot Z$

where V and I are the complex scalars in the voltage and current respectively and Z is the complex impedance.

This form of Ohm's law, with Z taking the place of R, generalizes the simpler form. When Z is complex, only the real part is responsible for dissipating heat.

In the general AC circuit, Z varies strongly with the frequency parameter s, and so also will the relationship between voltage and current.

For the common case of a steady *sinusoid*, the s parameter is taken to be $j\omega$, corresponding to a complex sinusoid $Ae^{j\omega t}$. The real parts of such complex current and voltage waveforms describe the actual sinusoidal currents and voltages in a circuit, which can be in different phases due to the different complex scalars.

LINEAR APPROXIMATIONS

Ohm's law is one of the basic equations used in the *analysis of electrical circuits*. It applies to both metal conductors and circuit components (*resistors*) specifically made for this behaviour. Both are ubiquitous in electrical engineering. Materials and components that obey Ohm's law are described as "ohmic" which means they produce the same value for resistance ($R = V/I$) regardless of the value of V or I which is applied and whether the applied voltage or current is DC (*direct current*) of either positive or negative polarity or AC (*alternating current*).

In a true ohmic device, the same value of resistance will be calculated from $R = V/I$ regardless of the value of the applied voltage V. That is, the ratio of V/I is constant, and when current is plotted as a function of voltage the curve is *linear* (a straight line). If voltage is forced to some value V, then that voltage V divided by measured current I will equal R. Or if the current is forced to some value I, then the measured voltage V divided by that current I is also R. Since the plot of I *versus* V is a straight line, then it is also true that for any set of two different voltages V_1 and V_2 applied across a given device of resistance R, producing currents $I_1 = V_1/R$ and $I_2 = V_2/R$, that the ratio/is also a constant equal to R. The operator "delta" (Δ) is used to represent a difference in a quantity, so we can write $\Delta V = V_1 - V_2$ and $\Delta I = I_1 - I_2$. Summarizing, for any truly ohmic device having resistance R, $V/I = \Delta V/\Delta I = R$ for any applied voltage or current or for the difference between any set of applied voltages or currents.

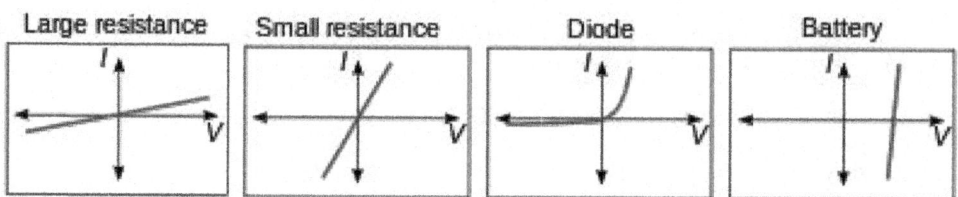

Fig. : The *I–V curves* of four devices : Two *resistors,* a *diode,* and a *battery.* The two resistors follow Ohm's law : The plot is a straight line through the origin. The other two devices do *not* follow Ohm's law.

There are, however, components of electrical circuits which do not obey Ohm's law; that is, their relationship between current and voltage (their *I–V curve*) is *non-linear* (or non-ohmic). An example is the *p-n junction diode.* The current only increases significantly if the applied voltage is positive, not negative. The ratio *V/I* for some point along the non-linear curve is sometimes called the *static,* or *chordal,* or *DC,* resistance, but as seen the value of total *V* over total *I* varies depending on the particular point along the non-linear curve which is chosen. This means the "DC resistance" V/I at some point on the curve is not the same as what would be determined by applying an AC signal having peak amplitude ΔV volts or ΔI amps centered at that same point along the curve and measuring $\Delta V/\Delta I$. However, in some diode applications, the AC signal applied to the device is small and it is possible to analyze the circuit in terms of the *dynamic, small-signal,* or *incremental* resistance, defined as the one over the slope of the V–I curve at the average value (DC operating point) of the voltage (that is, one over the *derivative* of current with respect to voltage). For sufficiently small signals, the dynamic resistance allows the Ohm's law small signal resistance to be calculated as approximately one over the slope of a line drawn tangentially to the V-I curve at the DC operating point.

TEMPERATURE EFFECTS

Ohm's law has sometimes been stated as, "for a conductor in a given state, the electromotive force is proportional to the current produced." That is, that the resistance, the ratio of the applied *electromotive force* (or voltage) to the current, "does not vary with the current strength." The qualifier "in a given state" is usually interpreted as meaning "at a constant temperature," since the resistivity of materials is usually temperature dependent. Because the conduction of current is related to *Joule heating* of the conducting body, according to *Joule's first law*, the temperature of a conducting body may change when it carries a current. The dependence of resistance on temperature therefore makes resistance depend upon the current in a typical experimental setup, making the law in this form difficult to directly verify. *Maxwell* and others worked out several methods to test the law experimentally in 1876, controlling for heating effects.

RELATION TO HEAT CONDUCTIONS

Ohm's principle predicts the flow of electrical charge (*i.e.* current) in electrical conductors when subjected to the influence of voltage differences; *Jean-Baptiste-Joseph Fourier*'s principle predicts the flow of *heat* in heat conductors when subjected to the influence of temperature differences.

The same equation describes both phenomena, the equation's variables taking on different meanings in the two cases. Specifically, solving a heat conduction (Fourier) problem with *temperature* (the driving "force") and *flux of heat* (the rate of flow of the driven "quantity", *i.e.* heat energy) variables also solves an analogous *electrical conduction* (Ohm) problem having *electric potential* (the driving "force") and *electric current* (the rate of flow of the driven "quantity", *i.e.* charge) variables.

The basis of Fourier's work was his clear conception and definition of *thermal conductivity*. He assumed that, all else being the same, the flux of heat is strictly proportional to the gradient of temperature. Although undoubtedly true for small temperature gradients, strictly proportional behaviour will be lost when real materials (*e.g.* ones having a thermal conductivity that is a function of temperature) are subjected to large temperature gradients.

A similar assumption is made in the statement of Ohm's law : other things being alike, the strength of the current at each point is proportional to the gradient of electric potential. The accuracy of the assumption that flow is proportional to the gradient is more readily tested, using modern measurement methods, for the electrical case than for the heat case.

OTHER VERSIONS

Ohm's law, in the form above, is an extremely useful equation in the field of electrical/electronic engineering because it describes how voltage, current and resistance are interrelated on a "macroscopic" level, that is, commonly, as circuit elements in an *electrical circuit*. Physicists who study the electrical properties of matter at the microscopic level use a closely related and more general *vector* equation, sometimes also referred to as Ohm's law, having variables that are closely related to the V, I, and R *scalar* variables of Ohm's law, but which are each functions of position within the conductor. Physicists often use this continuum form of Ohm's Law :

$E = \rho J$

where "E" is the *electric field* vector with units of volts per meter (analogous to "V" of Ohm's law which has units of volts), "J" is the *current density* vector with units of amperes per unit area (analogous to "I" of Ohm's law which has units of amperes), and "ρ" (Greek "rho") is the *resistivity* with units of ohm meters (analogous to "R" of Ohm's law which has units of ohms). The above equation is sometimes written as $J = \sigma E$ where "σ" (Greek "sigma") is the *conductivity* which is the reciprocal of ρ.

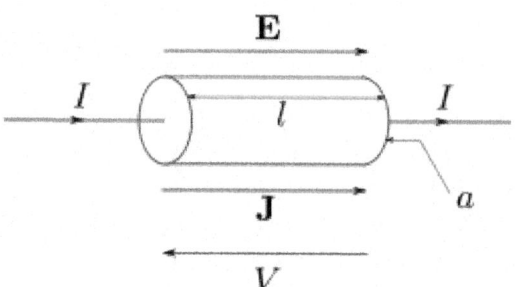

Fig. : Current flowing through a uniform cylindrical conductor (such as a round wire) with a uniform field applied.

The potential difference between two points is defined as :

$$\Delta V = - \int E \cdot dl$$

with *dl* the element of path along the integration of electric field vector E. If the applied E field is uniform and oriented along the length of the conductor, then defining the voltage V in the usual convention of being opposite in direction to the field, and with the understanding that the voltage V is measured differentially across the length of the conductor allowing us to drop the Δ symbol, the above vector equation reduces to the scalar equation :

$$V = El \quad \text{or} \quad E = \frac{V}{l}.$$

Since the E field is uniform in the direction of wire length, for a conductor having uniformly consistent resistivity ρ, the current density J will also be uniform in any cross-sectional area and oriented in the direction of wire length, so we may write :

$$J = \frac{I}{a}.$$

Substituting the above 2 results (for *E* and *J* respectively) into the continuum form shown at the beginning of this section :

$$\frac{V}{l} = \frac{I}{a}\rho \quad \text{or} \quad V = I\rho\frac{l}{a}.$$

The *electrical resistance* of a uniform conductor is given in terms of *resistivity* by :

$$R = \rho\frac{l}{a}$$

where l is the length of the conductor in *SI* units of meters, a is the cross-sectional area (for a round wire $a = \pi r^2$ if r is radius) in units of meters squared, and ρ is the resistivity in units of ohm ·meters.

After substitution of R from the above equation into the equation preceding it, the continuum form of Ohm's law for a uniform field (and uniform current density) oriented along the length of the conductor reduces to the more familiar form :

$V = IR.$

A perfect crystal lattice, with low enough thermal motion and no deviations from periodic structure, would have no *resistivity*, but a real metal has *crystallographic defects*, impurities, multiple *isotopes*, and thermal motion of the atoms. Electrons *scatter* from all of these, resulting in resistance to their flow.

The more complex generalized forms of Ohm's law are important to *condensed matter physics*, which studies the properties of *matter* and, in particular, its *electronic structure*. In broad terms, they fall under the topic of *constitutive equations* and the theory of *transport coefficients*.

Magnetic Effects

If an external **B**-field is present and the conductor is not at rest but moving at velocity **v**, then an extra-term must be added to account for the current induced by the *Lorentz force* on the charge carriers.

$\mathbf{J} = \sigma(\mathbf{E} + \mathbf{v} \times \mathbf{B})$

In the *rest frame* of the moving conductor this term drops out because **v** = 0. There is no contradiction because the electric field in the rest frame differs from the E-field in the lab frame : **E** ′ = **E** + **v** × **B**. Electric and magnetic fields are relative.

If the current **J** is alternating because the applied voltage or E-field varies in time, then reactance must be added to resistance to account for self-inductance. The reactance may be strong if the frequency is high or the conductor is coiled.

Chapter 3

ELECTRO-CHEMICAL CELL

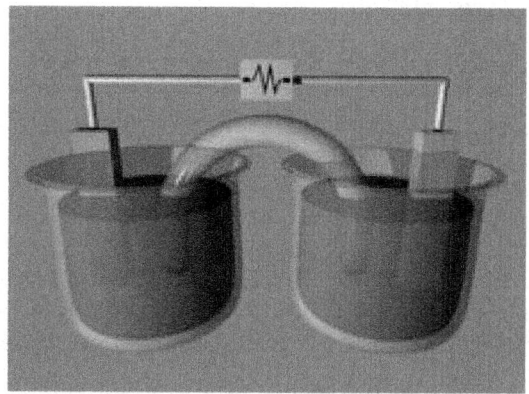

Fig.: A demonstration electro-chemical cell setup resembling the *Daniell cell*. The two half-cells are linked by a salt bridge carrying ions between them. Electrons flow in the external circuit.

An **electro-chemical cell** is a device capable of either deriving *electrical* energy from *chemical reactions* or facilitating chemical reactions through the introduction of electrical energy. A common example of an electro-chemical cell is a standard 1.5-volt "*battery*". (Actually a single "*Galvanic cell*"; a battery properly consists of multiple cells, connected in either *parallel or series* pattern.)

HALF-CELLS

An electro-chemical cell consists of two half-cells. Each *half-cell* consists of an *electrode* and an *electrolyte*. The two half-cells may use the same electrolyte, or they may use different electrolytes. The chemical reactions in the cell may involve the electrolyte, the electrodes, or an external substance (as in *fuel cells* that may use hydrogen gas as a reactant). In a full electro-chemical cell, species from one half-cell lose electrons (*oxidation*) to their *electrode* while species from the other half-cell gain electrons (*reduction*) from their electrode. A *salt bridge* (*e.g.*, filter paper soaked in KNO_3) is often employed to provide ionic contact between two half-cells with different electrolytes, to prevent the solutions from mixing and causing unwanted side reactions. As electrons flow from one half-cell to the other, a difference in charge is established. If no salt bridge were used, this charge difference would prevent further flow of electrons. A salt bridge allows the flow of ions to maintain a balance in charge between the oxidation and reduction vessels while keeping the contents of each separate. Other devices for achieving separation of solutions are porous pots and gelled solutions. A porous pot is used in the *Bunsen cell*.

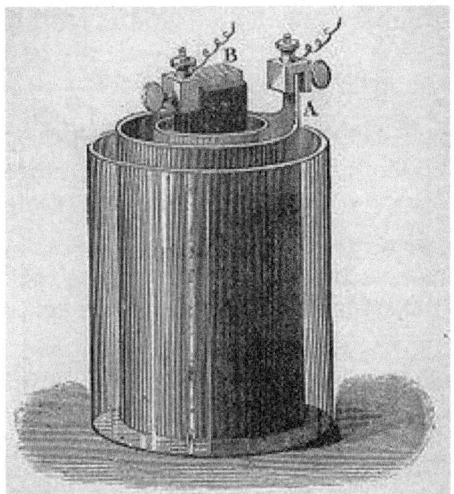

Fig. : The *Bunsen cell*, invented by *Robert Bunsen*.

EQUILIBRIUM REACTION

Each half-cell has a characteristic voltage. Different choices of substances for each half-cell give different potential differences. Each reaction is undergoing an *equilibrium* reaction between different *oxidation states* of the ions : When equilibrium is reached, the cell cannot provide further voltage. In the half-cell that is undergoing oxidation, the closer the equilibrium lies to the ion/atom with the more positive oxidation state the more potential this reaction will provide. Like-wise, in the reduction reaction, the closer the equilibrium lies to the ion/atom with the more *negative* oxidation state the higher the potential.

CELL POTENTIAL

The cell potential can be predicted through the use of *electrode potentials* (the voltages of each half-cell). These half-cell potentials are derived from the assignment of 0 volts to the standard hydrogen electrode (SHE). The difference in voltage between electrode potentials gives a prediction for the potential measured. When calculating the difference in voltage, one must first manipulate the half-cell reactions to obtain a balanced oxidation-reduction equation.

1. Reverse the reduction reaction with the smallest potential (to create an oxidation reaction/overall positive cell potential)
2. Half-reactions must be multiplied by integers to achieve electron balance.

An important note with this is that the cell potential does not change when the reaction is multiplied.

Cell potentials have a possible range of about zero to 6 volts. Cells using water-based electrolytes are usually limited to cell potentials less than about 2.5 volts, because the very powerful oxidizing and reducing agents that would be required to produce a higher cell potential tend to react with the water.

MASS TRANSPORT

In the electrode kinetics section we have seen that the rate of reaction can be influenced by the cell potential difference. However, the rate of transport to the surface can also effect or even dominate the overall reaction rate.

We have already seen that a typical electrolysis reaction involves the transfer of charge between an electrode and a species in solution. This whole process due to the interfacial nature of the electron transfer reactions typically involves a series of steps.

This is an exponential relationship, so we would predict from the electron transfer model that as the voltage is increased the reaction rate and therefore, the current will increase exponentially. This would mean that it is possible to pass unlimited quantities of current.

$$i_c = nF\,Ak_{red}\,[O]_o$$

Clearly for a fixed electrode area (A) the reaction can be controlled by two factors. First the rate constant k_{red} and second the surface concentration of the reactant ($[O]_o$). If the rate constant is large, such that any reactant close to the interface is immediately converted into products then the current will be controlled by the amount of fresh reactant reaching the interface from the bulk solution above. Thus movement of reactant in and out of the interface is important in predicting the current flowing.

There are three forms of mass transport which can influence an electrolysis reaction :

- Diffusion
- Convection
- Migration.

In order to predict the current flowing at any particular time in an electrolysis measurement we will need to have a quantitative model for each of these processes to complement the model for the electron transfer step(s).

Diffusion

Diffusion occurs in all solutions and arises from local uneven concentrations of reagents. Entropic forces act to smooth out these uneven distributions of concentration and are therefore, the main driving force for this process. One example of this can be seen in the animation below. Two materials are held separately in a single container separated by a barrier. When the barrier is removed the two reagents can mix and this processes on the microscopic scale is essentially random. For a large enough sample statistics can be used to predict how far material will move in a certain time-and this is oftern referred to as a random walk model.

Animation

Diffusion is particularly significant in an electrolysis experiment since the conversion reaction only occurs at the electrode surface. Consequently there will be a lower reactant concentration at the electrode than in bulk solution. Similarly a higher concentration of product will exist near the electrode than further out into solution.

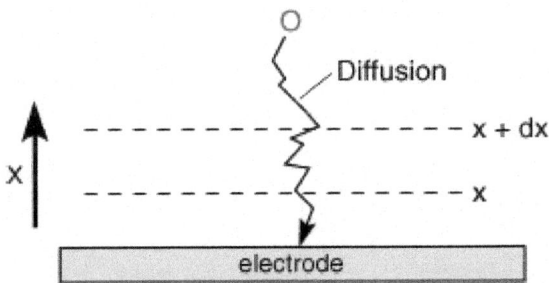

The rate of movement of material by diffusion can be predicted mathematically and *Fick* proposed two laws to quantify the processes. The first law :

$$J_o = -D_o \left(\frac{\partial C_o}{\partial x} \right)$$

relates the diffusional flux J_o (*i.e.* the rate of movement of material by diffusion) to the concentration gradient and the diffusion coefficient D_o. The negative sign simply signifies that material moves down a concentration gradient *i.e.* from regions of high to low concentration. However, in many measurements we need to know how the concentration of material varies as a function of time and this can be predicted from the first law. The result is Fick's second law :

$$\frac{\partial C_o}{\partial t} = D_o \left(\frac{\partial^2 C_o}{\partial x^2} \right)$$

In this case we consider diffusion normal to an electrode surface (x direction). The rate of change of the concentration ($[O]$) as a function of time (t) can be seen to be related to the change in the concentration gradient. So the steeper the change in concentration the greater the rate of diffusion. In practice diffusion is often found to be the most significant transport process for many electrolysis reactions.

Fick's second law is an important relationship since it permits the prediction of the variation of concentration of different species as a function of time within the electro-chemical cell. In order to solve these expressions analytical or computational models are usually employed.

Convection

Convection results from the action of a force on the solution. This can be a pump, a flow of gas or even gravity. There are two forms of convection the first is termed natural convection and is present in any solution. This natural convection is generated by small thermal or density differences and acts to mix the solution in a random and therefore unpredictable manner. In the case of electro-chemical measurements these effects tend to case problems if the measurement time for the experiment exceeds 20 seconds.

It is possible to drown out the natural convection effects from an electro-chemical experiment by deliberately introducing convection into the cell. This form of convection is termed *forced convection*. It is typically several orders of magnitude greater than any natural convection effects and therefore effectively removes the random aspect from the experimental measurements. This of course is only true if the convection is introduced in a well defined and quantitative manner.

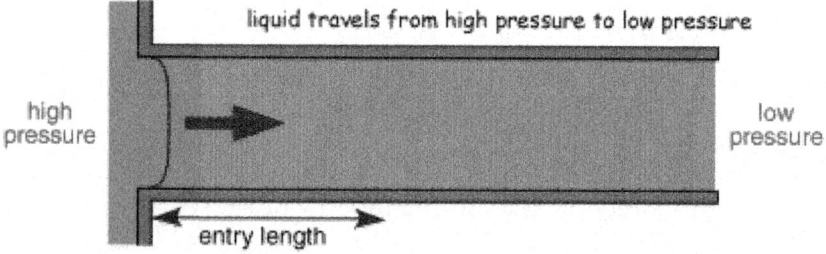

The figure above shows the cross-section of liquid flowing through a pipe. Solution is introduced from the right handside and pumped through the pipe. If the flow is controlled, after a small lead in length, the profile will become stable with no mixing in the lateral direction, this is termed laminar flow.

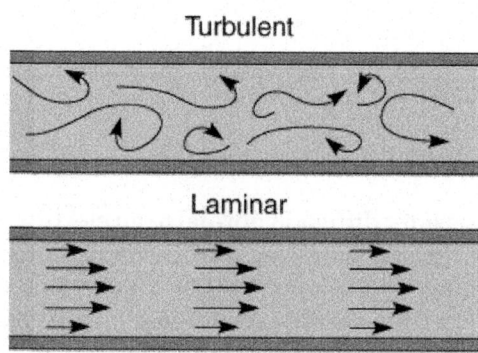

If however the solution is pumped through the cell at a high rate then the transport can become turbulent, where the solution movement is essentially a random and unpredictable. In order to predict this change over between turbulent and laminar behaviour, work was carried out by Hagan in the mid 1840's and later by Reynolds who was the first to put forward a predicitive model. In this particular example there is a maximum velocity in the centre and minimum velocity at the side walls. For laminar flow conditions the mass transport equation for (1 dimensional) convection is predicted by :

$$\frac{\partial C_o}{\partial t} = -v_x \left(\frac{\partial C_o}{\partial x} \right)$$

where v_x is the velocity of the solution which can be calculated in many situations be solving the appropriate form of the Navier-Stokes equations. An analogous form exists for the three dimensional convective transport. When an electro-chemical cell possesses forced convection we must be able to solve the electrode kinetic, diffusion and convection steps, to be able to predict the current flowing. This can be a difficult problem to solve even for modern computers and yet we still have one final form of mass transport to address!

Migration

The final form of mass transport we need to consider is migration. This is essentially an electrostatic effect which arises due the application of a voltage on the electrodes. This effectively creates a charged interface (the electrodes). Any charged species near that interface will either be attracted or repelled from it by electrostatic forces. The migratory flux induced can be described mathematically (in 1 dimension) using :

$$\frac{\partial C_o}{\partial t} = -u C_o \left(\frac{\partial \phi}{\partial x} \right)$$

However due to ion solvation effects and diffuse layer interactions in solution, migration is notoriously difficult to calculate accurately for real solutions. Consequently most voltammetric measurements are performed in solutions which contain a background electrolyte-this material is a salt (*e.g.* KCl) that does not undergo electrolysis itself but helps to shield the reactants from migratory effects. By adding a large quantity of the electrolyte (relative to the reactants) it is possible to ensure that the electrolysis reaction is not significantly effected by migration. The purpose of introducing a background electrolyte into a solution is not however solely to remove migration effects as it also acts as a conductor to help the passage of current through the solution.

Mass Transport in Electro-chemical Cells

To gain a quantitative model of the current flowing at the electrode we must account for the electrode kinetics, the 3 dimensional diffusion, convection and migration, of all the species involved. This is currently beyond the capacity of even the fastest computers and will be for some time. However, as we will discover later electro-chemical cells and experimental conditions can be employed to cheat the mass transport equations. We can effectively remove much (but not all) of the mass transport complexity by carefully designing and controlling the electro-chemical experiment.

DIFFUSION

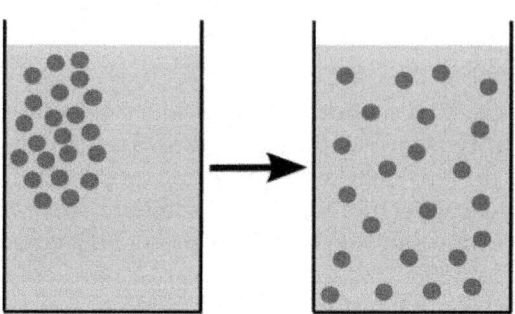

A diffusion process in science. Some particles are *dissolved* in a glass of water. At first, the particles are all near one corner of the glass. If the particles all randomly move around ("diffuse") in the water, then the particles will eventually become distributed randomly and uniformly, and organized (but diffusion will still continue to occur, just that there will be no net *flux*).

Diffusion is one of several *transport phenomena* that occur in nature. A distinguishing feature of diffusion is that it results in mixing or mass transport, without requiring bulk motion. Thus, diffusion should not be confused with *convection*, or *advection*, which are other transport mechanisms that utilize bulk motion to move particles from one place to another. In *Latin*, "diffundere" means "to spread out".

There are two ways to introduce the notion of *diffusion* : either a *phenomenological approach* starting with *Fick's laws of diffusion* and their mathematical consequences, or a physical and atomistic one, by considering the *random walk of the diffusing particles*.

In the phenomenological approach, according to Fick's laws, the diffusion *flux* is proportional to the negative *gradient* of concentrations. It goes from regions of higher concentration to regions of lower concentration. Some time later, various generalizations of Fick's laws were developed in the frame of *thermodynamics* and *non-equilibrium thermodynamics*.

From the *atomistic point of view*, diffusion is considered as a result of the random walk of the diffusing particles. In *molecular diffusion*, the moving molecules are self-propelled by thermal energy. Random walk of small particles in suspension in a fluid was discovered in 1827 by *Robert Brown*. The theory of the *Brownian motion* and the atomistic backgrounds of diffusion were developed by *Albert Einstein*.

The concept of diffusion is widely used in : *physics* (particle diffusion), *chemistry, biology, sociology, economics*, and *finance* (diffusion of people, ideas and of price values). The concept of diffusion is typically applied to any subject-matter involving random walks in *ensembles* of individuals.

History of Diffusion in Physics

In the scope of time, diffusion in solids was used long before the theory of diffusion was created. For example, *Pliny the Elder* had previously described the *cementation process* which produces steel from the element iron (Fe) through carbon diffusion. Another example is well known for many centuries, the diffusion of colours of *stained glass* or *earthenware* and *Chinese ceramics*.

In modern science, the first systematic experimental study of diffusion was performed by *Thomas Graham*. He studied diffusion in gases, and the main phenomenon was described by him in 1831–1833 :

"...gases of different nature, when brought into contact, do not arrange themselves according to their density, the heaviest undermost, and the lighter uppermost, but they spontaneously diffuse, mutually and equally, through each other, and so remain in the intimate state of mixture for any length of time."

The measurements of Graham contributed to *James Clerk Maxwell* deriving, in 1867, the coefficient of diffusion for CO_2 in air. The error rate is less than 5%.

In 1855, *Adolf Fick*, the 26-years old anatomy demonstrator from Zürich, proposed *his law of diffusion*. He used Graham's research, stating his goal as "the development of a fundamental law, for the operation of diffusion in a single element of space". He asserted a deep analogy between diffusion and conduction of heat or electricity, creating a formalism that is similar to *Fourier's law for heat conduction* and *Ohm's law* for electrical current.

Robert Boyle demonstrated diffusion in solids in the 17th century by penetration of Zinc into a copper coin. Nevertheless, diffusion in solids was not systematically studied till the second part of the 19th century. *William Chandler Roberts-Austen*, the well-known British metallurgist, and former assistant of Thomas Graham, studied systematically solid state diffusion on the example of gold in lead in 1896. :

"... My long connection with Graham's researches made it almost a duty to attempt to extend his work on liquid diffusion to metals."

In 1858, *Rudolf Clausius* introduced the concept of the *mean free path*. In the same year, *James Clerk Maxwell* developed the first atomistic theory of transport processes in gases. The modern atomistic theory of diffusion and *Brownian motion* was developed by *Albert Einstein, Marian Smoluchowski* and *Jean-Baptiste Perrin*. *Ludwig Boltzmann*, in the development of the atomistic backgrounds of the macroscopic transport processes, introduced the *Boltzmann equation*, which has served mathematics and physics with a source of transport process ideas and concerns for more than 140 years.

In 1920–1921 *George de Hevesy* measured *self-diffusion* using radio-isotopes. He studied self-diffusion of radioactive isotopes of lead in liquid and solid lead.

Yakov Frenkel (sometimes, Jakov/Jacov Frenkel) proposed, and elaborated in 1926, the idea of diffusion in crystals through local defects (vacancies and interstitial atoms). He concluded, the diffusion process in condensed matter is an ensemble of elementary jumps and quasi-chemical interactions of particles and defects. He introduced several mechanisms of diffusion and found rate constants from experimental data.

Some time later, *Carl Wagner* and *Walter H. Schottky* developed Frenkel's ideas about mechanisms of diffusion further. Presently, it is universally recognized that atomic defects are necessary to mediate diffusion in crystals.

Henry Eyring, with co-authors, applied his theory of *absolute reaction rates* to Frenkel's quasi-chemical model of diffusion. The analogy between *reaction kinetics* and diffusion leads to various non-linear versions of Fick's law.

Basic Models of Diffusion

Diffusion Flux

Each model of diffusion expresses the **diffusion flux** through concentrations, densities and their derivatives. Flux is a vector J. The transfer of a *physical quantity* N through a small *area* ΔS with normal v per time Δt is :

$$\Delta N = (J, v)\Delta S\Delta t + o(\Delta S\,\Delta t),$$

where (J, v) is the *inner product* and $o(\cdots)$ is the *little-o notation*. If we use the notation of *vector area* $\Delta S = v\Delta S$ then :

$$\Delta N = (J, \Delta S)\Delta t + t + o(\Delta S\Delta t).$$

The *dimension* of the diffusion flux is [flux]=[quantity]/([time] ·[area]). The diffusing physical quantity N may be the number of particles, mass, energy, electric charge, or any other scalar *extensive quantity*. For its density, n, the diffusion equation has the form :

$$\frac{\partial n}{\partial t} = -\nabla \cdot J + W,$$

where W is intensity of any local source of this quantity (the rate of a chemical reaction, for example). For the diffusion equation, the **no-flux boundary conditions** can be formulated as $(J(x), v(x)) = 0$ on the boundary, where v is the normal to the boundary at point x.

Fick's Law and Equations

Fick's first law : the diffusion flux is proportional to the negative of the concentration gradient :

$$J = -D\nabla n, \; J_i = -D\frac{\partial n}{\partial x_i}.$$

The corresponding diffusion equation (Fick's second law) is :

$$\frac{\partial n(x,t)}{\partial t} = \nabla \cdot (D\nabla n(x,t)) = D\Delta n(x,t),$$

where Δ is the *Laplace operator*,

$$\Delta n(x,\ t) = \sum_i \frac{\partial^2 n(x,t)}{\partial x_i^2}.$$

Onsager's Equations for Multi-component Diffusion and Thermodiffusion

Fick's law describes diffusion of an admixture in a medium. The concentration of this admixture should be small and the gradient of this concentration should be also small. The driving force of diffusion in Fick's law is the anti-gradient of concentration, $-\nabla n$.

In 1931, *Lars Onsager* included the multi-component transport processes in the general context of linear non-equilibrium thermodynamics. For multi-component transport,

$$J_i = \sum_j L_{ij} X_j,$$

where J_i is the flux of the *i*th physical quantity (component) and X_j is the *j*th *thermodynamic force*.

The thermodynamic forces for the transport processes were introduced by Onsager as the space gradients of the derivatives of the *entropy* density s (he used the term "force" in quotation marks or "driving force") :

$$X_i = \text{grad } \frac{\partial s(n)}{\partial n_i},$$

where n_i are the "thermodynamic co-ordinates". For the heat and mass transfer one can take $n_0 = u$ (the density of internal energy) and n_i is the concentration of the *i*th component. The corresponding driving forces are the space vectors :

$$X_0 = \text{grad}\frac{1}{T}, \ \ X_i = -\text{grad}\frac{\mu_i}{T}(i > 0), \ \text{because } ds = \frac{1}{T}du - \sum_{i\geq1}\frac{\mu_i}{T}dn_i$$

where T is the absolute temperature and μ_i is the chemical potential of the *i*th component. It should be stressed that the separate diffusion equations describe the mixing or mass transport without bulk motion. Therefore, the terms with variation of the total pressure are neglected. It is possible for diffusion of small admixtures and for small gradients.

For the linear Onsager equations, we must take the thermodynamic forces in the linear approximation near equilibrium :

$$X_i = \sum_{k\geq0} \frac{\partial^2 s(n)}{\partial n_i \partial n_k}\bigg|_{n=n^*} \text{grad} n_k,$$

where the derivatives of s are calculated at equilibrium n^*. The matrix of the *kinetic coefficients* L_{ij} should be symmetric (*Onsager reciprocal relations*) and *positive definite (for the entropy growth)*.

The transport equations are :

$$\frac{\partial n_i}{\partial t} = -\text{div} J_i = -\sum_{j\geq0} L_{ij} \text{div}\, X_j = \sum_{k\geq0}\left[-\sum_{j\geq0} L_{ij} \frac{\partial^2 s(n)}{\partial n_j \partial n_k}\bigg|_{n=n^*}\right]\Delta n_k.$$

Here, all the indexes i, j, $k=0,1,2,...$ are related to the internal energy (0) and various components. The expression in the square brackets is the matrix D_{ik} of the diffusion, thermodiffusion ($i>0$, $k=0$ or $k>0$, $i=0$) and *thermal conductivity* coefficients.

Under *isothermal conditions* T=const. The relevant thermodynamic potential is the free energy (or the *free entropy*). The thermodynamic driving forces for the isothermal diffusion are antigradients of chemical potentials, $-(1/T)\nabla\mu_j$, and the matrix of diffusion coefficients is :

$$D_{ik} = \frac{1}{T}\sum_{j\geq1} L_{ij} \frac{\partial\mu_j(n,T)}{\partial n_k}\bigg|_{n=n^*}.$$

There is intrinsic arbitrariness in the definition of the thermodynamic forces and kinetic coefficients because they are not measurable separately and only their combinations $\sum_j L_{ij} X_j$ can be measured. For example, in the original work of Onsager the thermodynamic forces include additional multiplier T, whereas in the *Course of Theoretical Physics* this multiplier is omitted but the sign of the thermodynamic forces is opposite. All these changes are supplemented by the corresponding changes in the coefficients and do not effect the measurable quantities.

Nondiagonal Diffusion Must be Non-linear

The formalism of linear irreversible thermodynamics (Onsager) generates the systems of linear diffusion equations in the form

$$\frac{\partial n_i}{\partial t} = \sum_j D_{ij}\Delta c_j.$$

If the matrix of diffusion coefficients is diagonal then this system of equations is just a collection of decoupled Fick's equations for various components. Assume that diffusion is non-diagonal, for example, $D_{12}\neq0$, and consider the state with $c_2 = \cdots = c_n = 0$. At this state, $\partial n_2/\partial t = D_{12}\Delta n_1$. If $D_{12}\Delta n_1(x) < 0$ at some points then $n_2(x)$ becomes negative at these points in a short time. Therefore, linear non-diagonal diffusion does not preserve positivity of concentrations. Non-diagonal equations of multi-component diffusion must be non-linear.

Einstein's Mobility and Teorell Formula

The *Einstein relation (kinetic theory)* connects the diffusion coefficient and the mobility (the ratio of the particle's terminal *drift velocity* to an applied *force*)

$$D = \mu\, k_B T$$

where D is the *diffusion constant*; μ is the "mobility"; k_B is *Boltzmann's constant*; T is the *absolute temperature*.

Below, to combine in the same formula the chemical potential μ and the mobility, we use for mobility the notation m.

The mobility – based approach was further applied by T. Teorell. In 1935, he studied the diffusion of ions through a membrane. He formulated the essence of his approach in the formula :

the flux is equal to mobility×concentration×force per gram ion.

This is the so-called *Teorell formula*.

The force under isothermal conditions consists of two parts :

1.   Diffusion force caused by concentration gradient : $-RT\dfrac{1}{n}\nabla n = -RT\nabla(\ln(n/n^{eq}))$.

2.   Electrostatic force caused by electric potential gradient : $q\nabla\varphi$.

Here R is the gas constant, T is the absolute temperature, n is the concentration, the equilibrium concentration is marked by a superscript "eq", q is the charge and φ is the electric potential.

The simple but crucial difference between the Teorell formula and the Onsager laws is the concentration factor in the Teorell expression for the flux. In the Einstein–Teorell approach, If for the finite force the concentration tends to zero then the flux also tends to zero, whereas the Onsager equations violate this simple and physically obvious rule.

The general formulation of the Teorell formula for non-perfect systems under isothermal conditions is :

$$J = m\exp\left(\frac{\mu - \mu_0}{RT}\right)(-\nabla\mu + (\text{external force per gram particle})),$$

where μ is the *chemical potential*, μ_0 is the standard value of the chemical potential. The expression $a = \exp\left(\dfrac{\mu - \mu_0}{RT}\right)$ is the so-called *activity*. It measures the "effective concentration" of a species in a non-ideal mixture. In this notation, the Teorell formula for the flux has a very simple form :

$$J = ma\,(-\nabla\mu + (\text{external force per gram particle})).$$

The standard derivation of the activity includes a normalization factor and for small concentrations $a = n/n^{\ominus} + o(n/n^{\ominus})$, where $n^{\ominus}$ is the standard concentration. Therefore, this formula for the flux describes the flux of the normalized dimensionless quantity, $n/n^{\ominus}$,

$$\frac{\partial(n/n^{\ominus})}{\partial t} = \nabla\cdot[ma(\nabla\mu - (\text{external force per gram particle}))].$$

Teorell Formula for Multi-component Diffusion

The Teorell formula with combination of Onsager's definition of the diffusion force gives :

$$J_i = m_i a_i \sum_j L_{ij} X_j,$$

where m_i is the mobility of the ith component, a_i is its activity, L_{ij} is the matrix of the coefficients, X_j is the thermodynamic diffusion force, $X_j = -\nabla \dfrac{\mu_j}{T}$. For the iso-

thermal perfect systems, $X_j = -R \dfrac{\nabla n_j}{n_j}$. Therefore, the Einstein-Teorell approach gives the following multi-component generalization of the Fick's law for multi-component diffusion :

$$\frac{\partial n_i}{\partial t} = \sum_j \nabla \cdot \left(D_{ij} \frac{n_i}{n_j} \nabla n_j \right).$$

where D_{ij} is the matrix of coefficients. The *Chapman-Enskog formulas for diffusion in gases* include exactly the same terms. Earlier, such terms were introduced in the *Maxwell–Stefan diffusion* equation.

Jumps on the Surface and in Solids

Diffusion of reagents on the surface of a *catalyst* may play an important role in heterogeneous catalysis. The model of diffusion in the ideal monolayer is based on the jumps of the reagents on the nearest free places. This model was used for CO on Pt oxidation under low gas pressure.

The system includes several reagents $A_1, A_2, \ldots A_m$ on the surface. Their surface concentrations are $c_1, c_2, \ldots c_m$. The surface is a lattice of the adsorption places. Each reagent molecule fills a place on the surface. Some of the places are free. The concentration of the free paces is $z = c_0$. The sum of all c_i (including free places) is constant, the density of adsorption places b.

The jump model gives for the diffusion flux of $A_i (i = 1,...,n)$:

$$J_i = -D_i[z \nabla c_i - c_i \nabla z].$$

The corresponding diffusion equation is :

$$\frac{\partial c_i}{\partial t} = -\text{div } J_i = D_i[z\Delta c_i - c_i\Delta z].$$

Due to the conservation law, $z = b - \sum_{i=1}^{n} c_i$, and we have the system of m diffusion equations. For one component we get Fick's law and linear equations because $(b - c)\nabla c - c\nabla(b - c) = b\nabla c$. For two and more components the equations are non-linear.

If all particles can exchange their positions with their closest neighbours then a simple generalization gives :

$$J_i = -\sum_j D_{ij}[c_j\nabla c_i - c_i\nabla c_j]$$

$$\frac{\partial c_i}{\partial t} = \sum_j D_{ij}[c_j\Delta c_i - c_i\Delta c_j]$$

where $D_{ij} = D_{ji} \geq 0$ is a symmetric matrix of coefficients which characterize the intensities of jumps. The free places (vacancies) should be considered as special "particles" with concentration c_0.

Various versions of these jump models are also suitable for simple diffusion mechanisms in solids.

Diffusion in Porous Media

For diffusion in porous media the basic equations are :

$$J = -D\nabla n^m$$

$$\frac{\partial n}{\partial t} = D\Delta n^m,$$

where D is the diffusion coefficient, n is the concentration, $m>0$ (usually $m>1$, the case $m=1$ corresponds to Fick's law).

For diffusion of gases in porous media this equation is the formalisation of *Darcy's law* : the velocity of a gas in the porous media is :

$$v = -\frac{k}{\mu}\nabla p$$

where k is the *permeability* of the medium, μ is the *viscosity* and p is the pressure. The flux $J=nv$ and for $p \sim n^\gamma$ Darcy's law gives the equation of diffusion in porous media with $m=\gamma+1$.

For underground water infiltration the *Boussinesq approximation* gives the same equation with $m=2$.

For plasma with the high level of radiation the *Zeldovich*-Raizer equation gives $m>4$ for the heat transfer.

Diffusion in Physics

Elementary Theory of Diffusion Coefficient in Gases

The diffusion coefficient D is the coefficient in the *Fick's first law* $J = -D\partial n/\partial x$, where J is the diffusion flux (*amount of substance*) per unit area per unit time, n (for ideal mixtures) is the concentration, x is the position [length].

Let us consider two gases with molecules of the same diameter d and mass m (*self-diffusion*). In this case, the elementary mean free path theory of diffusion gives for the diffusion coefficient :

$$D = \frac{1}{3}\ell v_T = \frac{2}{3}\sqrt{\frac{k_B^3}{\pi^3 m}}\frac{T^{3/2}}{Pd^2},$$

where k_B is the *Boltzmann constant*, T is the *temperature*, P is the *pressure*, ℓ is the *mean free path*, and v_T is the mean thermal speed :

$$\ell = \frac{k_B T}{\sqrt{2}\pi d^2 P}, v_T = \sqrt{\frac{8k_B T}{\pi m}}.$$

We can see that the diffusion coefficient in the mean free path approximation grows with T as $T^{3/2}$ and decreases with P as $1/P$. If we use for P the *ideal gas law* $P = RnT$ with the total concentration n, then we can see that for given concentration n the diffusion coefficient grows with T as $T^{1/2}$ and for given temperature it decreases with the total concentration as $1/n$.

For two different gases, A and B, with molecular masses m_A, m_B and molecular diameters d_A, d_B, the mean free path estimate of the diffusion coefficient of A in B and B in A is :

$$D_{AB} = \frac{1}{3}\sqrt{\frac{k_B^3}{\pi^3}}\sqrt{\frac{1}{2m_A} + \frac{1}{2m_B}}\frac{4T^{3/2}}{P(d_A + d_B)^2},$$

The Theory of Diffusion in Gases Based on Boltzmann's Equation

In Boltzmann's kinetics of the mixture of gases, each gas has its own distribution function, $f_i(x, c, t)$, where t is the time moment, x is position and c is velocity of molecule of the ith component of the mixture. Each component has its mean velocity $C_i(x, t) = \frac{1}{n_i}\int_c cf(x,c,t)dc$. If the velocities $C_i(x, t)$ do not concide then there exists *diffusion*.

In the *Chapman-Enskog* approximation, all the distribution functions are expressed through the densities of the conserved quantities :

- Individual concentrations of particles, $n_i(x, t) = \int_c f_i(x,c,t)dc$ (particles per volume),

- Density of moment $\sum_i m_i n_i C_i(x,t)$ (m_i is the ith particle mass),

- Density of kinetic energy $\sum_i\left(n_i\frac{m_i C_i^2(x,t)}{2} + \int_c \frac{m_i(c_i - C_i(x,t))^2}{2}f_i(x,c,t)dc\right)$.

The kinetic temperature T and pressure P are defined in 3D space as :

$$\frac{3}{2}k_B T = \frac{1}{n}\int_c \frac{m_i(c_i - C_i(x,t))^2}{2}f_i(x,c,t)dc ; P = k_B nT,$$

where $n = \sum_i n_i$ is the total density.

For two gases, the difference between velocities, $C_1 - C_2$ is given by the expression :

$$C_1 - C_2 = -\frac{n^2}{n_1 n_2} D_{12} \left\{ \nabla \left(\frac{n_1}{n} \right) + \frac{n_1 n_2 (m_2 - m_1)}{n(m_1 n_1 + m_2 n_2)} \nabla P - \frac{m_1 n_1 m_2 n_2}{P(m_1 n_1 + m_2 n_2)} (F_1 - F_2) + k_T \nabla T \right\},$$

where F_i is the force applied to the molecules of the ith component and k_T is the thermo-diffusion ratio.

The coefficient D_{12} is positive. This is the diffusion coefficient. Four terms in the formula for $C_1 - C_2$ describe four main effects in the diffusion of gases :

1. $\nabla \left(\dfrac{n_1}{n} \right)$ describes the flux of the first component from the areas with the high

   ratio n_1/n to the areas with lower values of this ratio (and, analogously the flux of the second component from high n_2/n to low n_2/n because $n_2/n = 1 - n_1/n$);

2. $\dfrac{n_1 n_2 (m_2 - m_1)}{n(m_1 n_1 + m_2 n_2)} \nabla P$ describes the flux of the heavier molecules to the areas with

   higher pressure and the lighter molecules to the areas with lower pressure, this is *barodiffusion*;

3. $\dfrac{m_1 n_1 m_2 n_2}{P(m_1 n_1 + m_2 n_2)} (F_1 - F_2)$ describes diffusion caused by the difference of the

   forces applied to molecules of different types. For example, in the Earth's gravitational field, the heavier molecules should go down, or in electric field the charged molecules should move, until this effect is not equilibrated by the sum of other terms. This effect should not be confused with barodiffusion caused by the pressure gradient.

4. $k_T \dfrac{1}{T} \nabla T$ describes *thermodiffusion*, the diffusion flux caused by the temperature

   gradient.
   All these effects are called *diffusion* because they describe the differences between velocities of different components in the mixture. Therefore, these effects cannot be described as a *bulk* transport and differ from advection or convection.
   In the first approximation,

- $D_{12} = \dfrac{3}{2n(d_1 + d_2)^2} \left[\dfrac{kT(m_1 + m_2)}{2\pi m_1 m_2} \right]^{1/2}$ for rigid spheres;

- $D_{12} = \dfrac{3}{8nA_1(v)\Gamma \left(3 - \dfrac{2}{v-1} \right)} \left[\dfrac{kT(m_1 + m_2)}{2\pi m_1 m_2} \right]^{1/2} \left(\dfrac{2kT}{\kappa_{12}} \right)^{\frac{2}{v-1}}$ for repulsing force $\kappa_{12} r^v$.

The number $A_1(v)$ is defined by quadratures (formulas (3.7), (3.9), Ch. 10 of the classical Chapman and Cowling book)

We can see that the dependence on T for the rigid spheres is the same as for the simple mean free path theory but for the power repulsion laws the exponent is different. Dependence on a total concentration n for a given temperature has always the same character, $1/n$.

In applications to gas dynamics, the diffusion flux and the bulk flow should be joined in one system of transport equations. The bulk flow describes the mass transfer. Its velocity V is the mass average velocity. It is defined through the momentum density and the mass concentrations :

$$V = \frac{\sum_i \rho_i C_i}{\rho}.$$

where $\rho_i = m_i n_i$ is the mass concentration of the ith species, $\rho = \sum_i \rho_i$ is the mass

density.

By definition, the diffusion velocity of the ith component is $v_i = C_i - V$, $\sum_i \rho_i v_i = 0$. The mass transfer of the ith component is described by the *continuity*

equation

$$\frac{\partial \rho_i}{\partial t} + \nabla(\rho_i V) + \nabla(\rho_i v_i) = W_i,$$

where W_i is the net mass production rate in chemical reactions, $\sum_i W_i = 0$.

In these equations, the term $\nabla(\rho_i V)$ describes advection of the ith component and the term $\nabla(\rho_i v_i)$ represents diffusion of this component.

In 1948, *Wendell H. Furry* proposed to use the *form* of the diffusion rates found in kinetic theory as a framework for the new phenomenological approach to diffusion in gases. This approach was developed further by F.A. Williams and S.H. Lam. For the diffusion velocities in multi-component gases (N components) they used :

$$v_i = -\left(\sum_{j=1}^{N} D_{ij} d_j + D_i^{(T)} \nabla(\ln T) \right);$$

$$d_j = \nabla X_j + (X_j - Y_j)\nabla(\ln P) + g_j;$$

$$g_j = \frac{\rho}{P}\left(Y_j \sum_{k=1}^{N} Y_k(f_k - f_j) \right).$$

Here, D_{ij} is the diffusion coefficient matrix, $D_i^{(T)}$ is the thermal diffusion coefficient, f_i is the body force per unite mass acting on the ith species, $X_i = P_i/P$ is the partial pressure fraction of the ith species (and P_i is the partial pressure), $Y_i = \rho_i/\rho$ is the mass fraction of the ith species, and $\sum_i X_i = \sum_i Y_i = 1.$.

Separation of Diffusion from Convection in Gases

While Brownian motion of multi-molecular mesoscopic particles (like pollen grains studied by Brown) is observable under an optical microscope, molecular diffusion can only be probed in carefully controlled experimental conditions. Since Graham experiments, it is well known that avoiding of convection is necessary and this may be a non-trivial task.

Under normal conditions, molecular diffusion dominates only on length scales between nanometer and millimeter. On larger length scales, transport in liquids and gases is normally due to another *transport phenomenon, convection,* and to study diffusion on the larger scale, special efforts are needed.

Therefore, some often cited examples of diffusion are *wrong* : If cologne is sprayed in one place, it will soon be smelled in the entire room, but a simple calculation shows that this can't be due to diffusion. Convective motion persists in the room because the temperature inhomogeneity. If ink is dropped in water, one usually observes an inhomogeneous evolution of the spatial distribution, which clearly indicates convection (caused, in particular, by this dropping).

In contrast, *heat conduction* through solid media is an everyday occurrence (*e.g.* a metal spoon partly immersed in a hot liquid). This explains why the diffusion of heat was explained mathematically before the diffusion of mass.

DOUBLE LAYER (INTERFACIAL)

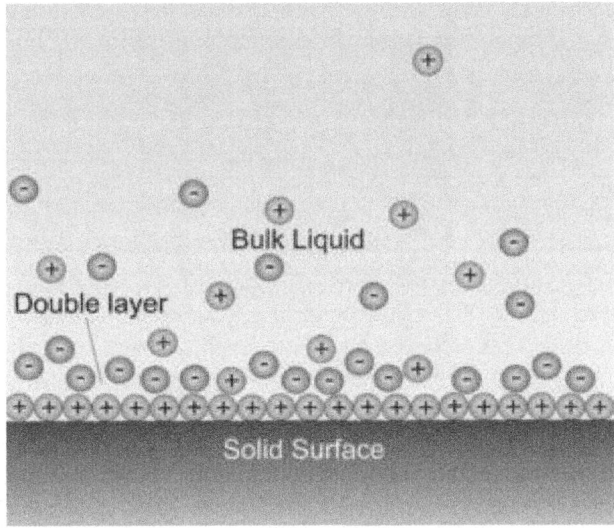

Fig. : Schematic of double layer in a liquid at contact with a negatively-charged solid. Depending on the nature of the solid, there may be another double layer (unmarked on the drawing) inside the solid.

A **double layer** (**DL**, also called an **electrical double layer, EDL**) is a structure that appears on the surface of an object when it is exposed to a fluid. The object might be a solid particle, a gas bubble, a liquid *droplet*, or a *porous body*. The DL refers to two parallel layers of charge surrounding the object. The first layer, the *surface charge* (either positive or negative), comprises ions *adsorbed* onto the object due to chemical interactions. The second layer is composed of ions attracted to the surface charge *via* the *coulomb force*, electrically *screening* the first layer. This second layer is loosely associated with the object. It is made of free ions that move in the fluid under the influence of *electric attraction* and *thermal motion* rather than being firmly anchored. It is thus called the "diffuse layer".

Interfacial DL is most apparent in systems with a large surface area to volume ratio, such as *colloid* or porous bodies with particles or pores (respectively) on the scale of micrometres to nanometres. However, DL is important to other phenomena, such as the *electro-chemical* behaviour of *electrodes*.

The DL plays a fundamental role in many everyday substances. For instance, milk exists only because fat droplets are covered with a DL that prevent their *coagulation* into butter. DLs exist in practically all *heterogeneous* fluid-based systems, such as blood, paint, ink and ceramic and cement *slurry*.

The DL is closely related to *electrokinetic phenomena* and *electroacoustic phenomena*.

Development of the Double Layer Model

Helmholtz

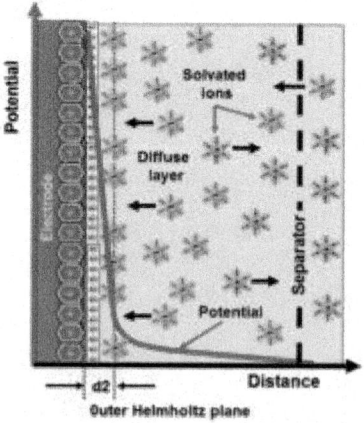

Fig.: Simplified illustration of the potential development in the area and in the further course of a Helmholtz double layer.

When an *electronic* conductor is brought in contact with a solid or liquid *ionic* conductor (electrolyte), a common boundary (*interface*) among the two *phases* appears. *Hermann von Helmholtz* was the first to realize that *charged* electrodes immersed in electrolytic solutions repel the *coions* of the charge while attracting counterions to their surfaces. Two layers of opposite *polarity* form at the interface between electrode and electrolyte. In 1853 he showed that an electrical double layer (DL), that is essentially a *molecular* dielectric, stored charge electrostatically. Below the electrolyte's decomposition voltage the stored charge is linearly dependent on the voltage applied.

This early model predicted a constant *differential capacitance* independent from the charge density depending on the *dielectric constant* of the electrolyte *solvent* and the thickness of the double-layer.

This model, while a good foundation for the description of the interface, does not consider important factors including diffusion/mixing of ions in solution, the possibility of *adsorption* onto the surface and the interaction between solvent *dipole moments* and the electrode.

Gouy-chapman

Louis Georges Gouy in 1910 and *David Leonard Chapman* in 1913 both observed that capacitance was not a constant and that it depended on the applied potential and the ionic concentration. The "Gouy-Chapman model" made significant improvements by introducing a *diffuse* model of the DL. In this model the charge distribution of ions as a function of distance from the metal surface allows *Maxwell–Boltzmann statistics* to be applied. Thus the *electric potential decreases exponentially* away from the surface of the fluid bulk.

Stern

Gouy-Chapman model fails for highly charged DLs. In 1924 *Otto Stern* suggested combining Helmholtz with Gouy-Chapman. In Stern's model, some ions adhere to the electrode as suggested by Helmholtz, giving an internal Stern layer, while some form a Gouy-Chapman diffuse layer.

The Stern layer accounted for ions' finite size and consequently ions' closest approach to the electrode is on the order of the ionic radius. The Stern model had its own limitations, effectively modelling ions as point charges, assuming all significant interactions in the *diffuse layer* are *Coulombic*, assuming *dielectric permittivity* to be constant throughout the double layer and that fluid viscosity is constant above the slipping plane.

Grahame

D. C. Grahame modified Stern in 1947. He proposed that some ionic or uncharged species can penetrate the Stern layer, although the closest approach to the electrode is normally occupied by solvent molecules. This could occur if ions lose their solvation shell as they approach the electrode. He called ions in direct contact with the electrode "specifically adsorbed ions". This model proposed the existence of three regions. The inner Helmholtz plane (IHP) plane passes through the centres of the specifically adsorbed ions. The outer Helmholtz plane (OHP) passes through the centres of solvated ions at the distance of their closest approach to the electrode. Finally the diffuse layer is the region beyond the OHP.

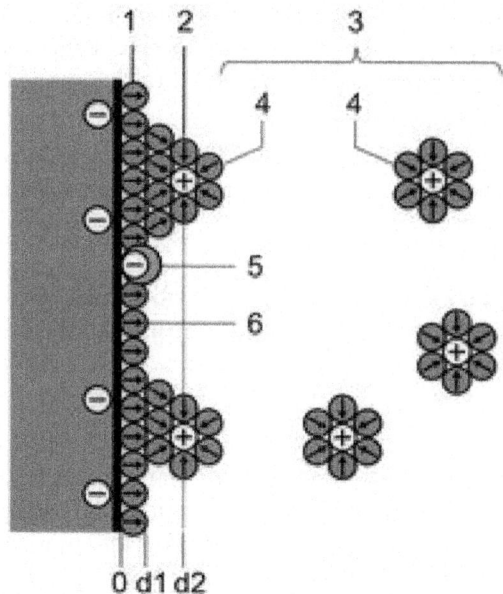

Fig. : Schematic representation of a double layer on an electrode (BMD) model. 1. Inner Helmholtz plane, (IHP), 2. Outer Helmholtz plane (OHP), 3. Diffuse layer, 4. Solvated ions (cations) 5. Specifically adsorbed ions (redox ion, which contributes to the pseudo-capacitance), 6. Molecules of the electrolyte solvent.

Bockris/Devanthan/Müller

In 1963 *J. O'M. Bockris*, M. A. V. Devanthan and *K. Alex Müller* proposed the BDM model of the double-layer that included the action of the solvent in the interface. They suggested that the attached molecules of the solvent, such as water, would have a fixed alignment to the electrode surface. This first layer of solvent molecules displays a strong orientation to the electric field depending on the charge. This orientation has great influence on the *permittivity* of the solvent that varies with field strength. The IHP passes through the centers of these molecules. Specifically adsorbed, partially solvated ions appear in this layer. The solvated ions of the electrolyte are outside the IHP. Through the centers of these ions pass the OHP. The diffuse layer is the region beyond the OHP. The BDM model now is most commonly used.

Trasatti/Buzzanca

Further research with double layers on ruthenium dioxide films in 1971 by Sergio Trasatti and Giovanni Buzzanca demonstrated that the electro-chemical behaviour of these electrodes at low voltages with specific adsorbed ions was like that of capacitors. The specific adsorption of the ions in this region of potential could also involve a partial charge transfer between the ion and the electrode. It was the first step towards understanding pseudo-capacitance.

Conway

Between 1975 and 1980 *Brian Evans Conway* conducted extensive fundamental and development work on *ruthenium oxide* electro-chemical capacitors. In 1991 he described the difference between 'Super-capacitor' and 'Battery' behaviour in electro-chemical energy storage. In 1999 he coined the term super-capacitor to explain the increased capacitance by surface redox reactions with faradaic charge transfer between electrodes and ions.

His "super-capacitor" stored electrical charge partially in the Helmholtz double-layer and partially as the result of faradaic reactions with "pseudo-capacitance" charge transfer of electrons and protons between electrode and electrolyte. The working mechanisms of pseudo-capacitors are redox reactions, intercalation and electrosorption.

Marcus

The physical and mathematical basics of electron charge transfer absent chemical bonds leading to pseudo-capacitance was developed by *Rudolph A. Marcus*. *Marcus Theory* explains the rates of electron transfer reactions — the rate at which an electron can move from one chemical species to another. It was originally formulated to address *outer sphere electron transfer* reactions, in which two chemical species change only in their charge, with an electron jumping. For redox reactions without making or breaking bonds, Marcus theory takes the place of *Henry Eyring*'s *transition state theory* which was derived for reactions with structural changes. Marcus received the *Nobel Prize in Chemistry* in 1992 for this theory.

Mathematical Description

There are detailed descriptions of the interfacial DL in many books on colloid and interface science and microscale fluid transport. There is also a recent IUPAC technical report on the subject of interfacial double layer and related *electrokinetic phenomena*.

As stated by Lyklema, "...the reason for the formation of a "relaxed" ("equilibrium") double layer is the non-electric affinity of charge-determining ions for a surface..." This process leads to the build up of an *electric surface charge*, expressed usually in C/m^2. This surface charge creates an electrostatic field that then affects the ions in the bulk of the liquid. This electrostatic field, in combination with the thermal motion of the ions, creates a counter charge, and thus screens the electric surface charge. The net electric charge in this screening diffuse layer is equal in magnitude to the net surface charge, but has the opposite polarity. As a result the complete structure is electrically neutral.

The diffuse layer, or at least part of it, can move under the influence of *tangential stress*. There is a conventionally introduced slipping plane that separates mobile fluid from fluid that remains attached to the surface. Electric potential at this plane is called *electrokinetic potential* or *zeta potential*. It is also denoted as ζ-potential.

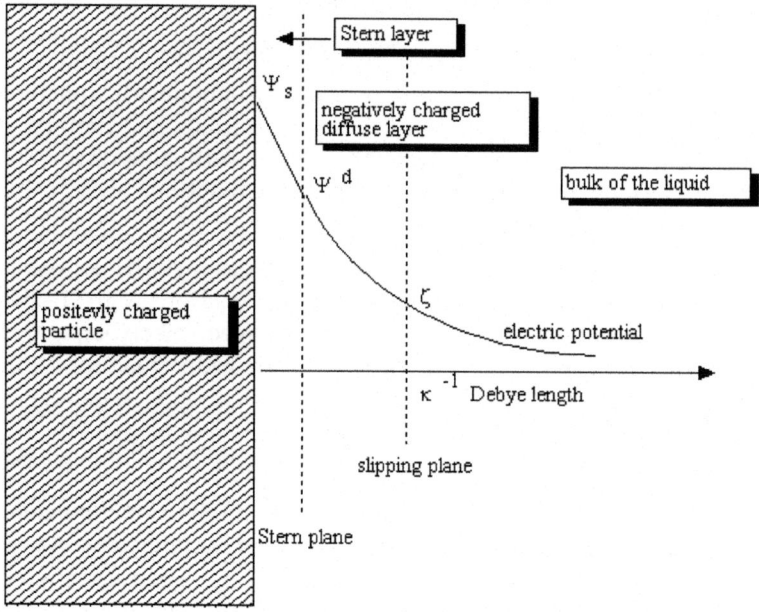

Fig. : Detailed illustration of interfacial DL.

The electric potential on the external boundary of the Stern layer *versus* the bulk electrolyte is referred to as *Stern potential*. Electric potential difference between the fluid bulk and the surface is called the electric surface potential.

Usually *zeta potential* is used for estimating the degree of DL charge. A characteristic value of this electric potential in the DL is 25 mV with a maximum value around 100 mV (up to several volts on electrodes). The chemical composition of the sample at which the ζ-potential is 0 is called the *point of zero charge* or the *iso-electric point*. It is usually determined by the solution pH value, since protons and hydroxyl ions are the charge-determining ions for most surfaces.

Zeta potential can be measured using *electrophoresis, electroacoustic phenomena, streaming potential,* and *electroosmotic flow*.

The characteristic thickness of the DL is the *Debye length*, κ^{-1}. It is reciprocally proportional to the square root of the ion concentration C. In aqueous solutions it is typically on the scale of a few nanometers and the thickness decreases with increasing concentration of the electrolyte.

The electric field strength inside the DL can be anywhere from zero to over 10^9 V/m. These steep electric potential gradients are the reason for the importance of the DLs.

The theory for a flat surface and a symmetrical electrolyte is usually referred to as the Gouy-Chapman theory. It yields a simple relationship between electric charge in the diffuse layer σ^d and the Stern potential Ψ^d :

$$\sigma^d = -\sqrt{8\varepsilon_0\varepsilon_m CRT} \sin h\frac{F\Psi^d}{2RT}$$

There is no general analytical solution for mixed electrolytes, curved surfaces or even spherical particles. There is an asymptotic solution for spherical particles with low charged DLs. In the case when electric potential over DL is less than 25 mV, the so-called Debye-Huckel approximation holds. It yields the following expression for electric potential Ψ in the spherical DL as a function of the distance r from the particle center :

$$\Psi(r) = \Psi^d\frac{a}{r}\exp(-\kappa(r-a))$$

There are several asymptotic models which play important roles in theoretical developments associated with the interfacial DL.

The first one is "thin DL". This model assumes that DL is much thinner than the colloidal particle or capillary radius. This restricts the value of the Debye length and particle radius as following :

$\kappa a \gg 1$

This model offers tremendous simplifications for many subsequent applications. Theory of *electrophoresis* is just one example. The theory of *electroacoustic phenomena* is another example.

The thin DL model is valid for most aqueous systems because the Debye length is only a few nanometers in such cases. It breaks down only for nano-colloids in solution with ionic strengths close to water.

The opposing "thick DL" model assumes that the Debye length is larger than particle radius :

$\kappa a < 1$

This model can be useful for some nano-colloids and non-polar fluids, where the Debye length is much larger.

The last model introduces "overlapped DLs". This is important in concentrated dispersions and emulsions when distances between particles become comparable with the Debye length.

Electrical Double Layers

The **electrical double layer (EDL)** is a structure which describes the variation of *electric potential* near a surface, and has a significant influence on the behaviour of *colloids* and other surfaces in contact with *solutions* or solid-state *fast ion conductors*.

The primary difference between a DL on an electrode and one on an interface is the mechanisms of *surface charge* formation. With an electrode, it is possible to regulate the surface charge by applying an external electric potential. This application, however, is impossible in colloidal and porous DLs, because for colloidal particles, one does not have access to the interior of the particle to apply a potential difference.

EDLs are analogous to the *double layer* in *plasma*.

Differential Capacitance

EDLs have an additional parameter defining their characterization : *differential capacitance*. Differential capacitance, denoted as C, is described by the equation below :

$$C = \frac{d\sigma}{d\Psi}$$

where σ is the *surface charge* and ψ is the *electric surface potential*.

Chapter 4

VOLTAMMETRIC TECHNIQUES

VOLTAMMETRY

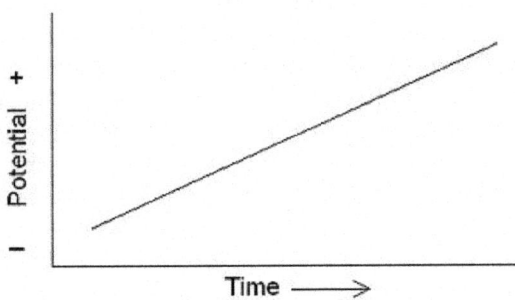

Fig. : Linear potential sweep.

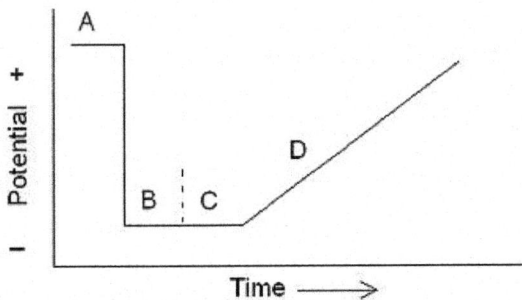

Fig. : Potential as a function of time for *anodic stripping voltammetry.*

Voltammetry is a category of *electro-analytical methods* used in *analytical chemistry* and various industrial processes. In voltammetry, information about an *analyte* is obtained by measuring the current as the potential is varied.

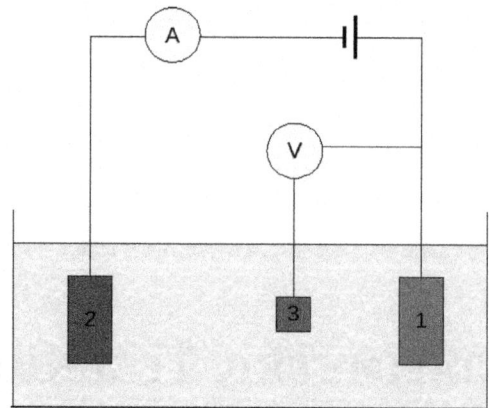

Fig. : Three-electrode setup : (1) working electrode; (2) auxiliary electrode; (3) reference electrode.

Three Electrodes System

Voltammetry experiments investigate the half cell reactivity of an *analyte*. Voltammetry is the study of current as a function of applied potential. These curves I = f(E) are called voltammograms. The potential is varied arbitrarily either step by step or continuously, and the actual current value is measured as the dependent variable. The opposite, *i.e.*, *amperometry*, is also possible but not common. The shape of the curves depends on the speed of potential variation (nature of driving force) and on whether the solution is stirred or quiescent (mass transfer). Most experiments control the *potential* (*volts*) of an electrode in contact with the analyte while measuring the resulting *current* (*amperes*).

To conduct such an experiment requires at least two electrodes. The *working electrode*, which makes contact with the analyte, must apply the desired potential in a controlled way and facilitate the transfer of charge to and from the analyte. A second electrode acts as the other half of the cell. This second electrode must have a known potential with which to gauge the potential of the working electrode, furthermore it must balance the charge added or removed by the working electrode. While this is a viable setup, it has a number of shortcomings. Most significantly, it is extremely difficult for an electrode to maintain a constant potential while passing current to counter redox events at the working electrode.

To solve this problem, the roles of supplying electrons and providing a reference potential are divided between two separate electrodes. The *reference electrode* is a half cell with a known reduction potential. Its only role is to act as reference in measuring and controlling the working electrodes potential and at no point does it pass any current. The *auxiliary electrode* passes all the current needed to balance the current observed at the working electrode. To achieve this current, the auxiliary will often swing to extreme potentials at the edges of the solvent window, where it oxidizes or reduces the solvent or supporting electrolyte. These electrodes, the *working*, *reference*, and *auxiliary* make up the modern three electrode system.

There are many systems which have more electrodes, but their design principles are generally the same as the three electrode system. For example, the *rotating ring-disk electrode* has two distinct and separate working electrodes, a disk and a ring, which can be used to scan or hold potentials independently of each other. Both of these electrodes are balanced by a single reference and auxiliary combination for an overall four electrode design. More complicated experiments may add working electrodes as required and at times reference or auxiliary electrodes.

In practice it can be very important to have a working electrode with known dimensions and surface characteristics. As a result, it is common to clean and polish working electrodes regularly. The auxiliary electrode can be almost anything as long as it doesn't react with the bulk of the analyte solution and conducts well. The reference is the most complex of the three electrodes; there are a variety of standards used and it is worth investigating elsewhere. For non-aqueous work, *IUPAC* recommends the use of the *ferrocene/ferrocenium* couple as an internal standard. In most voltammetry experiments, a bulk *electrolyte* (also known as a *supporting electrolyte*) is used to minimize solution resistance. It is possible to run an experiment without a bulk electrolyte, but the added resistance greatly reduces the accuracy of the results. With *room temperature ionic liquids*, the solvent can act as the electrolyte.

Theory

Data analysis requires the consideration of kinetics in addition to thermodynamics, due to the temporal component of voltammetry. Idealized theoretical electrochemical thermodynamic relationships such as the *Nernst equation* are modelled without a time component. While these models are insufficient alone to describe the dynamic aspects of voltammetry, models like the *Tafel equation* and *Butler-Volmer equation* lay the groundwork for the modified voltammetry relationships that relate theory to observed results.

History

The beginning of voltammetry was facilitated by the discovery of *polarography* in 1922 by the Nobel Prize winning chemist *Jaroslav Heyrovský*. Early voltammetric techniques had many problems, limiting their viability for everyday use in analytical chemistry. In 1942 Hickling built the first three electrodes potentiostat. The 1960s and 1970s saw many advances in the theory, instrumentation, and the introduction of computer added and controlled systems. These advancements improved sensitivity and created new analytical methods. Industry responded with the production of cheaper *potentiostat*, electrodes, and cells that could be effectively used in routine analytical work.

Applications

Voltammetric sensors A number of voltammetric systems are produced commercially for the determination of specific species that are of interest in industry and research. These devices are sometimes called *electrodes* but are, in fact, complete voltammetric cells and are better referred to as *sensors*. These sensors can be employed for the analysis of various organic and inorganic analytes in various matrices.

The oxygen electrode The determination of dissolved oxygen in a variety of aqueous environments, such as sea water, blood, sewage, effluents from chemical plants, and soils is of tremendous importance to industry, bio-medical and environmental research, and clinical medicine. One of the most common and convenient methods for making such measurements is with the *Clark oxygen sensor*, which was patented by *L.C. Clark, Jr.* in 1956.

TYPES OF VOLTAMMETRY

Linear Sweep Voltammetry

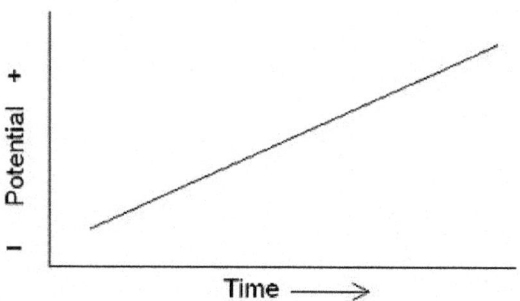

Fig. : Linear potential sweep.

Linear sweep voltammetry is a *voltammetric method* where the current at a *working electrode* is measured while the potential between the working electrode and a *reference electrode* is swept linearly in time. Oxidation or reduction of species is registered as a peak or trough in the current signal at the potential at which the species begins to be oxidized or reduced.

Experimental Method

The experimental setup for linear sweep voltammetry utilizes a potentiostat and a three-electrode setup to deliver a potential to a solution and monitor its change in current. The three-electrode setup consists of a working electrode, an auxiliary electrode, and a reference electrode. The potentiostat delivers the potentials through the three-electrode setup. A potential, E, is delivered through the working electrode. The slope of the potential *vs.* time graph is called the scan rate and can range from mV/s to 1,000,000 V/s. At higher scan rates the current is found to increase which improves the signal to noise ratio. Therefore higher scan rates lead to better signal to noise ratios.

The working electrode is where the oxidation/reduction reactions occur. The equation below gives an example of an oxidation occurring at the surface of the working electrode. ES is the standard reduction potential of A. As E approaches ES the current on the surface increases and when E=ES then the concentration of [A] = [A-] at the surface. As the molecules on the surface of the working electrode or oxidized/reduced they move away from the surface and new molecules come into contact with the surface of the working electrode. This flow of molecules to and from the working electrode causes the current.

$A + e^- = A^-$ E_s = 0.00V

Oxidation of molecule A at the surface of the working electrode

The auxiliary and reference electrode work in unison to balance out the charge added or removed by the working electrode. The auxiliary electrode balances the working electrode, but in order to know how much potential it has to add or remove it relies on the reference electrode. The reference electrode has a known reduction potential. The auxiliary electrode tries to keep the reference electrode at a certain reduction potential and to do this it has to balance the working electrode.

Characterization

Linear sweep voltammetry can identify unknown species and determine the concentration of solutions. E1/2 can be used to identify the unknown species while the height of the limiting current can determine the concentration. The sensitivity of current changes *vs.* voltage can be increased by increasing the scan rate. Higher potentials per second result in more oxidation/reduction of a species at the surface of the working electrode.

Variations

For reversible reactions cyclic voltammetry can be used to find information about the forward reaction and the reverse reaction. Like linear sweep voltammetry, cyclic voltammetry applies a linear potential over time and at a certain potential the potentiostat will reverse the potential applied and sweep back to the beginning point. Cyclic voltammetry provides information about the oxidation and reduction reactions.

Applications

While cyclic voltammetry is applicable to most cases where linear sweep voltammetry is used, there are some instances where linear sweep voltammetry is more useful. In cases where the reaction is irreversible cyclic voltammetry will not give any additional data that linear sweep voltammetry would give us. In one example, linear voltammetry was used to examine direct methane production *via* a biocathode. Since the production of methane from CO_2 is an irreversible reaction, cyclic voltammetry did not present any distinct advantage over linear sweep voltammetry. This group found that the biocathode produced higher current densities than a plain carbon cathode and that methane can be produced from a direct electrical current without the need of hydrogen gas.

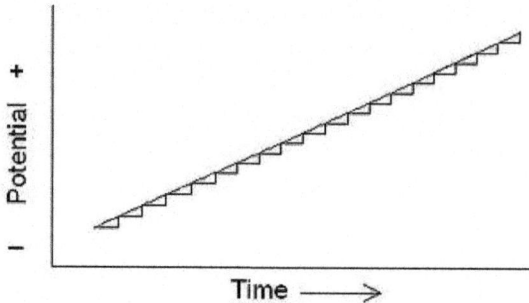

Fig. : Staircase potential sweep set against a linear potential sweep (blue).

Staircase Voltammetry

Staircase voltammetry is a derivative of *linear sweep voltammetry*. In linear sweep voltammetry the current at a *working electrode* is measured while the potential between the working electrode and a *reference electrode* is swept linearly in time. Oxidation or reduction of species is registered as a peak or trough in the current signal at the potential at which the species begins to be oxidized or reduced. In staircase voltammetry the potential sweep is a series of stair steps. The current is measured at the end of each potential change, right before the next, so that the contribution to the current signal from the capacitive charging current is reduced.

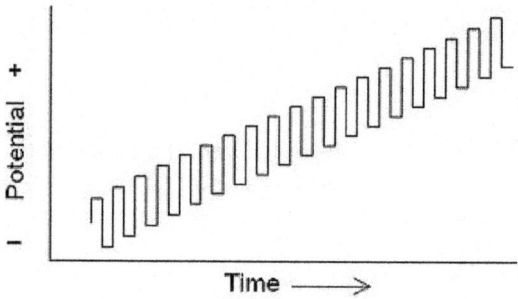

Fig. : Squarewave potential sweep.

Squarewave Voltammetry

Squarewave voltammetry (SWV) is a further improvement of *staircase voltammetry* which is itself a derivative of *linear sweep voltammetry*. In linear sweep voltammetry the current at a *working electrode* is measured while the potential between the working electrode and a *reference electrode* is swept linearly in time. In squarewave voltammetry, a *squarewave* is superimposed on the potential staircase sweep. Oxidation or reduction of species is registered as a peak or trough in the current signal at the potential at which the species begins to be oxidized or reduced. In staircase voltammetry the potential sweep is a series of stair steps. The current is measured at the end of each potential change, right before the next, so that the contribution to the current signal from the capacitive charging current is minimized. The *differential* current is then plotted as a function of potential, and the reduction or oxidation of species is measured as a peak or trough. Due to the lesser contribution of capacitative charging current the detection limits for SWV are on the order of nanomolar concentrations.

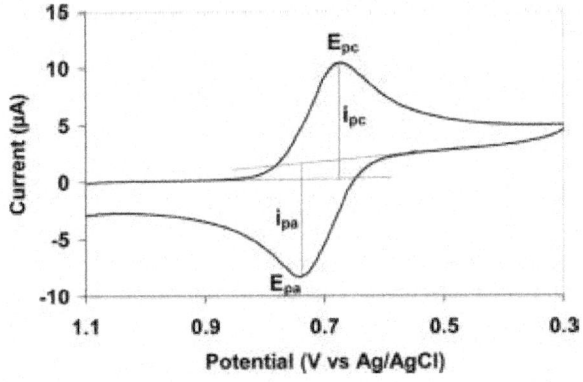

Fig. : Typical cyclic voltammogram where i_{pc} and i_{pa} show the peak cathodic and anodic current respectively for a reversible reaction.

Cyclic Voltammetry

Cyclic voltammetry or CV is a type of *potentiodynamic electro-chemical* measurement. In a cyclic voltammetry experiment the working electrode potential is ramped linearly versus time like *linear sweep voltammetry*. Cyclic voltammetry takes the experiment a step further than linear sweep voltammetry which ends when it reaches a set potential. When cyclic voltammetry reaches a set potential, the working electrode's potential ramp is inverted. This inversion can happen multiple times during a single experiment. The current at the *working electrode* is plotted *versus* the applied voltage to give the cyclic voltammogram trace. Cyclic voltammetry is generally used to study the electro-chemical properties of an *analyte* in solution.

Experimental Method

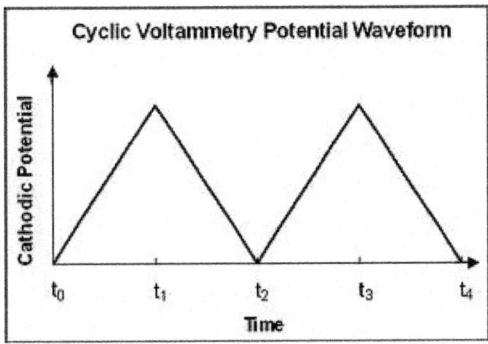

Fig. : Cyclic voltammetry waveform.

In cyclic voltammetry, the electrode potential ramps linearly versus time as shown. This ramping is known as the experiment's scan rate (V/s). The potential is applied between the reference electrode and the working electrode and the current is measured between the working electrode and the counter electrode. These data are then plotted as current (*i*) *vs.* potential (*E*). As the waveform shows, the forward scan produces a current peak for any analytes that can be reduced (or oxidized depending on the initial scan direction) through the range of the potential scanned. The current will increase as the potential reaches the reduction potential of the analyte, but then falls off as the concentration of the analyte is depleted close to the electrode surface. If the *redox couple* is reversible then when the applied potential is reversed, it will reach the potential that will reoxidize the product formed in the first reduction reaction, and produce a current of reverse polarity from the forward scan. This oxidation peak will usually have a similar shape to the reduction peak. As a result, information about the redox potential and electro-chemical reaction rates of the compounds are obtained.

For instance if the electronic transfer at the surface is fast and the current is limited by the *diffusion* of species to the electrode surface, then the *current* peak will be proportional to the *square root* of the scan rate. This relationship is described by the *Cottrell equation*. The CV experiment then samples only a small portion of the solution, the material within the *diffusion layer*.

Characterization

The utility of cyclic voltammetry is highly dependent on the analyte being studied. The analyte has to be redox active within the experimental potential window. It is also highly desirable for the analyte to display a reversible wave. A reversible wave is when an analyte is reduced or oxidized on a forward scan and is then reoxidized or rereduced in a predictable way on the return scan.

Even reversible couples contain polarization *overpotential* and thus display a hysteresis between absolute potential between the reduction (E_{pc}) and oxidation peak (E_{pa}). This overpotential emerges from a combination of analyte diffusion rates and the intrinsic activation barrier of transferring electrons from an electrode to analyte. A theoretical description of polarization overpotential is in part described by the *Butler-Volmer equation* and *Cottrell equation*. Conveniently in an ideal system the relationships reduces to, $|E_{pc} - E_{pa}| = \dfrac{57\text{ mV}}{n}$, for an *n* electron process.

Reversible couples will display ratio of the peak currents passed at reduction (i_{pc}) and oxidation (i_{pa}) that is near unity ($1 = i_{pa}/i_{pc}$). This ratio can be perturbed for reversible couples in the presence of a *following chemical reaction*, stripping wave, or nucleation event.

When such reversible peaks are observed *thermodynamic information* in the form of half cell potential $E^0_{1/2}$ can be determined. When waves are semi-reversible such as when i_{pa}/i_{pc} is less than or greater than 1, it can be possible to determine even more information especially kinetic processes like following *chemical reaction*.

When waves are non-reversible it is impossible to determine what their thermodynamic $E^0_{1/2}$ is with cyclic voltammetry. This $E^0_{1/2}$ can be determined, however it often requires equal quantities of the analyte in both oxidation states. When a wave is non-reversible cyclic voltammetry can not determine if the wave is at its thermodynamic potential or shifted to a more extreme potential by some form of *overpotential*. The couple could be irreversible because of a following chemical process, a common example for *transition metals* is a shift in the geometry of the *co-ordination* sphere. If this is the case, then higher scan rates may show a reversible wave. It is also possible that the wave is irreversible due to a physical process most commonly some form of *precipitation* as discussed below. Some speculation can be made in regards to irreversible waves however they are generally outside the scope of cyclic voltammetry.

Experimental Setup

The method uses *reference electrode, working electrode,* and *counter electrode* which in combination are sometimes referred to as a *three-electrode setup. Electrolyte* is usually added to the test solution to ensure sufficient conductivity. The combination of the solvent, electrolyte and specific working electrode material determines the range of the potential.

Electrodes are static and sit in unstirred solutions during cyclic voltammetry. This "still" solution method results in cyclic voltammetry's characteristic diffusion controlled peaks. This method also allows a portion of the *analyte* to remain after reduction or oxidation where it may display further redox activity. Stirring the solution between cyclic voltammetry traces is important as to supply the electrode surface with fresh analyte for each new experiment. The solubility of an analyte can change drastically with its overall charge. Since cyclic voltammetry usually alters the charge of the analyte it is common for reduced or oxidized analyte to *precipitate* out onto the electrode. This layering of analyte can insulate the electrode surface, display its own redox activity in subsequent scans, or at the very least alter the electrode surface. For this and other reasons it is often necessary to clean electrodes between scans.

Common materials for *working electrodes* include *glassy carbon, platinum,* and *gold.* These electrodes are generally encased in a rod of inert insulator with a disk exposed at one end. A regular working electrode has a radius within an order of magnitude of 1 mm. Having a controlled surface area with a defined shape is important for interpreting cyclic voltammetry results.

To run cyclic voltammetry experiments at high scan rates a regular working electrode is insufficient. High scan rates create peaks with large currents and increased resistances which result in distortions. *Ultra-microelectrodes* can be used to minimize the current and resistance.

The *counter electrode*, also known as the auxiliary or second electrode, can be any material which conducts easily and won't react with the bulk solution. Reactions occurring at the counter electrode surface are unimportant as long as it continues to conduct current well. To maintain the observed current the counter electrode will often oxidize or reduce the solvent or bulk electrolyte.

Reference electrodes are a complex subject and worth investigating elsewhere.

Variations

In some experiments an electro-active species is fixed to the surface of the electrode, for instance in microparticle voltammetry.

Potentiodynamic techniques also exist that add low-amplitude ac perturbation to a potential ramp and measure variable response in a single frequency (ac voltammetry) or in many frequencies simultaneously (*potentiodynamic electro-chemical impedance spectroscopy*). The response in alternating current is two-dimensional–it is characterised by *amplitude* and *phase*. The amplitude and phase depend differently on frequency for constituents of ac response attributed to different processes (charge transfer, diffusion, double layer charging, etc.). *Frequency response* analysis enables simultaneous monitoring of the various processes that contribute to the potentiodynamic ac response of electro-chemical system.

Distinctions

Cyclic voltammetry is not a *hydrodynamic technique*. In a hydrodynamic technique flow is achieved at the electrode surface by stirring the solution, pumping the solution, or rotating the electrode as is the case with *rotating disk electrodes* and *rotating ring-disk electrodes*. These techniques target steady state conditions which appear the same scanned from the positive or the negative, thus limiting them to *linear sweep voltammetry*.

Applications

Cyclic voltammetry (CV) has become an important and widely used electro-analytical technique in many areas of chemistry. It is widely used to study a variety of redox processes, for obtaining stability of reaction products, the presence of intermediates in oxidation-reduction reactions, reaction and electron transfer kinetics, and the reversibility of a reaction. CV can also be used to determine the electron stoichiometry of a system, the diffusion coefficient of an analyte, and the formal reduction potential, which can be used as an identification tool. In addition, because concentration is proportional to current in a reversible, Nernstian system, concentration of an unknown solution can be determined by generating a calibration curve of current *vs.* concentration.

This latter application is gaining interest in the field of cellular biology where it is used to measure concentration of various chemicals in the cells of organisms, including living ones.

Electro-chemical Cell Set-up

Anodic stripping voltammetry usually incorporates three electrodes, a *working electrode*, *auxiliary electrode* (sometimes called the counter electrode), and *reference electrode*. The *solution* being analyzed usually has an *electrolyte* added to it. For most standard tests, the working electrode is a bismuth or *mercury* film electrode (in a disk or planar strip configuration). The mercury film forms an *amalgam* with the analyte of interest, which upon oxidation results in a sharp peak, improving resolution between analytes. The mercury film is formed over a *glassy carbon* electrode. A mercury drop electrode has also been used for much the same reasons. In cases where the analyte of interest has an oxidizing potential above that of mercury, or where a mercury electrode would be otherwise unsuitable, a solid, inert metal such as *silver, gold,* or *platinum* may also be used.

Steps

Anodic stripping voltammetry usually incorporates 4 steps if the working electrode is a mercury film or mercury drop electrode and the solution incorporates stirring. The solution is stirred during the first two steps at a repeatable rate. The first step is a cleaning step; in the cleaning step, the potential is held at a more oxidizing potential than the analyte of interest for a period of time in order to fully remove it from the electrode. In the second step, the potential is held at a lower potential, low enough to reduce the analyte and deposit it on the electrode. After the second step, the stirring is stopped, and the electrode is kept at the lower potential. The purpose of this third step is to allow the deposited material to distribute more evenly in the mercury. If a solid inert electrode is used, this step is unnecessary. The last step involves raising the working electrode to a higher potential (anodic), and stripping (oxidizing) the analyte. As the analyte is oxidized, it gives off electrons which are measured as a current.

Stripping analysis is an analytical technique that involves (i) preconcentration of a metal phase onto a solid electrode surface or into Hg (liquid) at negative potentials and (ii) selective oxidation of each metal phase species during an anodic potential sweep.

Stripping analysis has the following properties :

- Very sensitive and reproducible (RSD<5%) method for trace metal ion analysis in aqueous media.
- Concentration limits of detection for many metals are in the low ppb to high ppt range and this compares favourably with AAS or ICP analysis.
- Field deployable instrumentation that is inexpensive.
- Approximately 12-15 metal ions can be analyzed for by this method.
- The stripping peak currents and peak widths are a function of the size, coverage and distribution of the metal phase on the electrode surface (Hg or alternate)

Sensitivity

Anodic stripping voltammetry can detect µg/L concentrations of analyte. This method has an excellent detection limit (typically 10^{-9}–10^{-10} M)

Applications

Stripping voltammetry (Anodic, Cathodic and Adsorptive) have been employed for analysis of organic molecules as well as metal ions. Carbon paste, glassy carbon paste, glassy carbon etc. electrodes when modified are termed as chemically modified electrodes. Chemically modified electrodes have been employed for the analysis of organic molecules (*viz.*, Paracetamol, aspirin, caffeine, phenol, catechol, resorcinol, hydroquinone, dopamine, L-dopa, epinephrine, nor epinephrine, methyl parathion, ethyl parathion, venlafaxine, desvenlafaxine, imipramine, trimipramine, desipramine etc.) as well as metal ions (bismuth, antimony etc.).

Cathodic Stripping Voltammetry

Cathodic stripping voltammetry is a voltammetric method for *quantitative* determination of specific ionic species. It is similar to the *trace* analysis method *anodic stripping voltammetry*, except that for the plating step, the potential is held at an oxidizing potential, and the *oxidized* species are stripped from the electrode by sweeping the potential positively. This technique is used for ionic species that form *insoluble salts* and will deposit on or near the *anodic*, working electrode during deposition. The stripping step can be either *linear, staircase, squarewave*, or pulse.

Adsorptive Stripping Voltammetry

Adsorptive stripping voltammetry is similar to *anodic stripping voltammetry* and *cathodic stripping voltammetry* except that the preconcentration step is not controlled by *electrolysis*. The preconcentration step in adsorptive stripping voltammetry is accomplished by *adsorption* on the *working electrode* surface, or by reactions with chemically modified electrodes.

Polarography

Polarography is a sub-class of *voltammetry* where the *working electrode* is a *dropping mercury electrode* (DME) or a static mercury drop electrode (SMDE), which are useful for their wide *cathodic ranges* and renewable surfaces. It was invented by *Jaroslav Heyrovský*, for which he won the Nobel prize in 1959.

Theory of Operation

Polarography is a voltammetric measurement whose response is determined by combined diffusion/convection mass transport. Simple principle of polarography is the study of solutions or of electrode processes by means of *electrolysis* with two *electrodes*, one polarizable and one unpolarizable, the former formed by mercury

regularly dropping from a *capillary tube*. Polarography is a specific type of measurement that falls into the general category of linear-sweep voltammetry where
the electrode potential is altered in a linear fashion from the initial potential to
the final potential. As a linear sweep method controlled by convection/diffusion
mass transport, the current *vs.* potential response of a polarographic experiment
has the typical *sigmoidal shape*. What makes polarography different from other
linear sweep voltammetry measurements is that polarography makes use of the
dropping mercury electrode (DME) or the static mercury drop electrode.

Fig. : Heyrovský's Polarograph.

A plot of the current *vs.* potential in a polarography experiment shows the
current oscillations corresponding to the drops of Hg falling from the capillary.
If one connected the maximum current of each drop, a sigmoidal shape would
result. The limiting current (the plateau on the sigmoid), called the diffusion current because diffusion is the principal contribution to the flux of electro-active
material at this point of the Hg drop life.

Limitations

There are various limitations in particular for the classical polarography experiment for quantitative analytical measurements. Because the current is continuously
measured during the growth of the Hg drop, there is a substantial contribution
from capacitive current. As the Hg flows from the capillary end, there is initially
a large increase in the surface area. As a consequence, the initial current is dominated by capacitive effects as charging of the rapidly increasing interface occurs.
Toward the end of the drop life, there is little change in the surface area which
diminishes the contribution of capacitance changes to the total current. At the same
time, any redox process which occurs will result in faradaic current that decays

approximately as the square root of time (due to the increasing dimensions of the Nernst diffusion layer). The exponential decay of the capacitive current is much more rapid than the decay of the faradaic current; hence, the faradaic current is proportionally larger at the end of the drop life. Unfortunately, this process is complicated by the continuously changing potential that is applied to the *working electrode* (the Hg drop) throughout the experiment. Because the potential is changing during the drop lifetime (assuming typical experimental parameters of a 2 mV/s scan rate and a 4 s drop time, the potential can change by 8 mV from the beginning to the end of the drop), the charging of the interface (capacitive current) has a continuous contribution to the total current, even at the end of the drop when the surface area is not rapidly changing. As such, the typical signal to noise of a polarographic experiment allows detection limits of only approximately 10^{-5} or 10^{-6} M.

Fig. : Heyrovský's Polarograph and DME.

Improvements

Dramatically better discrimination against the capacitive current can be obtained using the tast and pulse polarographic techniques. These have been developed with introduction of analog and digital electronic potentiostats. A first major improvement is obtained, if the current is only measured at the end of each drop lifetime (tast polarography). An even greater enhancement has been the introduction of differential pulse polarography. Here, the current is measured before the beginning and before the end of short potential pulses. The latter are superimposed to the linear potential-time-function of the voltammetric scan. Typical amplitudes of these pulses range between 10 and 50 mV, whereas pulse duration is 20 to 50 ms. The difference between both current values is that taken as the analytical signal.

This technique results in a 100 to 1000-fold improvement of the detection limit, because the capacitive component is effectively suppressed.

Qualitative Information

Qualitative information can also be determined from the half-wave potential of the polarogram (the current *vs.* potential plot in a polarographic experiment). The value of the half-wave potential is related to the standard potential for the redox reaction being studied.

This technique and especially the DPASV one can be used for environmental analysis, and especially for marine study for characterisation of Organic matter and metals interactions.

Rotated Electrode Voltammetry

Rotated electrode voltammetry, a sub-class of *voltammetry*, is technique in which the working *electrode* usually a *rotating disk electrode* (RDE) or *rotating ring-disk electrode* (RRDE) is rotated at a very high rate (2000 to 10,000 (?) *RPM*). This asserts some control over the *mass transport* process which brings the *analyte* to the electrode surface. This technique was made popular through the work of the Russian electro-chemist, *Benjamin Levich*. This technique is useful for studying the *kinetics* and *electro-chemical reaction mechanism* for a *half reaction*. This technique is part of a larger class of electro-analytical methods known as hydrodynamic voltammetry.

Differential Pulse Voltammetry

Differential Pulse *Voltammetry*, DPV, (or Differential Pulse *Polarography*, DPP) is often used to make *electro-chemical* measurements. It can be considered as a derivative of *linear sweep voltammetry* or *staircase voltammetry*, with a series of regular voltage pulses superimposed on the potential linear sweep or stairsteps. The current is measured immediately before each potential change, and the current difference is plotted as a function of potential. By sampling the current just before the potential is changed, the effect of the charging current can be decreased.

By contrast, in *normal pulse voltammetry* the current resulting from a series of ever larger potential pulses is compared with the current at a constant 'baseline' voltage. Another type of pulse voltammetry is *squarewave voltammetry*, which can be considered a special type of differential pulse voltammetry in which equal time is spent at the potential of the ramped baseline and potential of the superimposed pulse.

Electro-chemical Cell Set-up

The system of this measurement is usually the same as that of standard *voltammetry*. The *potential* between the *working electrode* and the *reference electrode* is changed as a *pulse* from an initial potential to an inter-level potential and remains at the inter-level potential for about 5 to 100 milliseconds; then it changes to the final potential, which is different from the initial potential. The pulse is repeated, changing the final potential, and a constant difference is kept between the initial and the inter-level potential. The value of the current between the working electrode and *auxiliary electrode* before and after the pulse are sampled and their differences are plotted versus potential

Uses

These measurements can be used to study the *redox* properties of extremely small amounts of chemicals because of the following two features :

1. In these measurements, the effect of the charging current can be minimized, so high *sensitivity* is achieved.
2. *Faradaic current* is extracted, so electrode reactions can be analyzed more precisely.

Characteristics

Differential pulse voltammetry has these characteristics :

1. Reversible reactions show *symmetrical* peaks, and irreversible reactions show asymmetrical peaks.
2. The peak potential is equal to $E_{1/2}{}^{r}-\Delta E$ in reversible reactions, and the peak *current* is proportional to the concentration.
3. The detection limit is about 10^{-8} M.

Chronoamperometry

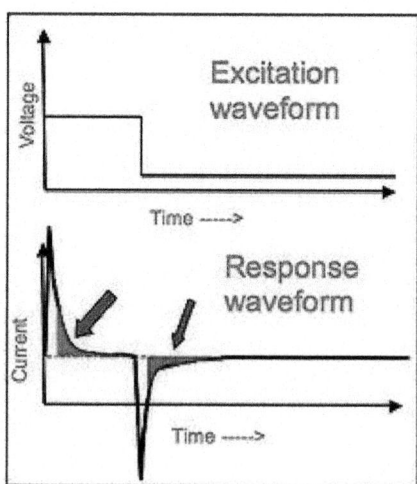

Fig. : Double-pulsed chronoamperometry waveform showing integrated region for charge determination.

Chronoamperometry is an electro-chemical technique in which the potential of the *working electrode* is stepped and the resulting current from faradic processes occurring at the electrode (caused by the potential step) is monitored as a function of time. Limited information about the identity of the electrolyzed species can be obtained from the ratio of the peak oxidation current *versus* the peak reduction current. However, as with all pulsed techniques, chronoamperometry generates high charging currents, which decay exponentially with time as any RC circuit. The Faradaic current–which is due to electron transfer events and is most often the current component of interest–decays as described in the *Cottrell equation*. In most electro-chemical cells this decay is much slower than the charging decay–cells with no supporting electrolyte are notable exceptions. Most commonly investigated with a *three electrode system*. Since the current is integrated over relatively longer time intervals, chronoamperometry gives a better signal to noise ratio in comparison to other amperometric technique.

Example

Anthracene in deoxygenated *dimethylformamide* (DMF) will be reduced (An + e⁻-> An⁻) at the electrode surface that is at a certain negative *potential*. The *reduction* will be *diffusion-limited,* thereby causing the *current* to drop in time (proportional to the diffusion gradient that is formed by diffusion).

You can do this experiment several times increasing electrode potentials from low to high. (In between the experiments, the solution should be stirred.) When you measure the current i(t) at a certain fixed time point τ after applying the voltage, you will see that at a certain moment the current i(τ) does not rise anymore; you have reached the mass-transfer-limited region. This means that anthracene arrives as fast as diffusion can bring it to the electrode.

Neopolarogram

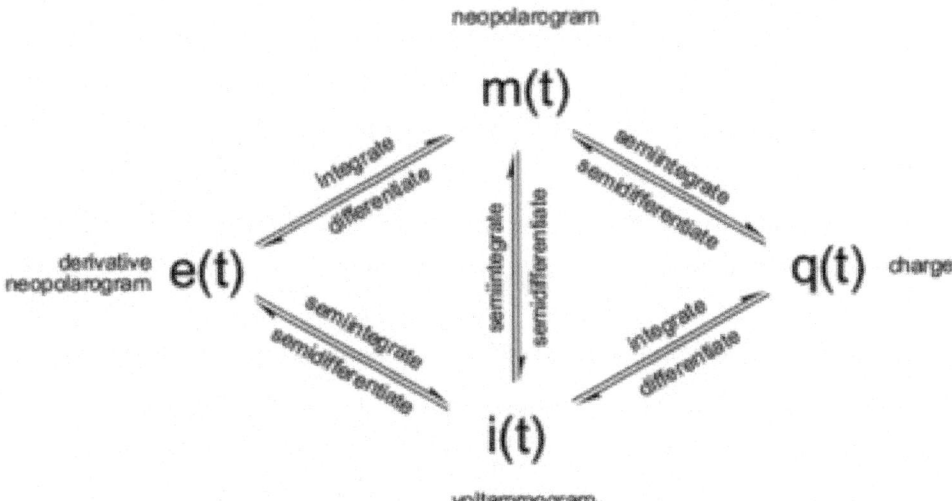

Fig. : Mathematical relation of current, charge and fractional derivatives.

The term **neopolarogram** refers to mathematical derivatives of *polarograms* or *cyclic voltammograms* that in effect deconvolute diffusion and electro-chemical kinetics. This is achieved by analog or digital implementations of *fractional calculus*. The implementation of fractional derivative calculations by means of numerical methods is straight forward. The G1-(*Grünwald–Letnikov derivative*) and the RL0-algorithms (*Riemann–Liouville integral*) are recursive methods to implement a numerical calculation of fractional differ integrals. Yet *differ integrals* are faster to compute in discrete fourier space using *FFT*.

Applications

The graphs below show the behaviour of fractional derivatives calculated by different algorithms for *ferrocene* in *acetonitrile* at 100mV/s, the *reference electrode* is 0.1M Ag+/Ag in acetonitrile (+0.04V *vs.* Fc).

1st Derivative of the "Semi-derivative" or 1.5th Order Derivative in Voltammetry

1.5th order derivative of a voltammogram hits the abscissa exactly at the point where the formal potential of the electrode reaction is found.

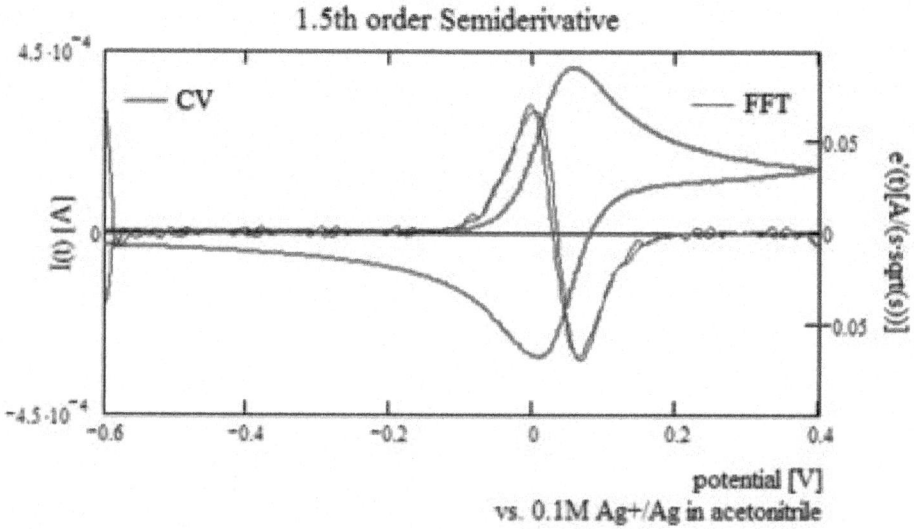

Fig. : Typical 1.5th order semi-derivative for a reversible reaction, *Ferrocene* has a formal potential of 40mV *vs.* ATE1.

"Semi-derivative" or Numerical Grünberg-Letnikov Derivative in Voltammetry

The G1 algorithm produces a numerical derivative that has the shape of a *bell curve*, this derivative obeys to certain laws, for example the G1 derivative of a cyclic voltammogram is mirrored at the *abscissa* as long as the electro-chemical reaction is diffusion controlled, the planar diffusion approximation can be applied to the electrode geometry and ohmic drop distortion is minimal. The *FWHM* of the curve is approximately 100 mV for a system that behaves in the described manor. The maximum is found at the value of the formal potential, this is quivalent to the 1.5th order semi-derivative hitting the abszissa at this potential. Moreover the semi-derivative scales linearly with the scanrate, while the current scales linearly with the square root of the scanrate (Randles-Sevcik equation). Plotting the semi-derivatives produced at different scanrates gives a *family of curves* that are linearly related by the scanrate quotient in an ideal system.

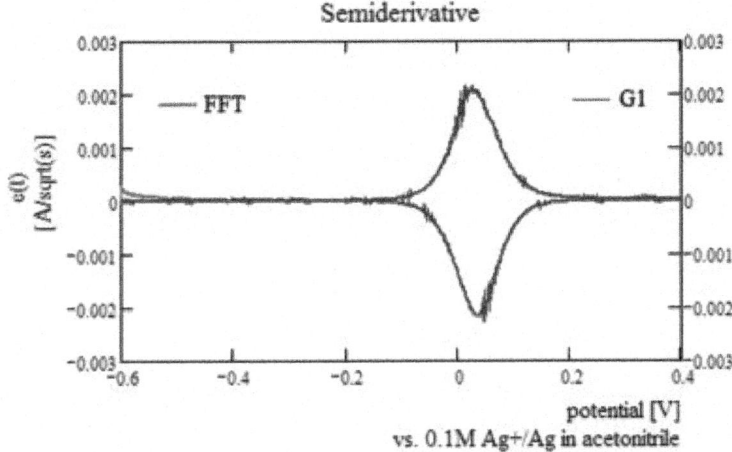

Fig. : Typical semi-derivative for a reversible reaction, recursive algorithms and FFT methods yield equivalent results.

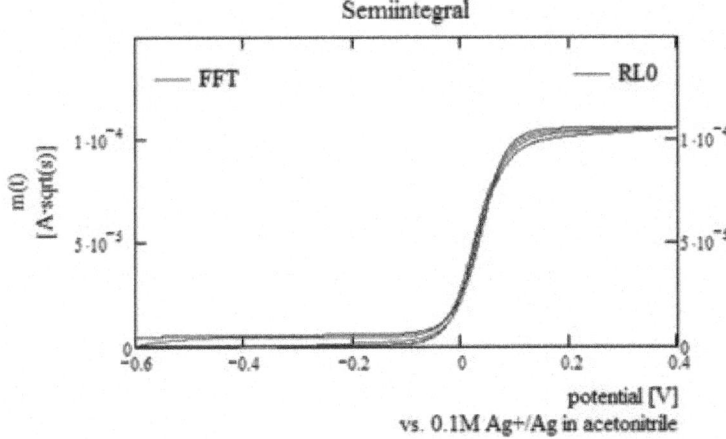

Fig. : Typical semi-integral for a reversible reaction, recursive algorithms and FFT methods yield slightly different results due to non-perfect periodicity of cyclic voltammetry data.

"Semi-integral" or Numerical Riemann-Liouville Integral in Voltammetry

The shape of the semi-integral can be used as an easy method to measure the amount of *ohmic drop* of an electro-chemical cell in *cyclic voltammetry*. Essentially the semi-integral of a cyclic voltammogram at a planar electrode (an electrode that obeys to the rules of planar diffusion) has the shape of a *sigmoid* while the original data is gauss-sigmoid convoluted. This enables the operator to optimize parameters necessary for *positive feedback compensation* in an easy manor. If ohmic drop distortion is present the two sigmoids for the forward and the backward scan are far away from congruence, the ohmic drop can be calculated from the deviation from congruence in these cases. In the example shown slight distortion is present, yet this does not have adverse effects on data quality.

Merits of FFT Techniques

The implementation differ-integral calculation using fast fourier transform has certain benefits because it is easily combined with low pass quadratic filtering methods. This is very useful when cyclic voltammograms are recorded in high resistivity solvents like *tetrahydrofuran* or *toluene*, where feedback oscillations are a frequent problem.

ELECTRO-CATALYST

An **electro-catalyst** is a *catalyst* that participates in *electro-chemical reactions*. Catalyst materials modify and increase the rate of chemical reactions without being consumed in the process. Electro-catalysts are a specific form of catalysts that function at *electrode* surfaces or may be the electrode surface itself. An electro-catalyst can be *heterogeneous* such as a platinum surface or *nanoparticles*, or *homogeneous* like a *co-ordination complex* or *enzyme*. The electro-catalyst assists in transferring electrons between the electrode and reactants, and/or facilitates an intermediate chemical transformation described by an overall *half-reaction*.

Context

There are multiple ways for many transformations to occur. For example, *hydrogen* and *oxygen* can be combined to form water through a *free radical mechanism* commonly referred to as *combustion*. Useful energy can be obtained from the thermal heat of this reaction through an *internal combustion engine* with an upper efficiency of 60% (for compression ratio of 10 and specific heat ratio of 1.4) based on the *Otto thermodynamic cycle*. It is also possible to combine the hydrogen and oxygen through redox mechanism as in the case of a *fuel cell*. In this process, the reaction is broken into two half-reactions which occur at separate electrodes. In this situation the reactant's energy is directly converted to electricity.

Half-reaction	$E°$ (V)
$H_2(g) \rightleftharpoons 2H^+ + 2e^-$	$\equiv 0$
$O_2(g) + 4H^+ + 4e^- \rightleftharpoons 2H_2O$	+1.23

This process is not governed by the same thermodynamic cycles as combustion engines, it is governed by the total energy available to do work as described by the *Gibbs free energy*. In the case of this reaction, that limit is 83% efficient at 298K. This half-reaction pair and many others don't achieve their theoretical limit in practical application due to lack of an effective electro-catalyst.

One of the greatest drawbacks to *galvanic cells*, like *fuel cells* and various forms of *electrolytic cells*, is that they can suffer from high activation barriers. The energy diverted to overcome these activation barriers is transformed into heat. In most exothermic combustion reactions this heat would simply propagate the reaction catalytically. In a redox reaction, this heat is a useless by-product lost to the system. The extra energy required to overcome kinetic barriers is usually described in terms of low *faradaic efficiency* and high *overpotentials*. In the example above, each of the two *electrodes* and its associated *half-cell* would require its own specialized electro-catalyst.

Half-reactions involving *multiple steps*, multiple electron transfers, and the evolution or consumption of gases in their overall chemical transformations, will often have considerable kinetic barriers. Furthermore, there is often more than one possible reaction at the surface of an electrode. For example, during the *electrolysis of water*, the anode can oxidize water through a two electron process to *hydrogen peroxide* or a four electron process to oxygen. The presence of an electro-catalyst could facilitate either of the reaction pathways.

Like other catalysts, an electro-catalyst lowers the *activation energy* for a reaction without altering the *reaction equilibrium*. Electro-catalysts go a step further than other catalysts by lowering the excess energy consumed by a redox reaction's activation barriers.

Ethanol-powered Fuel Cells

An electro-catalyst of *platinum* and *rhodium* on carbon backed tin-dioxide nano-particles can break *carbon bonds* at room temperature with only *carbon dioxide* as a by-product, so that *ethanol* can be oxidized into the necessary hydrogen ions and electrons required to create electricity.

ADSORPTION

Adsorption is the *adhesion* of *atoms*, *ions*, or *molecules* from a gas, liquid, or dissolved solid to a *surface*. This process creates a film of the *adsorbate* on the surface of the *adsorbent*. This process differs from *absorption*, in which a *fluid* (the *absorbate*) *permeates* or is *dissolved* by a liquid or solid (the *absorbent*). Adsorption is a surface-based process while absorption involves the whole volume of the material. The term *sorption* encompasses both processes, while *desorption* is the reverse of it. Adsorption is a *surface phenomenon*.

IUPAC Definition

Increase in the concentration of a substance at the interface of a condensed and a liquid or gaseous layer owing to the operation of surface forces.

Note 1 : Adsorption of proteins is of great importance when a material is in contact with blood or body fluids. In the case of blood, *albumin*, which is largely predominant, is generally adsorbed first, and then re-arrangements occur in favour of other minor proteins according to surface affinity against mass law selection (*Vroman effect*).

Note 2 : Adsorbed molecules are those that are resistant to washing with the same solvent medium in the case of adsorption from solutions. The washing conditions can thus modify the measurement results, particularly when the interaction energy is low.

Similar to *surface tension*, adsorption is a consequence of *surface energy*. In a *bulk material*, all the bonding requirements (be they *ionic, covalent,* or *metallic*) of the constituent *atoms* of the material are filled by other atoms in the material. However, atoms on the surface of the adsorbent are not wholly surrounded by other adsorbent atoms and therefore can attract adsorbates. The exact nature of the bonding depends on the details of the species involved, but the adsorption process is generally classified as *physisorption* (characteristic of weak *van der Waals forces*) or *chemisorption* (characteristic of covalent bonding). It may also occur due to electrostatic attraction.

Adsorption is present in many natural, physical, biological, and chemical systems, and is widely used in industrial applications such as *activated charcoal*, capturing and using waste heat to provide cold water for air conditioning and other process requirements (*adsorption chillers*), *synthetic resins*, increase storage capacity of *carbide-derived carbons*, and *water purification*. Adsorption, *ion exchange*, and *chromatography* are sorption processes in which certain adsorbates are selectively transferred from the fluid phase to the surface of insoluble, rigid particles suspended in a vessel or packed in a column. Lesser known, are the pharmaceutical industry applications as a means to prolong neurological exposure to specific drugs or parts thereof.

The word "adsorption" was coined in 1881 by German physicist *Heinrich Kayser*.

Isotherms

Adsorption is usually described through *isotherms*, that is, the amount of adsorbate on the adsorbent as a function of its pressure (if gas) or concentration (if liquid) at constant temperature. The quantity adsorbed is nearly always normalized by the mass of the adsorbent to allow comparison of different materials.

HENRY ADSORPTION CONSTANT

The **Henry adsorption constant** is the constant appearing in the **linear adsorption isotherm**, which formally resembles *Henry's law*; therefore, it is also called **Henry's adsorption isotherm**. This is the simplest *adsorption isotherm* in that the amount of the surface adsorbate is represented to be proportional to the *partial pressure* of the adsorptive gas :

$$X = K_H P$$

where :

- X-surface coverage,
- P-partial pressure,
- K_H-Henry's adsorption constant.

For solutions, concentrations, or *activities*, are used instead of the partial pressures.

The linear isotherm can be used to describe the initial part of many practical isotherms. It is typically taken as valid for low surface coverages, and the adsorption energy being independent of the coverage (lack of inhomogeneities on the surface).

The Henry adsorption constant can be defined as :

$$K_H = \lim_{Q \to 0} \frac{Q_s}{Q(z)},$$

where :

- $Q(z)$ is the number density at free phase,
- Q_s is the surface number density.

Application at a Permeable Wall

If a solid body is modelled by a constant field and the structure of the field is such that it has a penetrable core, then

$$K_H = \int_{-\infty}^{x'} [\exp(-\beta u) - \exp(-\beta u_0)]dx - \int_{x'}^{\infty} [1 - \exp(-\beta u)]dx.$$

Here x' is the position of the dividing surface, $u = u(x)$ is the external force field, simulating a solid, u_0 is the field value deep in the solid, $\beta = 1/k_B T$, k_B is the Boltzmann constant, and T is the temperature.

Introducing "the surface of zero adsorption"

$$x_0 = - \int_{-\infty}^{0} \tilde{\theta}(x)dx + \int_{0}^{\infty} \tilde{\varphi}(x)dx,$$

where :

$$\tilde{\theta} = \frac{\exp(-\beta u) - \exp(-\beta u_0)}{1 - \exp(-\beta u_0)}$$

and

$$\tilde{\varphi} = \frac{1 - \exp(-\beta u)}{1 - \exp(-\beta u_0)},$$

we get

$$K_H(x') = [x' - x_0(T)][1 - \exp(-\beta u_0)]$$

and the problem of K_H determination is reduced to the calculation of x_0.

Taking into account that for Henry *absorption* constant we have :

$$k_H = \lim_{Q \to 0} \frac{Q(z')}{Q(z)} = \exp(-\beta u_0),$$

where $Q(z')$ is the number density inside the solid, we arrive at the parametric dependence :

$$K_H = \int_{-\infty}^{x'} \left[k_H^{\tilde{u}(x)} - k_H \right] dx - \int_{x'}^{\infty} \left[1 - k_H^{\tilde{u}(x)} \right] dx$$

where :

$$\tilde{u}(x) = \frac{u(x)}{u_0}.$$

Application at an Impermeable Wall

If a solid body is modelled by a constant hard-core field, then :

$$K_H = \int_{-\infty}^{x'} \exp(-\beta u)dx - \int_{x'}^{\infty} [1 - \exp(-\beta u)] \, dx,$$

or

$$K_H(x') = x - x_0(T),$$

where :

$$x_0 = -\int_{-\infty}^{0} \theta(x)dx + \int_{0}^{\infty} \varphi(x)dx.$$

Here

$$\theta = \exp(-\beta u)$$

$$\varphi = 1 - \exp(-\beta u).$$

For the hard solid potential

$$x_0 = x_{step},$$

where x_{step} is the position of the potential discontinuity. So, in this case :

$$K_H(x') = x' - x_{step}.$$

Choice of the Dividing Surface

The choice of the dividing surface, strictly speaking, is arbitrary, however, it is very desirable to take into account the type of external potential $u(x)$. Otherwise, these expressions are at odds with the generally accepted concepts and common sense.

First, x' must lie close to the transition layer (*i.e.*, the region where the number density varies), otherwise it would mean the attribution of the bulk properties of one of the phase to the surface.

Second. In the case of weak adsorption, for example, when the potential is close to the step wise, it is logical to choose x' close to x_0. (In some cases, choosing $x_0 \pm R$, where R is particle radius, excluding the "dead" volume.)

In the case of pronounced adsorption it is advisable to choose x' close to the right border of the transition region. In this case all particles from the transition layer will be attributed to the solid, and K_H is always positive. Trying to put $x' = x_0$ in this case will lead to a strong shift of x' to the solid body domain, which is clearly unphysical.

Conversely, if $u_0 < 0$ (fluid on the left), it is advisable to choose x' lying on the left side of the transition layer. In this case the surface particles once again refer to the solid and K_H is back positive.

Thus, we can always avoid the "negative adsorption" for one-component systems.

FREUNDLICH EQUATION

The **Freundlich equation** or **Freundlich adsorption isotherm**, an *adsorption isotherm*, is a curve relating the *concentration* of a *solute* on the surface of an *adsorbent* to the concentration of the solute in the liquid with which it is in contact. In 1909, Freundlich gave an empirical expression representing the isothermal variation of Adsorption of a quantity of gas adsorbed by unit mass of solid adsorbent with pressure. This equation is known as Freundlich Adsorption Isotherm or Freundlich Adsorption equation. There are basically two well established types of adsorption isotherm : the Freundlich adsorption isotherm and the *Langmuir adsorption isotherm*. Here the amount of mass that is adsorbed is plotted against the temperature which gives an idea about the variation of *adsorption* with temperature.

Freundlich Adsorption Isotherm

The Freundlich Adsorption Isotherm is mathematically expressed as :

$$\frac{x}{m} = Kp^{1/n}$$

It is also written as :

$$\log \frac{x}{m} = \log K + \frac{1}{n}\log p$$

or

$$\frac{x}{m} = Kc^{1/n}$$

It is also written as :

$$\log \frac{x}{m} = \log K + \frac{1}{n} \log c$$

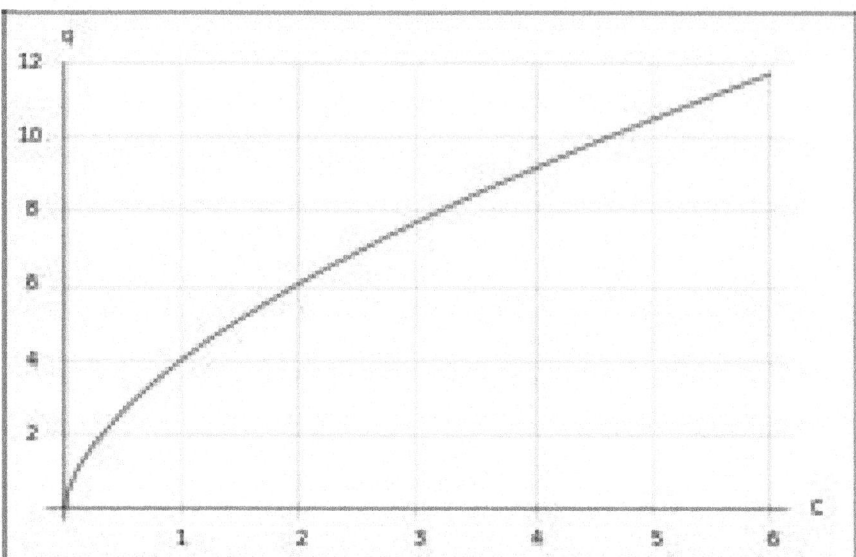

Fig. : Example of the Freundlich isotherm, showing the amount adsorbed, q (*e.g.*, in mol/kg), as a function of equilibrium concentration in the solution, c (*e.g.*, in mol/L). The graph is for the values of the constants of K=4 and 1/n=0.6.

where :

   x = *mass* of *adsorbate*

   m = mass of *adsorbent*

   p = *Equilibrium pressure* of adsorbate

   c = Equilibrium *concentration* of adsorbate in solution.

   K and n are constants for a given adsorbate and adsorbent at a particular temperature.

   At high pressure $1/n = 0$, hence extent of adsorption becomes independent of pressure.

   It is used in cases where the actual identity of the solute is not known, such as adsorption of colored material from sugar, vegetable oil etc.

Limitation of Freundlich Adsorption Isotherm

Experimentally it was determined that extent of adsorption varies directly with pressure until saturation pressure Ps is reached. Beyond that point rate of adsorption saturates even after applying higher pressure. Thus Freundlich Adsorption Isotherm failed at higher pressure.

Langmuir

Irving Langmuir was the first to derive a scientifically based adsorption isotherm in 1918. The model applies to gases adsorbed on solid surfaces. It is a semi-empirical isotherm with a kinetic basis and was derived based on statistical thermodynamics. It is the most common isotherm equation to use due to its simplicity and its ability to fit a variety of adsorption data. It is based on four assumptions :

1. All of the adsorption sites are equivalent and each site can only accommodate one molecule.

2. The surface is energetically homogeneous and adsorbed molecules do not interact.

3. There are no phase transitions.

4. At the maximum adsorption, only a monolayer is formed. Adsorption only occurs on localized sites on the surface, not with other adsorbates.

These four assumptions are seldom all true : there are always imperfections on the surface, adsorbed molecules are not necessarily inert, and the mechanism is clearly not the same for the very first molecules to adsorb to a surface as for the last. The fourth condition is the most troublesome, as frequently more molecules will adsorb to the monolayer; this problem is addressed by the *BET isotherm* for relatively flat (non-*microporous*) surfaces. The Langmuir isotherm is nonetheless the first choice for most models of adsorption, and has many applications in surface kinetics (usually called *Langmuir–Hinshelwood kinetics*) and *thermodynamics*.

Langmuir suggested that adsorption takes place through this mechanism : $Ag + S \rightleftharpoons AS$, where A is a gas molecule and S is an adsorption site. The direct and inverse rate constants are k and k_{-1}. If we define surface coverage, θ, as the fraction of the adsorption sites occupied, in the equilibrium we have :

$$K = \frac{k}{k_{-1}} = \frac{\theta}{(1-\theta)P}$$

or

$$\theta = \frac{KP}{1+KP}$$

where P is the partial pressure of the gas or the molar concentration of the solution. For very low pressures $\theta \approx KP$ and for high pressures $\theta \approx 1$.

θ is difficult to measure experimentally; usually, the adsorbate is a gas and the quantity adsorbed is given in moles, grams, or gas volumes at *standard temperature and pressure* (STP) per gram of adsorbent. If we call v_{mon} the STP volume of adsorbate required to form a monolayer on the adsorbent (per gram of adsorbent), $\theta = \dfrac{v}{v_{mon}}$ and we obtain an expression for a straight line :

$$\frac{1}{v} = \frac{1}{Kv_{mon}}\frac{1}{P} + \frac{1}{v_{mon}}$$

Through its slope and y-intercept we can obtain v_{mon} and K, which are constants for each adsorbent/adsorbate pair at a given temperature. v_{mon} is related to the number of adsorption sites through the *ideal gas law*. If we assume that the number of sites is just the whole area of the solid divided into the cross-section of the adsorbate molecules, we can easily calculate the surface area of the adsorbent. The surface area of an adsorbent depends on its structure; the more pores it has, the greater the area, which has a big influence on *reactions on surfaces*.

If more than one gas adsorbs on the surface, we define θ_E as the fraction of empty sites and we have :

$$\theta_E = \frac{1}{1+\sum_{i=1}^{n} K_i P_i}$$

Also, we can define θ_j as the fraction of the sites occupied by the j-th gas :

$$\theta_j = \frac{K_j P_j}{1+\sum_{i=1}^{n} K_i P_i}$$

where i is each one of the gases that adsorb.

BET THEORY

Brunauer–Emmett–Teller (BET) theory aims to explain the physical *adsorption of gas molecules* on a *solid surface* and serves as the basis for an important analysis technique for the measurement of the specific surface area of a material. In 1938, Stephen Brunauer, *Paul Hugh Emmett*, and *Edward Teller* published the first article about the BET theory in the *Journal of the American Chemical Society*.

Concept

The concept of the theory is an extension of the *Langmuir theory*, which is a theory for *monolayer* molecular adsorption, to multi-layer adsorption with the following hypotheses : (a) gas molecules physically adsorb on a solid in layers infinitely; (b) there is no interaction between each adsorption layer; and (c) the Langmuir theory can be applied to each layer. The resulting *BET equation* is :

$$\frac{1}{v[(p_0/p)-1]} = \frac{c-1}{v_m c}\left(\frac{p}{p_0}\right)+\frac{1}{v_m c}, \tag{1}$$

where p and p_0 are the *equilibrium* and the *saturation pressure* of adsorbates at the temperature of adsorption, v is the adsorbed gas quantity (for example, in volume units), and v_m is the *monolayer* adsorbed gas quantity. c is the *BET constant*,

$$c = \exp\left(\frac{E_1-E_L}{RT}\right), \tag{2}$$

where E_1 is the heat of adsorption for the first layer, and E_L is that for the second and higher layers and is equal to the heat of *liquefaction*.

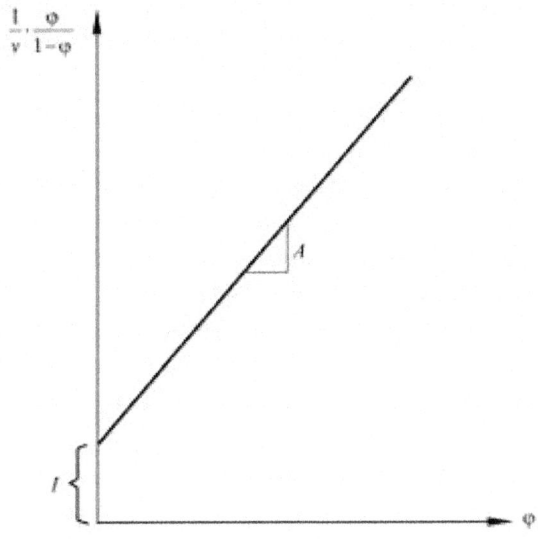

Fig. : BET plot.

Equation (1) is an *adsorption isotherm* and can be plotted as a straight line with $1/v[p_0/p) - 1$ on the y-axis and $\phi = p/p_0$ on the x-axis according to experimental results. This plot is called a *BET plot*. The linear relationship of this equation is maintained only in the range of $0.05 < p/p_0 < 0.35$. The value of the slope A and the y-intercept I of the line are used to calculate the monolayer adsorbed gas quantity v_m and the BET constant c. The following equations can be used :

$$v_m = \frac{1}{A+I} \tag{3}$$

$$c = 1 + \frac{A}{I}. \tag{4}$$

The BET method is widely used in *surface* science for the calculation of *surface areas* of *solids* by physical adsorption of gas molecules. The total surface area S_{total} and the *specific surface area* S_{BET} are given by :

$$S_{total} = \frac{(v_m N s)}{V}, \tag{5}$$

$$S_{BET} = \frac{S_{total}}{a}, \tag{6}$$

where v_m is in units of volume which are also the units of the molar volume of the adsorbate gas, N is *Avogadro's number*, s the adsorption cross-section of the adsorbing species, V the molar volume of the adsorbate gas, and a the mass of the adsorbent.

Derivation

The BET theory can be derived similar to the *Langmuir theory*, but by considering multi-layered gas molecule adsorption, where it is not required for a layer to be completed before an upper layer formation starts. Furthermore, the authors made five assumptions :

1. Adsorptions occur only on well-defined sites of the sample surface (one per molecule)

2. The only molecular interaction considered is the following one : a molecule can act as a single adsorption site for a molecule of the upper layer.

3. The uppermost molecule layer is in equilibrium with the gas phase, *i.e.* similar molecule adsorption and desorption rates.

4. The desorption is a kinetically-limited process, *i.e.* a heat of adsorption must be provided :

   - These phenomenon are homogeneous, *i.e.* same heat of adsorption for a given molecule layer.

   - It is E_1 for the first layer, *i.e.* the heat of adsorption at the solid sample surface

   - The other layers are assumed similar and can be represented as condensed species, *i.e.* liquid state. Hence, the heat of adsorption is E_L is equal to the heat of liquefaction.

5. At the saturation pressure, the molecule layer number tends to infinity (*i.e.* equivalent to the sample being surrounded by a liquid phase)

Let us consider a given amount of solid sample in a controlled atmosphere. Let θ_i be the fractional coverage of the sample surface covered by a number i of successive molecule layers. Let us assume that the adsorption rate $R_{ads,i-1}$ for molecules on a layer $(i-1)$ (*i.e.* formation of a layer i) is proportional to both its fractional surface θ_{i-1} and to the pressure P, and that the desorption rate $R_{des,i}$ on a layer i is also proportional to its fractional surface θ_i :

$$R_{ads,i-1} = k_i P \Theta_{i-1}$$
$$R_{des,i} = k_{-i} \Theta_i,$$

where k_i and k_{-i} are the kinetic constants (depending on the temperature) for the adsorption on the layer $(i-1)$ and desorption on layer i, respectively. For the adsorptions, these constant are assumed similar whatever the surface. Assuming an Arrhenius law for desorption, the related constants can be expressed as :

$$k_i = \exp(-E_i / RT),$$

where E_i is the heat of adsorption, equal to E_1 at the sample surface and to E_L otherwise.

Example

Cement Paste

By application of the BET theory it is possible to determine the inner surface of hardened *cement* paste. If the quantity of adsorbed water vapour is measured at different levels of relative humidity a BET plot is obtained. From the slope A and y-intersection I on the plot it is possible to calculate v_m and the BET constant c. In case of cement paste hardened in water ($T = 97°C$), the slope of the line is $A = 24.20$ and the y-intersection $I = 0.33$; from this follows :

$$v_m = \frac{1}{A+I} = 0.0408,$$

$$c = 1 + \frac{A}{I} = 73.6.$$

From this the specific BET surface area S_{BET} can be calculated by use of the above mentioned equation (one water molecule covers $s = 0.114nm^2$). It follows thus $S_{BET} = 156m^2/g$ which means that hardened cement paste has an inner surface of 156 square meters per g of cement.

Activated Carbon

For example, *activated carbon*, which is a strong adsorbate and usually has an adsorption *cross-section s* of 0.16 nm² for *nitrogen* adsorption at *liquid nitrogen* temperature, is revealed from experimental data to have a large surface area around 3000 m² g⁻¹. Moreover, in the field of solid *catalysis*, the surface area of *catalysts* is an important factor in *catalytic activity*. Porous inorganic materials such as *mesoporous silica* and layer *clay minerals* have high surface areas of several hundred m² g⁻¹ calculated by the BET method, indicating the possibility of application for efficient catalytic materials.

Kisliuk

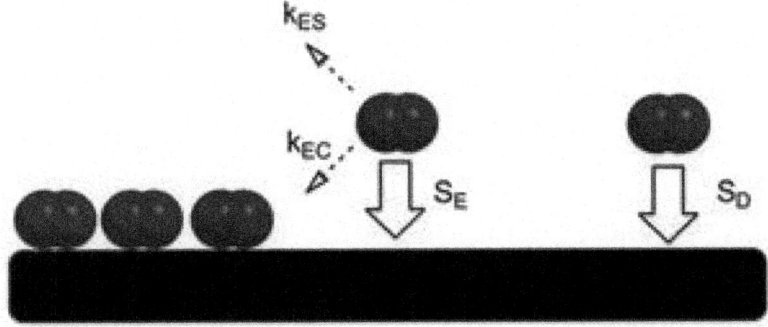

Fig. : Two adsorbate nitrogen molecules adsorbing onto a tungsten adsorbent from the precursor state around an island of previously adsorbed adsorbate (left) and *via* random adsorption.

In other instances, molecular interactions between gas molecules previously adsorbed on a solid surface form significant interactions with gas molecules in the gaseous phases. Hence, adsorption of gas molecules to the surface is more likely to occur around gas molecules that are already present on the solid surface, rendering the Langmuir adsorption isotherm ineffective for the purposes of modelling. This effect was studied in a system where nitrogen was the adsorbate and tungsten was the adsorbent by Paul Kisliuk in 1957. To compensate for the increased probability of adsorption occurring around molecules present on the substrate surface, Kisliuk developed the precursor state theory, whereby molecules would enter a precursor state at the interface between the solid adsorbent and adsorbate in the gaseous phase. From here, adsorbate molecules would either adsorb to the adsorbent or desorb into the gaseous phase. The probability of adsorption occurring from the precursor state is dependent on the adsorbate's proximity to other adsorbate molecules that have already been adsorbed. If the adsorbate molecule in the precursor state is in close proximity to an adsorbate molecule that has already formed on the surface, it has a sticking probability reflected by the size of the S_E constant and will either be adsorbed from the precursor state at a rate of k_{EC} or will desorb into the gaseous phase at a rate of k_{ES}. If an adsorbate molecule enters the precursor state at a location that is remote from any other previously adsorbed adsorbate molecules, the sticking probability is reflected by the size of the S_D constant.

These factors were included as part of a single constant termed a "sticking coefficient," k_E, described below :

$$k_E = \frac{S_E}{k_{ES}.S_D}.$$

As S_D is dictated by factors that are taken into account by the Langmuir model, S_D can be assumed to be the adsorption rate constant. However, the rate constant for the Kisliuk model (R') is different from that of the Langmuir model, as R' is used to represent the impact of diffusion on monolayer formation and is proportional to the square root of the system's diffusion coefficient. The Kisliuk adsorption isotherm is written as follows, where $\Theta_{(t)}$ is fractional coverage of the adsorbent with adsorbate, and t is immersion time :

$$\frac{d\theta_{(t)}}{dt} = R'(1-\theta)(1+k_E\theta).$$

Solving for $\Theta_{(t)}$ yields :

$$\theta_{(t)} = \frac{1-e^{-R'(1+k_E)t}}{1+k_E e^{-R'(1+k_E)t}}.$$

Adsorption Enthalpy

Adsorption constants are equilibrium constants, therefore they obey the *van 't Hoff equation* :

$$\left(\frac{\partial \ln K}{\partial \frac{1}{T}} \right)_{\theta} = -\frac{\Delta H}{R}.$$

As can be seen in the formula, the variation of K must be isosteric, that is, at constant coverage. If we start from the BET isotherm and assume that the entropy change is the same for liquefaction and adsorption we obtain :

$$\Delta H_{ads} = \Delta H_{liq} - RT \ln c,$$

that is to say, adsorption is more exothermic than liquefaction.

Adsorbents

Characteristics and General Requirements

Adsorbents are used usually in the form of spherical pellets, rods, moldings, or monoliths with hydrodynamic diameters between 0.5 and 10 mm. They must have high *abrasion* resistance, high *thermal stability* and small pore diameters, which results in higher exposed surface area and hence high capacity for adsorption. The adsorbents must also have a distinct pore structure that enables fast transport of the gaseous vapours.

Most industrial adsorbents fall into one of three classes :

- *Oxygen-containing compounds* : Are typically hydrophilic and polar, including materials such as *silica gel* and *zeolites*.
- *Carbon-based compounds* : Are typically hydrophobic and non-polar, including materials such as *activated carbon* and *graphite*.
- *Polymer-based compounds* : Are polar or non-polar functional groups in a porous polymer matrix.

Silica Gel

Silica gel is a chemically inert, non-toxic, polar and dimensionally stable (< 400°C or 750°F) amorphous form of SiO_2. It is prepared by the reaction between sodium silicate and acetic acid, which is followed by a series of after-treatment processes such as aging, pickling, etc. These after treatment methods results in various pore size distributions.

Silica is used for drying of process air (*e.g.* oxygen, natural gas) and adsorption of heavy hydrocarbons from natural gas.

Zeolites

Zeolites are natural or synthetic crystalline alumino-silicates, which have a repeating pore network and release water at high temperature. Zeolites are polar in nature.

They are manufactured by hydrothermal synthesis of sodium alumino-silicate or another silica source in an autoclave followed by ion exchange with certain cations (Na⁺, Li⁺, Ca²⁺, K⁺, NH₄⁺). The channel diameter of zeolite cages usually ranges from 2 to 9 $\mathring{A}$ (200 to 900 pm). The ion exchange process is followed by drying of the crystals, which can be pelletized with a binder to form macroporous pellets.

Zeolites are applied in drying of process air, CO_2 removal from natural gas, CO removal from reforming gas, air separation, catalytic cracking, and catalytic synthesis and reforming.

Non-polar (siliceous) zeolites are synthesized from aluminum-free silica sources or by dealumination of aluminum-containing zeolites. The dealumination process is done by treating the zeolite with steam at elevated temperatures, typically greater than 500°C (930°F). This high temperature heat treatment breaks the aluminum-oxygen bonds and the aluminum atom is expelled from the zeolite framework.

Activated Carbon

Activated carbon is a highly porous, amorphous solid consisting of microcrystallites with a graphite lattice, usually prepared in small pellets or a powder. It is non-polar and cheap. One of its main drawbacks is that it reacts with oxygen at moderate temperatures (over 300°C).

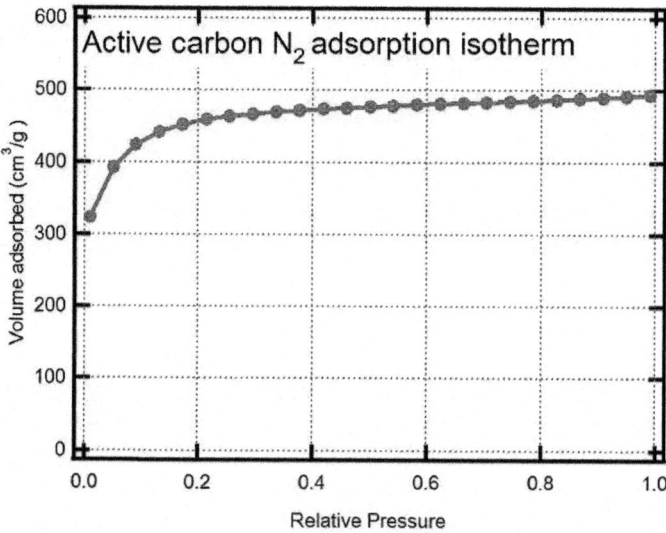

Fig. : Activated carbon nitrogen isotherm showing a marked microporous type I behaviour.

Activated carbon can be manufactured from carbonaceous material, including coal (bituminous, sub-bituminous, and lignite), peat, wood, or nutshells (*e.g.*, coconut). The manufacturing process consists of two phases, carbonization and activation. The carbonization process includes drying and then heating to separate by-products, including tars and other hydrocarbons from the raw material, as well as to drive off any gases generated. The process is completed by heating the material over 400°C (750°F) in an oxygen-free atmosphere that cannot support combustion. The carbonized particles are then "activated" by exposing them to an oxidizing agent, usually steam or carbon dioxide at high temperature. This agent burns off the pore blocking structures created during the carbonization phase and so, they develop a porous, three-dimensional graphite lattice structure. The size of the pores developed during activation is a function of the time that they spend in this stage. Longer exposure times result in larger pore sizes. The most popular aqueous phase carbons are bituminous based because of their hardness, abrasion resistance, pore size distribution, and low cost, but their effectiveness needs to be tested in each application to determine the optimal product.

Activated carbon is used for adsorption of organic substances and non-polar adsorbates and it is also usually used for waste gas (and waste water) treatment. It is the most widely used adsorbent since most of its chemical (*e.g.* surface groups) and physical properties (*e.g.* pore size distribution and surface area) can be tuned according to what is needed. Its usefulness also derives from its large micropore (and sometimes mesopore) volume and the resulting high surface area.

Protein Adsorption of Bio-materials

Protein adsorption is a process that has a fundamental role in the field of *bio-materials*. Indeed, bio-material surfaces in contact with biological media, such as blood or serum, are immediately coated by proteins. Therefore, living *cells* do not interact directly with the bio-material surface, but with the adsorbed proteins layer. This protein layer mediates the interaction between bio-materials and cells, translating bio-material physical and chemical properties into a "biological language". In fact, *cell membrane receptors* bind to protein layer bio-active sites and these receptor-protein binding events are transduced, through the cell membrane, in a manner that stimulates specific intra-cellular processes that then determine cell adhesion, shape, growth and differentiation. Protein adsorption is influenced by many surface properties such as surface *wettability*, surface chemical composition and surface nanometre-scale morphology.

Adsorption Chillers

Combining an adsorbent with a refrigerant, adsorption *chillers* use heat to provide a cooling effect. This heat, in the form of hot water, may come from any number of industrial sources including waste heat from industrial processes, prime heat from solar thermal installations or from the exhaust or water jacket heat of a piston engine or turbine.

Although there are similarities between *absorption* and adsorption refrigeration, the latter is based on the interaction between gases and solids. The adsorption chamber of the chiller is filled with a solid material (for example, zeolite, silica gel, alumina, active carbon and certain types of metal salts), which in its neutral state has adsorbed the refrigerant. When heated, the solid desorbs (releases) refrigerant vapour, which subsequently is cooled and liquefied. This liquid refrigerant then provides its cooling effect at the evaporator, by *ab*sorbing external heat and turning back into a vapour. In the final stage the refrigerant vapour is (re)adsorbed into the solid. As an adsorption chiller requires no moving parts, it is relatively quiet.

Portal Site Mediated Adsorption

Portal site mediated adsorption is a model for site-selective activated gas adsorption in metallic catalytic systems that contain a variety of different adsorption sites. In such systems, low-co-ordination "edge and corner" defect-like sites can exhibit significantly lower adsorption enthalpies than high-co-ordination (*basal plane*) sites. As a result, these sites can serve as "portals" for very rapid adsorption to the rest of the surface. The phenomenon relies on the common "spillover" effect, where certain adsorbed species exhibit high mobility on some surfaces. The model explains seemingly inconsistent observations of gas adsorption thermodynamics and kinetics in catalytic systems where surfaces can exist in a range of co-ordination structures, and it has been successfully applied to bimetallic catalytic systems where synergistic activity is observed.

In contrast to pure spillover, portal site adsorption refers to surface diffusion to adjacent adsorption sites, not to non-adsorptive support surfaces.

The model appears to have been first proposed for carbon monoxide on silica-supported platinum by Brandt *et. al.*. A similar, but independent model was developed by King and co-workers to describe hydrogen adsorption on silica-supported alkali promoted ruthenium, silver-ruthenium and copper-ruthenium bimetallic catalysts. The same group applied the model to CO hydrogenation (Fischer–Tropsch synthesis). Zupanc *et. al.* subsequently confirmed the same model for hydrogen adsorption on magnesia-supported caesium-ruthenium bimetallic catalysts. Trens *et. al.* have similarly described CO surface diffusion on carbon-supported Pt particles of varying morphology.

Adsorption Spillover

In the case catalytic or adsorbent systems where a metal species is dispersed upon a support (or carrier) material (often quasi-inert oxides, such as alumina or silica), it is possible for an adsorptive species to indirectly adsorb to the support surface under conditions where such adsorption is thermodynamically unfavourable. The presence of the metal serves as a lower-energy pathway for gaseous species to first adsorb to the metal and then diffuse on the support surface. This is possible because the adsorbed species attains a lower energy state once it has adsorbed to the metal, thus lowering the activation barrier between the gas phase species and the support-adsorbed species.

Hydrogen spillover is the most common example of an adsorptive spillover. In the case of hydrogen, adsorption is most often accompanied with dissociation of molecular hydrogen (H_2) to atomic hydrogen (H), followed by spillover of the hydrogen atoms present.

The spillover effect has been used to explain many observations in heterogeneous catalysis and adsorption.

Polymer Adsorption

Adsorption of molecules onto polymer surfaces is central to a number of applications, including development of non-stick coatings and in various bio-medical devices. Polymers may also be adsorbed to surfaces through *polyelectrolyte adsorption*.

Adsorption in Viruses

Adsorption is the first step in the *viral life cycle*. The next steps are penetration, uncoating, synthesis (transcription if needed, and translation), and release. The virus replication cycle, in this respect, is similar for all types of viruses. Factors such as transcription may or may not be needed if the virus is able to integrate its genomic information in the cell's nucleus, or if the virus can replicate itself directly within the cell's cytoplasm.

Chapter 5

CONTROLLED POTENTIAL ELECTROLYSIS

INTRODUCTION

The principle behind the Controlled Potential Electrolysis (CPE) experiment is very simple. If only the oxidized species is initially present, then the potential is set at a constant value sufficiently negative to cause rapid reduction and is maintained at this value until only the reduced species is present in solution. The total charge passed during the CPE experiment (Q) is calculated by integrating the current and is related to the number of electrons transferred per molecule (n) and the number of moles of the oxidized species initially present (N) through Faraday's law :

$Q = nFN$

where F is Faraday's constant ($96500 \, C \, mol^{-1}$). Therefore, if one of n or N is known, the other can be calculated. Hence, CPE has both analytical and synthetic applications, and is a standard technique on the epsilon.

SETTING UP A CONSTANT POTENTIAL ELECTROLYSIS EXPERIMENT

The parameter values for CPE are set using the Change Parameters dialog box in either the Experiment menu or the pop-up menu.

1. Potential values are entered in mV, and time values are entered in minutes or seconds (selected using T-Units)

2. The time resolution of the data is specified by the Sample Interval (*e.g.*, the default condition is that a data point is recorded every second).

3. There are two gain stages for the current-to-voltage converter. The default values of these stages that are used for a given current Full Scale value are determined by the software. However, they can be adjusted manually using the Filter/F.S. dialog box. This dialog box is also used to change the analog Noise Filter Value settings from the default values set by the software.

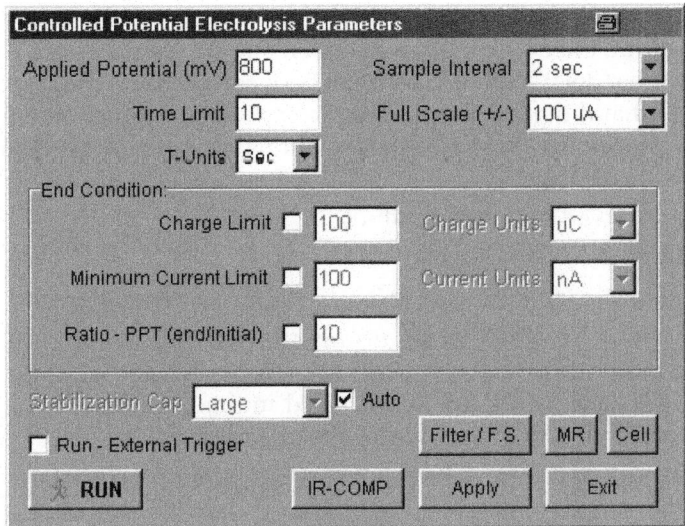

Fig. : Change Parameters dialog box for controlled potential eletrolysis.

4. A stabilizing capacitor (Stabilization Cap) between the auxiliary and reference electrodes is switched in during the CPE experiment. The Auto default capacitor (Large-0.1 uF) can be manually changed to Small (0.01 uF) or No Cap (no capacitor).

5. The default condition of the cell is that the cell is On (*i.e.*, the electronics are connected to the electrodes) during the experiment, and is Off between experiments. However, the potential can be switched On between experiments using the Cell dialog box. however, this option should be used with caution since connecting or disconnecting the electrodes when the cell is on can result in damage to the potentiostat, the cell, and/or the user !

6. A series of identical experiments on the same cell can be programmed using the MR (Multi-Run) option. However, if any of the optional End Conditions are checked, any MR settings are ignored.

7. Clicking Exit will exit the dialog box without saving any changes made to the parameter values. Any changes can be saved by clicking Apply before exiting.

8. The end of the CPE experiment can be set by the user in a number of ways. The most basic criterion is the Time Limit, which must be set by the user; that is, the experiment will end after a user-defined time period. However, there are three optional criteria that can also be set :

• *Charge Limit :* The absolute value of the charge limit should be specified.

• *Minimum Current Limit :* The absolute minimum current value should be specified.

• *Ratio-PPT (end/initial) :* The criterion is the ratio of the final current to the initial current in parts per thousand.

Any one of these three optional criteria can be used in addition to the Time Limit. However, it is important to note that the Time Limit always takes precedence; that is, if the Time Limit is attained before *e.g.*, the charge exceeds the Charge Limit, the experiment will end. It should also be noted that there will typically be a delay of 1 or 2 seconds between the time the selected criterion is exceeded and the termination of the experiment, due to the time required for data processing. If any of these optional criteria are used, the multi-run capability is disabled.

9. Range of allowed parameter values :
    *   Potential =–3275-+3275 mV
    *   Sample Interval = 0.05, 0.1, 0.2, 0.5, 1, 2, 5, 10, 30 and 60 sec
    *   Time Limit = 0–32000 in sec or min (the maximum value allowed for the Time Limit is also determined by the Sample Interval, since a maximum of 64000 data points can be recorded in one experiment (e.g, the shorter the Sample Interval, the shorter the experiment)).

10. Once the parameters have been set, the experiment can be started by clicking Run (either in this dialog box, in the Experiment menu, in the pop-up menu, on the Tool Bar, or using the F5 key).

11. Neither the Pre-Run Drops function nor the internal dummy cell are available for CPE.

The potential required for a CPE experiment is determined by the redox potential of the analyte (measured by *e.g.*, cyclic voltammetry). For a reduction, the ideal potential is ca. 200 mV more negative than the redox potential so that the rate of electrolysis is controlled by the rate of mass transport to the working electrode. However, it is not always possible to use a potential too far removed from the redox potential due to electrolysis of other electro-active materials (*e.g.*, electrolyte, solvent, or other components of the solution mixture).

The cell required for CPE is significantly different to that required for voltammetry experiments (in which only a very small fraction of the electro-active molecule of interest is electrolyzed). The rate of electrolysis is enhanced by using a working electrode with a large surface area (*e.g.*, platinum gauze, reticulated vitreous carbon or a mercury pool) and an auxiliary electrode with a large surface area (*e.g.*, platinum coil or gauze); in addition, the solution is stirred to increase the rate of mass transport to and from the working electrode. The auxiliary electrode must be isolated from the working electrode to prevent species that are electro-generated at the auxiliary electrode from interfering with electrolysis at the working electrode. However, care must be taken when choosing the material used to isolate the auxiliary electrode from the working electrode, since high resistance material may affect the efficiency of the electrolysis.

The output from a CPE experiment is a current *vs.* time plot.

The following parameters are displayed in a separate window during the experiment, and are updated every second :

*   Elapsed Time
*   Initial Current (the current at the start of the experiment)

- Current (the most recent current value)
- Current Ratio (ratio of the most recent current value to the initial current value in part per thousand)
- Charge (the accumulated charge)

At the end of the experiment, the following parameters are displayed in the Text Info box to the right of the display :

- Initial Current I
- End Current I (the current value at the end of the experiment, calculated as the average of the last 5 current values)
- End Current Ratio
- Charge
- Net Charge

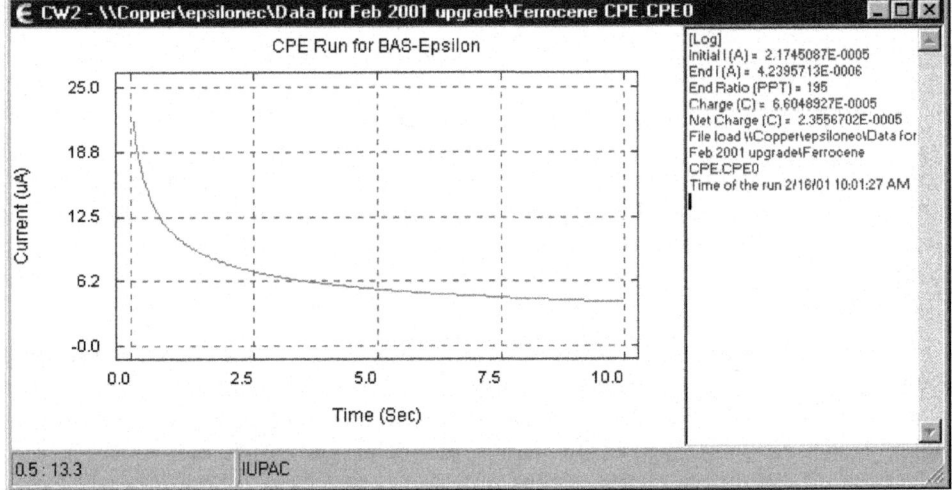

Fig. : Current *vs.* time plot for constant potential electrolysis.

Net Charge is calculated by subtracting the background charge from the total charge passed during the experiment (Charge). The background current is assumed to be constant during the experiment, and equal to End I. Therefore, the background charge is equal to the product of the background current and the experimental time. Once the Net Charge has been calculated, it can be used to calculate the number of electrons transferred or the amount of material electrolyzed (in moles) through Faraday's Law :

$Q = nFN$

where :

n = number of electrons transferred/molecule

F = Faraday's constant (96,485 C mol-1)

N = amount of material electrolyzed (mol).

Chapter 6

THE ELECTRO-CHEMICAL
BEHAVIOUR OF TRANISITION

VANADIUM

Vanadium is a chemical element with the symbol V and atomic number 23. It is a hard, silvery gray, ductile and malleable transition metal. The element is found only in chemically combined form in nature, but once isolated artificially, the formation of an oxide layer stabilizes the free metal somewhat against further oxidation.

Andrés Manuel del Río discovered compounds of vanadium in 1801 by analyzing a new lead-bearing mineral he called "brown lead," and presumed its qualities were due to the presence of a new element, which he named erythronium (Greek for "red") since, upon heating, most of its salts turned from their initial colour to red. Four years later, however, he was (erroneously) convinced by other scientists that erythronium was identical to chromium. Chlorides of vanadium were generated in 1830 by Nils Gabriel Sefström who thereby proved that a new element was involved, which he named "vanadium" after the Scandinavian goddess of beauty and fertility, Vanadís (Freyja). Both names were attributed to the wide range of colours found in vanadium compounds. Del Rio's lead mineral was later renamed vanadinite for its vanadium content. In 1867 Henry Enfield Roscoe obtained the pure element.

Vanadium occurs naturally in about 65 different minerals and in fossil fuel deposits. It is produced in China and Russia from steel smelter slag; other countries produce it either from the flue dust of heavy oil, or as a by-product of uranium mining. It is mainly used to produce specialty steel alloys such as high speed tool steels. The most important industrial vanadium compound, vanadium pentoxide, is used as a catalyst for the production of sulfuric acid.

Large amounts of vanadium ions are found in a few organisms, possibly as a toxin. The oxide and some other salts of vanadium have moderate toxicity.

Particularly in the ocean, vanadium is used by some life forms as an active center of enzymes, such as the vanadium bromoperoxidase of some ocean algae. Vanadium is probably a micro-nutrient in mammals, including humans, but its precise role in this regard is unknown.

History

Vanadium was originally discovered by Andrés Manuel del Río, a Spanish-Mexican mineralogist, in 1801. Del Río extracted the element from a sample of Mexican "brown lead" ore, later named vanadinite. He found that its salts exhibit a wide variety of colours, and as a result he named the element panchromium (Greek : παγχρώμιο "all colours"). Later, Del Río renamed the element erythronium (Greek : ερυθρός "red") as most of its salts turned red upon heating. In 1805, the French chemist Hippolyte Victor Collet-Descotils, backed by del Río's friend Baron Alexander von Humboldt, incorrectly declared that del Río's new element was only an impure sample of chromium. Del Río accepted Collet-Descotils' statement and retracted his claim.

In 1831, the Swedish chemist Nils Gabriel Sefström rediscovered the element in a new oxide he found while working with iron ores. Later that same year, Friedrich Wöhler confirmed del Río's earlier work. Sefström chose a name beginning with V, which had not been assigned to any element yet. He called the element vanadium after Old Norse Vanadís (another name for the Norse Vanr goddess Freyja, whose facets include connections to beauty and fertility), because of the many beautifully coloured chemical compounds it produces. In 1831, the geologist George William Featherstonhaugh suggested that vanadium should be renamed "rionium" after del Río, but this suggestion was not followed.

The isolation of vanadium metal proved difficult. In 1831, Berzelius reported the production of the metal, but Henry Enfield Roscoe showed that Berzelius had in fact produced the nitride, vanadium nitride (VN). Roscoe eventually produced the metal in 1867 by reduction of vanadium(II) chloride, VCl_2, with hydrogen. In 1927, pure vanadium was produced by reducing vanadium pentoxide with calcium. The first large-scale industrial use of vanadium in steels was found in the chassis of the Ford Model T, inspired by French race cars. Vanadium steel allowed for reduced weight while simultaneously increasing tensile strength.

Characteristics

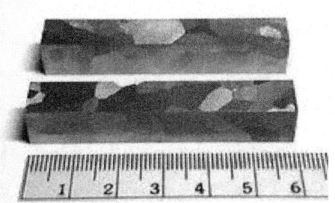

Fig. : High-purity (99.95%) vanadium cuboids, ebeam remelted and macro etched.

Vanadium is a medium hard, ductile, steel-blue metal. Some sources describe vanadium as "soft", perhaps because it is ductile, malleable and not brittle. Vanadium is harder than most metals and steels. It has good resistance to corrosion and it is stable against alkalis, sulfuric and hydrochloric acids. It is oxidized in air at about 933 K (660°C, 1220°F), although an oxide layer forms even at room temperature.

Compounds

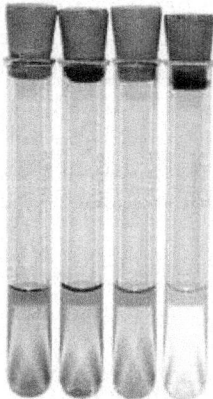

Fig. : From left : $[V(H_2O)_6]^{+2}$, $[V(H_2O)_6]^{+3}$, $[VO(H_2O)_5]^{+2}$ (blue) and $[VO(H_2O)_5]^{+3}$ (yellow).

The chemistry of vanadium is noteworthy for the accessibility of the four adjacent oxidation states 2-5. In aqueous solution, vanadium forms metal aquo complexes the colours are lilac $[V(H_2O)_6]^{2+}$, green $[V(H_2O)_6]^{3+}$, blue $[VO(H_2O)_5]^{2+}$, yellow VO_3^-. Vanadium(II) compounds are reducing agents, and vanadium(V) compounds are oxidizing agents. Vanadium (IV) compounds often exist as vanadyl derivatives which contain the VO^{2+} center.

Ammonium vanadate(V) (NH_4VO_3) can be successively reduced with elemental zinc to obtain the different colours of vanadium in these four oxidation states. Lower oxidation states occur in compounds such as $V(CO)_6$, $[V(CO)_6]^-$ and substituted derivatives.

The most commercially important compound is vanadium pentoxide. It is used as a catalyst for the production of sulfuric acid. This compound oxidizes sulfur dioxide (SO_2) to the trioxide (SO_3). In this redox reaction, sulfur is oxidized from +4 to +6, and vanadium is reduced from +5 to +4 :

$$V_2O_5 + SO_2 \rightarrow 2\ VO_2 + SO_3$$

The catalyst is regenerated by oxidation with air :

$$2\ VO_2 + O_2 \rightarrow V_2O_5$$

Similar oxidations are used in the production of maleic anhydride, phthalic anhydride, and several other bulk organic compounds.

Oxyanions

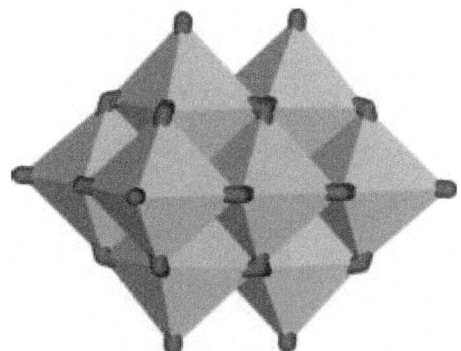

Fig. : The decavanadate structure.

In aqueous solution, vanadium (V) forms an extensive family of oxyanions. The interrelationships within this family are described by the predominance diagram, shows at least 11 species, depending on pH and concentration. The tetrahedral orthovanadate ion, VO_{3-4}, is the principal species present at pH 12-14. Analogies exist between between orthovanadate and orthophosphate owing to the similarity in size and charge of phosphorus (V) and vanadium (V). Orthovanadate VO_{3-4} is used in protein crystallography to study the bio-chemistry of phosphate. The tetrathiovanadate $[VS_4]^{3-}$ is analogous to the orthovanadate ion.

At lower pH's, the monomer $[HVO_4]^{2-}$ and dimer $[V_2O_7]^-$ are formed, with the monomer predominant at vanadium concentration of less than ca. $10^{-2}M$ (pV > 2; pV is equal to minus the logarithm of the total vanadium concentration/M). The formation of the divanadate ion is analogous to the formation of the dichromate ion. As the pH is reduced, further protonation and condensation to polyvanadates occur : at pH 4-6 $[H_2VO_4]^-$ is predominant at pV greater than ca. 4, while at higher concentrations trimers and tetramers are formed. Between pH 2-4 decavanadate predominates, its formation from orthovanadate is represented by this condensation reaction :

$$10\ [VO_4]^{3-} + 24\ H^+ \rightarrow [V_{10}O_{28}]^{6-} + 12\ H_2O$$

In decavanadate, each V(V) center is surrounded by six oxide ligands. Vanadic acid, H_3VO_4 exists only a very low concentrations because protonation of the tetrahedral species $[H_2VO_4]^-$ results in the preferential formation of the octahedral $[VO_2(H_2O)_4]^+$ species. In strongly acidic solutions, pH<2. $[VO_2(H_2O)_4]^+$ is the predominant species, while the oxide V_2O_5 precipitates from solution at high concentrations. The oxide is formally the inorganic anhydride of vanadic acid. The structures of many vanadate compounds have been characterized by X-ray crystallography.

The Pourbaix diagram for vanadium in water, which shows the redox potentials between various vanadium species in different oxidation states is also complex.

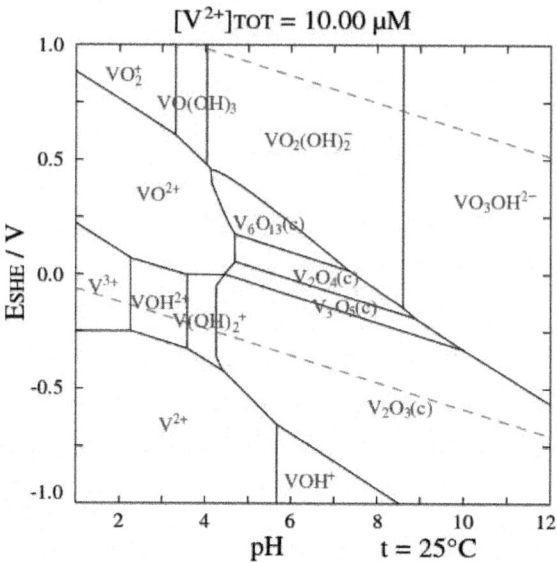

Fig. : The Pourbaix diagram for vanadium in water.

Vanadium (V) also forms various peroxo complexes, most notably in the active site of the vanadium-containing bromoperoxidase enzymes. The species $VO(O)_2(H_2O)_4^+$ is stable in acidic solutions. In alkaline solutions species with 2, 3 and 4 peroxide groups are known; the last forms violet salts with the formula $M_3V(O_2)_4 \cdot nH_2O$ (M = Li, Na, etc.), in which the vanadium has an 8-co-ordinate dodecahedral structure.

Halide Derivatives

Twelve binary halides, compounds with the formula VX_n, are known. VI_4, VCl_5, VBr_5, and VI_5 do not exist or are extremely unstable. In combination with other reagents, VCl_4 is used as a catalyst for polymerization of dienes. Like all binary halides, those of vanadium are Lewis acidic, especially those of V(IV) and V(V). Many of the halides form octahedral complexes with the formula VX_nL_{6-n} (X = halide; L = other ligand).

Many vanadium oxyhalides (formula VO_mX_n) are known. The oxytrichloride and oxytrifluoride, VOF_3) and $VOCl_3$) are the most widely studied. Akin to $POCl_3$, they are volatile, adopt a tetrahedral structures in the gas phase, and are Lewis acidic.

Co-ordination Compounds

Complexes of vanadium (II) and (III) are relatively exchange inert and reducing. Those of V(IV) and V(V) are oxidants. Vanadium ion is rather large and some complexes achieve co-ordination numbers greater than 6, as is the case in $[V(CN)_7]^{4-}$. The co-ordination chemistry of V^{4+} is dominated by the vanadyl center,

VO^{2+}, which binds four other ligands strongly and one weakly (the one trans to the vanadyl center). An example is vanadyl acetylacetonate ($V(O)(O_2C_5H_7)_2$). In this complex, the vanadium is 5-co-ordinate, square pyramidal, meaning that a sixth ligand, such as pyridine, may be attached, though the association constant of this process is small. Many 5-co-ordinate vanadyl complexes have a trigonal bypyramidal geometry, such as $VOCl_2(NMe_3)_2$. The co-ordination chemistry of V^{5+} is dominated by the polyoxovanadates, such as decavanadate.

Organometallic Compounds

Organometallic chemistry of vanadium is well developed, although they are mainly only academic significance. Vanadocene dichloride is a versatile start-ing reagent and even finds some applications in organic chemistry. Vanadium carbonyl, $V(CO)_6$, is a rare example of a paramagnetic metal carbonyl. Reduction yields $V(CO)_{-6}$ (isoelectronic with $Cr(CO)_6$), which may be further reduced with sodium in liquid ammonia to yield $V(CO)_{3-5}$ (isoelectronic with $Fe(CO)_5$).

Occurrence

Metallic vanadium is not found in nature, but is known to exist in about 65 different minerals. Economically significant examples include patronite (VS_4), vanadinite ($Pb_5(VO_4)_3Cl$), and carnotite ($K_2(UO_2)_2(VO_4)_2 \cdot 3H_2O$). Much of the world's vanadium production is sourced from vanadium-bearing magnetite found in ultramafic gab-bro bodies. Vanadium is mined mostly in South Africa, north-western China, and eastern Russia. In 2010 these three countries mined more than 98% of the 56,000 tonnes of produced vanadium.

Vanadium is also present in bauxite and in fossil fuel deposits such as crude oil, coal, oil shale and tar sands. In crude oil, concentrations up to 1200 ppm have been reported. When such oil products are burned, the traces of vanadium may initiate corrosion in motors and boilers. An estimated 110,000 tonnes of vanadium per year are released into the atmosphere by burning fossil fuels. Vanadium has also been detected spectro-scopically in light from the Sun and some other stars.

Production

Most vanadium is used as an alloy called ferro-vanadium as an additive to improve steels. Ferro-vanadium is produced directly by reducing a mixture of vanadium oxide, iron oxides and iron in an electric furnace. Vanadium-bearing magnetite iron ore is the main source for the production of vanadium. The vanadium ends up in pig iron produced from vanadium bearing magnetite. During steel pro-duction, oxygen is blown into the pig iron, oxidizing the carbon and most of the other impurities, forming slag. Depending on the ore used, the slag contains up to 25% of vanadium.

Vanadium metal is obtained *via* a multistep process that begins with the roasting of crushed ore with NaCl or Na_2CO_3 at about 850 °C to give sodium metavanadate. An aqueous extract of this solid is acidified to give "red cake", a polyvanadate salt, which is reduced with calcium metal. As an alternative for small-scale production, vanadium pentoxide is reduced with hydrogen or magnesium. Many other methods are also in use, in all of which vanadium is produced as a by-product of other processes. Purification of vanadium is possible by the crystal bar process developed by Anton Eduard van Arkel and Jan Hendrik de Boer in 1925. It involves the formation of the metal iodide, in this example vanadium (III) iodide, and the subsequent decomposition to yield pure metal.

$$2\,V + 3\,I_2 \rightleftharpoons 2\,VI_3$$

Applications

Alloys

Approximately 85% of vanadium produced is used as ferro-vanadium or as a steel additive. The considerable increase of strength in steel containing small amounts of vanadium was discovered in the beginning of the 20th century. Vanadium forms stable nitrides and carbides, resulting in a significant increase in the strength of the steel. From that time on vanadium steel was used for applications in axles, bicycle frames, crankshafts, gears, and other critical components. There are two groups of vanadium containing steel alloy groups. Vanadium high-carbon steel alloys contain 0.15% to 0.25% vanadium and high speed tool steels (HSS) have a vanadium content of 1% to 5%. For high speed tool steels, a hardness above HRC 60 can be achieved. HSS steel is used in surgical instruments and tools. Some powder metallurgic alloys can contain up to 18% per cent vanadium. The high content of vanadium carbides in those alloys increases the wear resistivity significantly. One application for those alloys are tools and knives.

Vanadium stabilizes the beta form of titanium and increases the strength and temperature stability of titanium. Mixed with aluminium in titanium alloys it is used in jet engines, high-speed airframes and dental implants. One of the common alloys is Titanium 6AL-4V, a titanium alloy with 6% aluminium and 4% vanadium.

Other Uses

Vanadium is compatible with iron and titanium, therefore vanadium foil is used in cladding titanium to steel. The moderate thermal neutron-capture cross-section and the short half-life of the isotopes produced by neutron capture makes vanadium a suitable material for the inner structure of a fusion reactor. Several vanadium alloys show super-conducting behaviour. The first A15 phase superconductor was a vanadium compound, V_3Si, which was discovered in 1952. Vanadium-gallium tape is used in super-conducting magnets (17.5 teslas or 175,000 gauss). The structure of the super-conducting A15 phase of V_3Ga is similar to that of the more common Nb_3Sn and Nb_3Ti.

The most common oxide of vanadium, vanadium pentoxide V_2O_5, is used as a catalyst in manufacturing sulfuric acid by the contact process and as an oxidizer in maleic anhydride production. Vanadium pentoxide is also used in making ceramics. Another oxide of vanadium, vanadium dioxide VO_2, is used in the production of glass coatings, which blocks infrared radiation (and not visible light) at a specific temperature. Vanadium oxide can be used to induce colour centers in corundum to create simulated alexandrite jewelry, although alexandrite in nature is a chrysoberyl. The possibility to use vanadium redox couples in both half-cells, thereby eliminating the problem of cross-contamination by diffusion of ions across the membrane is the advantage of vanadium redox rechargeable batteries. Vanadate can be used for protecting steel against rust and corrosion by electro-chemical conversion coating. Lithium vanadium oxide has been proposed for use as a high energy density anode for lithium ion batteries, at 745 Wh/L when paired with a lithium cobalt oxide cathode. It has been proposed by some researchers that a small amount, 40 to 270 ppm, of vanadium in Wootz steel and Damascus steel, significantly improves the strength of the material, although it is unclear what the source of the vanadium was. Lithium vanadium phosphate has been proposed for a new battery as well, and is very commercially applicable because phosphates are inexpensive and vanadium makes the battery very energy dense.

Biological Role

Vanadium plays a very limited role in biology, and is more important in marine environments than terrestrial ones.

Fig. : Active site of the enzyme vanadium bromoperoxidase, which produces the preponderance of organobromine compoorganobromine compounds.

Vanadoenzymes

A number of species of marine algae produce vanadium-containing vanadium bromoperoxidase as well as the closely related chloroperoxidase and iodoperoxidases. The bromoperoxidase produce an estimated 1–2 million tons of bromoform and 56,000 tons of bromomethane annually. Most naturally occurring organobromine compounds, accounting arise by the action of this enzyme. They catalyse the following reaction (R-H is hydrocarbon substrate) :

$$R\text{-}H + Br^- + H_2O_2 \rightarrow R\text{-}Br + H_2O + OH^-$$

A vanadium nitrogenase is used by some nitrogen-fixing micro-organisms, such as *Azotobacter*. In this role vanadium replaces more common molybdenum or iron, and gives the nitrogenase slightly different properties.

Vanadium Accumulation in Tunicates and Ascidians

Vanadium is essential to ascidians and tunicates, where it is stored in the highly acidified vacuoles of certain blood cell types, designated vanadocytes. Vanabins (vanadium binding proteins) have been identified in the cytoplasm of such cells. The concentration of vanadium in these ascidians' blood is up to ten million times higher than the concentration of vanadium in the seawater around them, the seawater contains 1 to 2 µg/l. The function of this vanadium concentration system, and these vanadium-containing proteins, is still unknown but the vanadocytes are later deposited just under the outer surface of the tunic where their presence may deter predation.

Fungi

Several species of macrofungi, namely *Amanita muscaria* and related species, accumulate vanadium (up to 500 mg/kg in dry weight). Vanadium is present in the co-ordination complex amavadin, in fungal fruit-bodies. However, the biological importance of the accumulation process is unknown. Toxin functions or peroxidase enzyme functions have been suggested.

Mammals and Birds

Deficiencies in vanadium result in reduced growth and impaired reproduction in rats and chickens. Vanadium is a relatively controversial dietary supplement, used primarily for increasing insulin sensitivity and body-building. Whether it works for the latter purpose has not been proven; some evidence suggests that athletes who take it are merely experiencing a placebo effect. Vanadyl sulfate may improve glucose control in people with type 2 diabetes. Decavanadate and oxovanadates appear to play a role in a variety of biochemical processes, such as those relating to oxidative stress.

Safety

All vanadium compounds should be considered toxic. Tetravalent $VOSO_4$ has been reported to be over 5 times more toxic than trivalent V_2O_3. The Occupational Safety and Health Administration has set an exposure limit of 0.05 mg/m^3 for vanadium pentoxide dust and 0.1 mg/m^3 for vanadium pentoxide fumes in workplace air for an 8-hour workday, 40-hour work week. The National Institute for Occupational Safety and Health has recommended that 35 mg/m^3 of vanadium be considered immediately dangerous to life and health. This is the exposure level of a chemical that is likely to cause permanent health problems or death.

Vanadium compounds are poorly absorbed through the gastrointestinal system. Inhalation exposures to vanadium and vanadium compounds result primarily in adverse effects on the respiratory system. Quantitative data are, however, insufficient to derive a sub-chronic or chronic inhalation reference dose. Other effects have been reported after oral or inhalation exposures on blood parameters, on liver, on neurological development in rats, and other organs.

There is little evidence that vanadium or vanadium compounds are reproductive toxins or teratogens. Vanadium pentoxide was reported to be carcinogenic in male rats and male and female mice by inhalation in an NTP study, although the interpretation of the results has recently been disputed. Vanadium has not been classified as to carcinogenicity by the United States Environmental Protection Agency.

Vanadium traces in diesel fuels present a corrosion hazard; it is the main fuel component influencing high temperature corrosion. During combustion, it oxidizes and reacts with sodium and sulfur, yielding vanadate compounds with melting points down to 530 °C, which attack the passivation layer on steel, rendering it susceptible to corrosion. The solid vanadium compounds also cause abrasion of engine components.

ORGANOVANADIUM CHEMISTRY

Organovanadium chemistry is the chemistry of organometallic compounds containing a carbon to vanadium (V) chemical bond. Organovanadium compounds are of some relevance to organic synthesis and to polymer chemistry as reagents and catalysts.

Oxidation states for vanadium are +2, +3, +4 and +5. Low valency vanadium is usually stabilized with carbonyl ligands. Oxo ligands for example in vanadyl ions are common when the valency increases. In most compounds outside the oxidation state of +5, vanadium is paramagnetic hampering NMR spectroscopy. Typical vanadium precursors are vanadium (III) chloride and its adduct with THF $VCl_3(THF)_3$ and vanadium tetrachloride.

Synthesis

Common ligands for vanadium are carbonyl, phosphine and cyclopentadienyl.

Vanadium carbonyl can be prepared from vanadium (II) chloride, vanadium oxytrichloride or vanadium acetylacetonate in reaction with carbon monoxide, pyridine and magnesium or zinc as reducing agent :

The hexacarbonylvanadate ion can be prepared in from VCl_3 :

$$4\,Na + VCl_3 + 6\,CO + 2\,diglyme \rightarrow [Na(diglyme)_2][V(CO)_6] + 3\,NaCl$$

Vanadocene dichloride is prepared from sodium cyclopentadienyl :

$$NaC_5H_5 + VCl_4 \rightarrow VCp_2Cl_2$$

Reduction of this compound gives the parent vanadocene $(Cp)_2V$:

$$VCp_2Cl_2 + LiAlH_4 \rightarrow V(Cp)_2$$

Vanadocene is the first stable metallocene in the transition metals. Titanocene and zirconocene only exist as the dichlorides titanocene dichloride and zirconocene dichloride. It reacts as a metal carbene to alkyne and nitrile ligands and is also used as a reducing agent.

Indene can also be a ligand :

$VCl_3(THF)_3 + Zn \rightarrow [VCl_2(THF)]_n$

$VCl_2(THF)]_n + Na(indenyl) \rightarrow V(C_9H_7)_2$

and benzene :

$VCl_4 + AlCl_3 + C_6H_6 \rightarrow [V(\eta^6C_6H_6)2][AlH_4]$

$[V(\eta^6C_6H_6)2][AlH_4] + H_2O \rightarrow V(\eta^6C_6H_6)_2$

Many η^1 alkyl and aryl complexes exist for example with mesitylene groups :

$VCl_3(THF)_3 + (mes)MgBr \rightarrow V(mes)_3(THF)$

$V(mes)_3(THF) + LiMes \rightarrow Li[V(mes)_4]$

$Li[V(mes)_4] + air \rightarrow V(mes)_4(THF)$

or norbornyl groups :

$VCl_4 + Li(norbornyl) \rightarrow V(norbornyl)_4$

Vanadium oxytrichloride is a starting material for vanadium(V) compounds :

$VOCl_3 + Li(mes) \rightarrow Li[VO(mes)_3]$

$Li[VO(mes)_3] + chloranil \rightarrow VO(mes)_3$

$VOCl_3 + ZnPh_2 \rightarrow VOPhCl_2$

Reactions

Vanadium compounds appear as catalysts and reagents in several specific reactions. Two main reactions are coupling reactions :

$VCl_3 + RLi + R'-CHO \rightarrow R-C(O)-R'$

and insertion reaction into the C-V bond for example in a sequence forming acetone :

$CpVCl_2(PPh_3)_2 + MeLi \rightarrow CpVMe_2(PPh_3)_2$

$CpVMe_2(PPh_3)_2 + CO \rightarrow CpVCO_2(PPh_3)_2 + Me-C(O)-Me$

Higher Group 5 Organometallics

In **organoniobium** (Nb) and **organotantalum** (Ta) chemistry the oxidation state +5 is favoured with convenient starting materials niobium (V) chloride and tantalum (V) chloride. In high oxidation states Nb and Ta compounds resemble that of group 4 elements being strongly oxophilic with strong resistance to reduction. In low oxidation states Nb and Ta compounds resemble group 6 elements with ease of formation of metal-metal bonds. Both group 5 & 6 elements form multiple bonds with non-metal ligands as in metal carbenes, nitrenes, oxides and sulfide for example in a Ta=C metal carbene :

$TaCp_3Cl + Ph_3P=CH2 \rightarrow CpTa_2$

Niobocene dichloride is a metallocene and $NbCl_5$ is a Lewis acid catalyst in carbonyl-ene reactions for example :

CHROMIUM

Chromium is a chemical element which has the symbol **Cr** and atomic number 24. It is the first element in Group 6. It is a steely-gray, lustrous, hard and brittle metal which takes a high polish, resists tarnishing, and has a high melting point. The name of the element is derived from the Greek word χρῶμα, *chrōma*, meaning colour, because many of its compounds are intensely coloured.

Chromium oxide was used by the Chinese in the Qin dynasty over 2,000 years ago to coat metal weapons found with the Terracotta Army. Chromium was discovered as an element after it came to the attention of the western world in the red crystalline mineral crocoite (lead (II) chromate), discovered in 1761 and initially used as a pigment. Louis Nicolas Vauquelin first isolated chromium metal from this mineral in 1797. Since Vauquelin's first production of metallic chromium, small amounts of native (free) chromium metal have been discovered in rare minerals, but these are not used commercially. Instead, nearly all chromium is commercially extracted from the single commercially viable ore chromite, which is iron chromium oxide ($FeCr_2O_4$). Chromite is also now the chief source of chromium for chromium pigments.

Chromium metal and ferro-chromium alloy are commercially produced from chromite by silicothermic or aluminothermic reactions, or by roasting and leaching processes. Chromium metal has proven of high value due to its high corrosion resistance and hardness. A major development was the discovery that steel could be made highly resistant to corrosion and discolouration by adding metallic chromium to form stainless steel. This application, along with chrome plating (electroplating with chromium) currently comprise 85% of the commercial use for the element, with applications for chromium compounds forming the remainder.

Trivalent chromium (Cr (III)) ion is possibly required in trace amounts for sugar and lipid metabolism, although the issue remains in debate. In larger amounts and in different forms, chromium can be toxic and carcinogenic. The most prominent example of toxic chromium is hexavalent chromium). Abandoned chromium production sites often require environmental cleanup.

Characteristics

Physical

Chromium is remarkable for its magnetic properties : it is the only elemental solid which shows anti-ferromagnetic ordering at room temperature. Above 38 °C, it transforms into a paramagnetic state.

Passivation

Chromium metal left standing in air is passivated by oxygen, forming a thin protective oxide surface layer. This layer is a spinel structure only a few atoms thick. It is very dense, and prevents the diffusion of oxygen into the underlying material. This barrier is in contrast to iron or plain carbon steels, where the oxygen migrates into the underlying material and causes rusting. The passivation can be enhanced by short contact with oxidizing acids like nitric acid. Passivated chromium is stable against acids. The opposite effect can be achieved by treatment with a strong reducing agent that destroys the protective oxide layer on the metal. Chromium metal treated in this way readily dissolves in weak acids.

Chromium, unlike metals such as iron and nickel, does not suffer from hydrogen embrittlement. However, it does suffer from nitrogen embrittlement, reacting with nitrogen from air and forming brittle nitrides at the high temperatures necessary to work the metal parts.

Occurrence

Chromium is the 22nd most abundant element in Earth's crust with an average concentration of 100 ppm. Chromium compounds are found in the environment, due to erosion of chromium-containing rocks and can be distributed by volcanic eruptions. The concentrations range in soil is between 1 and 300 mg/kg, in sea water 5 to 800 µg/liter, and in rivers and lakes 26 µg/liter to 5.2 mg/liter. Chromium is mined as chromite ($FeCr_2O_4$) ore. About two-fifths of the chromite ores and concentrates in the world are produced in South Africa, while Kazakhstan, India, Russia, and Turkey are also substantial producers. Untapped chromite deposits are plentiful, but geographically concentrated in Kazakhstan and southern Africa.

Although rare, deposits of native chromium exist. The Udachnaya Pipe in Russia produces samples of the native metal. This mine is a kimberlite pipe, rich in diamonds, and the reducing environment helped produce both elemental chromium and diamond.

The relation between Cr (III) and Cr (VI) strongly depends on pH and oxidative properties of the location, but in most cases, the Cr (III) is the dominating species, although in some areas the ground water can contain up to 39 µg/liter of total chromium of which 30 µg/liter is present as Cr (VI).

Chromium is a member of the transition metals, in group 6. Chromium(0) has an electronic configuration of $4s^13d^5$, owing to the lower energy of the high spin configuration. Chromium exhibits a wide range of possible oxidation states, where the +3 state is most stable energetically; the +3 and +6 states are most commonly observed in chromium compounds, whereas the +1, +4 and +5 states are rare.

The following is the Pourbaix diagram for chromium in pure water, perchloric acid or sodium hydroxide :

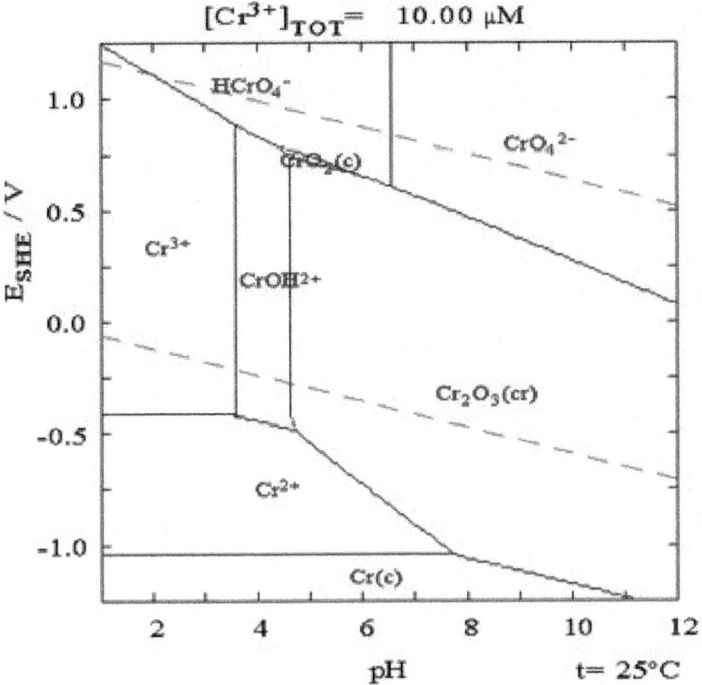

$[Cr^{3+}]_{TOT} = 10.00\ \mu M$

Chromium (III)

A large number of chromium (III) compounds are known. Chromium (III) can be obtained by dissolving elemental chromium in acids like hydrochloric acid or sulfuric acid. The Cr3+ ion has a similar radius (63 pm) to the Al3+ ion (radius 50 pm), so they can replace each other in some compounds, such as in chrome alum and alum. When a trace amount of Cr3+ replaces Al3+ in corundum (aluminium oxide, Al_2O_3), the red-coloured ruby is formed.

Chromium (III) ions tend to form octahedral complexes. The colours of these complexes is determined by the ligands attached to the Cr centre. The commercially available chromium (III) chloride hydrate is the dark green complex $[CrCl_2(H_2O)_4]$ Cl. Closely related compounds have different colours : pale green $[CrCl(H_2O)_5]$ Cl_2 and the violet $[Cr(H_2O)_6]Cl_3$. If water-free green chromium (III) chloride is dissolved in water then the green solution turns violet after some time, due to the substitution of water by chloride in the inner co-ordination sphere. This kind of reaction is also observed with solutions of chrome alum and other water-soluble chromium (III) salts.

Chromium (III) hydroxide ₃) is amphoteric, dissolving in acidic solutions to form $[Cr(H_2O)_6]^{3+}$, and in basic solutions to form $[Cr(OH)_6]^{3-}$. It is dehydrated by heating to form the green chromium (III) oxide , which is the stable oxide with a crystal structure identical to that of corundum.

Chromium(VI)

Chromium (VI) compounds are powerful oxidants at low or neutral pH. Most important are chromate anion (CrO_{2-4}) and dichromate ($Cr_2O_7^{2-}$) anions, which exist in equilibrium :

$$2 [CrO_4]^{2-} + 2 H^+ \rightleftharpoons [Cr_2O_7]^{2-} + H_2O$$

Chromium (VI) halides are known also and include the hexafluoride CrF_6 and chromyl chloride (CrO_2Cl_2).

Sodium chromate is produced industrially by the oxidative roasting of chromite ore with calcium or sodium carbonate. The dominant species is therefore, by the law of mass action, determined by the pH of the solution. The change in equilibrium is visible by a change from yellow (chromate) to orange (dichromate), such as when an acid is added to a neutral solution of potassium chromate. At yet lower pH values, further condensation to more complex oxyanions of chromium is possible.

Both the chromate and dichromate anions are strong oxidizing reagents at low pH : Cr

2O2–

7 + 14 H

3O+

+ 6 e⁻ → 2 Cr3+

+ 21 H

2O ($\varepsilon_0 = 1.33$ V)

They are, however, only moderately oxidizing at high pH :

CrO_{2-}

4 + 4 H

2O + 3 e⁻ → Cr(OH)

3 + 5 OH–

($\varepsilon_0 = -0.13$ V)

Chromium (VI) compounds in solution can be detected by adding an acidic hydrogen peroxide solution. The unstable dark blue chromium (VI) peroxide (CrO_5) is formed, which can be stabilized as an ether adduct CrO :

5·OR

2.

Chromic acid has the hypothetical formula H

2CrO

It is a vaguely described chemical, despite many well-defined chromates and dichromates being known. The dark red chromium (VI) oxide CrO_3, the acid anhydride of chromic acid, is sold industrially as "chromic acid". It can be produced by mixing sulfuric acid with dichromate, and is a strong oxidizing agent.

Chromium (V) and chromium (IV)

The oxidation state +5 is only realized in few compounds but are intermediates in many reactions involving oxidations by chromate. The only binary compound is the volatile chromium (V) fluoride (CrF_5). This red solid has a melting point of 30 °C and a boiling point of 117 °C. It can be synthesized by treating chromium metal with fluorine at 400 °C and 200 bar pressure. The peroxochromate (V) is another example of the +5 oxidation state. Potassium peroxochromate (K_3[Cr is made by reacting potassium chromate with hydrogen peroxide at low temperatures. This red brown compound is stable at room temperature but decomposes spontaneously at 150–170 °C.

Compounds of chromium (IV) (in the +4 oxidation state) are slightly more common than those of chromium (V). The tetrahalides, CrF_4, $CrCl_4$, and $CrBr_4$, can be produced by treating the trihalides with the corresponding halogen at elevated temperatures. Such compounds are susceptible to disproportionation reactions and are not stable in water.

Chromium (II)

Many chromium(II) compounds are known, including the water-stable chromium (II) chloride, $CrCl_2$, which can be made by reduction of chromium (III) chloride with zinc. The resulting bright blue solution is only stable at neutral pH. Many chromous carboxylates are also known, most famously, the red chromous acetate ($Cr_2(O_2CCH_3)_4$), which features a quadruple bond.

Chromium (I)

Most Cr (I) compounds are obtained by oxidation of electron-rich, octahedral Cr (0) complexes. Other Cr (I) complexes contain cyclopentadienyl ligands. As verified by X-ray diffraction, a Cr-Cr quintuple bond (length 183.51(4) pm) has also been described. Extremely bulky monodentate ligands stabilize this compound by shielding the quintuple bond from further reactions.

History

Weapons found in burial pits dating from the late 3rd century B.C. Qin Dynasty of the Terracotta Army near Xi'an, China have been analyzed by archaeologists. Although buried more than 2,000 years ago, the ancient bronze tips of crossbow bolts and swords found at the site showed unexpectedly little corrosion, possibly because the bronze was deliberately coated with a thin layer of chromium oxide. However, this oxide layer was not chromium metal or chrome plating as we know it.

Chromium minerals as pigments came to the attention of the west in the 18th century. On 26 July, 1761, Johann Gottlob Lehmann found an orange-red mineral in the Beryozovskoye mines in the Ural Mountains which he named *Siberian red lead*. Though misidentified as a lead compound with selenium and iron components, the mineral was in fact crocoite (*lead chromate*) with a formula of $PbCrO_4$.

In 1770, Peter Simon Pallas visited the same site as Lehmann and found a red lead mineral that had useful properties as a pigment in paints. The use of Siberian red lead as a paint pigment then developed rapidly. A bright yellow pigment made from crocoite also became fashionable.

In 1797, Louis Nicolas Vauquelin received samples of crocoite ore. He produced chromium trioxide (CrO_3) by mixing crocoite with hydrochloric acid. In 1798, Vauquelin discovered that he could isolate metallic chromium by heating the oxide in a charcoal oven, making him the discoverer of the element. Vauquelin was also able to detect traces of chromium in precious gemstones, such as ruby or emerald.

During the 1800s, chromium was primarily used as a component of paints and in tanning salts. At first, crocoite from Russia was the main source, but in 1827, a larger chromite deposit was discovered near Baltimore, United States. This made the United States the largest producer of chromium products till 1848 when large deposits of chromite were found near Bursa, Turkey.

Chromium is also known for its luster when polished. It is used as a protective and decorative coating on car parts, plumbing fixtures, furniture parts and many other items, usually applied by electroplating. Chromium was used for electroplating as early as 1848, but this use only became widespread with the development of an improved process in 1924.

Metal alloys now account for 85% of the use of chromium. The remainder is used in the chemical industry and refractory and foundry industries.

Production

Approximately 4.4 million metric tons of marketable chromite ore were produced in 2000, and converted into ~3.3 million tons of ferro-chrome with an approximate market value of 2.5 billion United States dollars. The largest producers of chromium ore have been South Africa (44%) India (18%), Kazakhstan (16%) Zimbabwe (5%), Finland (4%) Iran (4%) and Brazil (2%) with several other countries producing the rest of less than 10% of the world production.

The two main products of chromium ore refining are ferro-chromium and metallic chromium. For those products the ore smelter process differs considerably. For the production of ferro-chromium, the chromite ore ($FeCr_2O_4$) is reduced in large scale in electric arc furnace or in smaller smelters with either aluminium or silicon in an aluminothermic reaction.

For the production of pure chromium, the iron has to be separated from the chromium in a two step roasting and leaching process. The chromite ore is heated with a mixture of calcium carbonate and sodium carbonate in the presence of air. The chromium is oxidized to the hexavalent form, while the iron forms the stable Fe_2O_3. The subsequent leaching at higher elevated temperatures dissolves the chromates and leaves the insoluble iron oxide. The chromate is converted by sulfuric acid into the dichromate.

$$4\ FeCr_2O_4 + 8\ Na_2CO_3 + 7\ O_2 \rightarrow 8\ Na_2CrO_4 + 2\ Fe_2O_3 + 8\ CO_2$$
$$2\ Na_2CrO_4 + H_2SO_4 \rightarrow Na_2Cr_2O_7 + Na_2SO_4 + H_2O$$

The dichromate is converted to the chromium(III) oxide by reduction with carbon and then reduced in an aluminothermic reaction to chromium.

$$Na_2Cr_2O_7 + 2\ C \rightarrow Cr_2O_3 + Na_2CO_3 + CO$$
$$Cr_2O_3 + 2\ Al \rightarrow Al_2O_3 + 2\ Cr$$

Applications

Metallurgy

The strengthening effect of forming stable metal carbides at the grain boundaries and the strong increase in corrosion resistance made chromium an important alloying material for steel. The high-speed tool steels contain between 3 and 5% chromium. Stainless steel, the main corrosion-proof metal alloy, is formed when chromium is added to iron in sufficient concentrations, usually above 11%. For its formation, ferro-chromium is added to the molten iron. Also nickel-based alloys increase in strength due to the formation of discrete, stable metal carbide particles at the grain boundaries. For example, Inconel 718 contains 18.6% chromium. Because of the excellent high-temperature properties of these nickel superalloys, they are used in jet engines and gas turbines *in lieu* of common structural materials.

The relative high hardness and corrosion resistance of unalloyed chromium makes it a good surface coating, being still the most "popular" metal coating with unparalleled combined durability. A thin layer of chromium is deposited on pre-treated metallic surfaces by electroplating techniques. There are two deposition methods : Thin, below 1 µm thickness, layers are deposited by chrome plating, and are used for decorative surfaces. If wear-resistant surfaces are needed then thicker chromium layers are deposited. Both methods normally use acidic chromate or dichromate solutions. To prevent the energy-consuming change in oxidation state, the use of chromium (III) sulfate is under development, but for most applications, the established process is used.

In the chromate conversion coating process, the strong oxidative properties of chromates are used to deposit a protective oxide layer on metals like aluminium, zinc and cadmium. This passivation and the self-healing properties by the chromate stored in the chromate conversion coating, which is able to migrate to local defects, are the benefits of this coating method. Because of environmental and health regulations on chromates, alternative coating method are under development.

Anodizing of aluminium is another electro-chemical process, which does not lead to the deposition of chromium, but uses chromic acid as electrolyte in the solution. During anodization, an oxide layer is formed on the aluminium. The use of chromic acid, instead of the normally used sulfuric acid, leads to a slight difference of these oxide layers. The high toxicity of Cr (VI) compounds, used in the established chromium electroplating process, and the strengthening of safety and environmental regulations demand a search for substitutes for chromium or at least a change to less toxic chromium (III) compounds.

Dye and Pigment

The mineral crocoite (lead chromate $PbCrO_4$) was used as a yellow pigment shortly after its discovery. After a synthesis method became available starting from the more abundant chromite, chrome yellow was, together with cadmium yellow, one of the most used yellow pigments. The pigment does not photodegrade, but it tends to darken due to the formation of chromium (III) oxide. It has a strong colour, and was used for school buses in the US and for Postal Service (for example Deutsche Post) in Europe. The use of chrome yellow declined due to environmental and safety concerns and was replaced by organic pigments or alternatives free from lead and chromium. Other pigments based on chromium are, for example, the bright red pigment chrome red, which is a basic lead chromate ($PbCrO_4 \cdot Pb(OH)_2$). A very important chromate pigment, which was used widely in metal primer formulations, was zinc chromate, now replaced by zinc phosphate. A wash primer was formulated to replace the dangerous practice of pretreating aluminium aircraft bodies with a phosphoric acid solution. This used zinc tetroxychromate dispersed in a solution of polyvinyl butyral. An 8% solution of phosphoric acid in solvent was added just before application. It was found that an easily oxidized alcohol was an essential ingredient. A thin layer of about 10–15 μm was applied, which turned from yellow to dark green when it was cured. There is still a question as to the correct mechanism. Chrome green is a mixture of Prussian blue and chrome yellow, while the chrome oxide green is chromium (III) oxide.

Chromium oxides are also used as a green colour in glassmaking and as a glaze in ceramics. Green chromium oxide is extremely light-fast and as such is used in cladding coatings. It is also the main ingredient in IR reflecting paints, used by the armed forces, to paint vehicles, to give them the same IR reflectance as green leaves.

Synthetic Ruby and the First Laser

Natural rubies are corundum (aluminum oxide) crystals that are coloured red (the rarest type) due to chromium (III) ions (other colours of corundum gems are termed sapphires). A red-coloured artificial ruby may also be achieved by doping chromium (III) into artificial corundum crystals, thus making chromium a requirement for making synthetic rubies. Such a synthetic ruby crystal was the basis for the first laser, produced in 1960, which relied on stimulated emission of light from the chromium atoms in such a crystal.

Wood Preservative

Because of their toxicity, chromium(VI) salts are used for the preservation of wood. For example, chromated copper arsenate (CCA) is used in timber treatment to protect wood from decay fungi, wood attacking insects, including termites, and marine borers. The formulations contain chromium based on the oxide CrO_3 between 35.3% and 65.5%. In the United States, 65,300 metric tons of CCA solution have been used in 1996.

Tanning

Chromium (III) salts, especially chrome alum and chromium (III) sulfate, are used in the tanning of leather. The chromium (III) stabilizes the leather by cross-linking the collagen fibers. Chromium tanned leather can contain between 4 and 5% of chromium, which is tightly bound to the proteins. Although the form of chromium used for tanning is not the toxic hexavalent variety, there remains interest in management of chromium in the tanning industry such as recovery and reuse, direct/indirect recycling, use of less chromium or "chrome-less" tanning are practiced to better manage chromium in tanning.

Refractory Material

The high heat resistivity and high melting point makes chromite and chromium (III) oxide a material for high temperature refractory applications, like blast furnaces, cement kilns, molds for the firing of bricks and as foundry sands for the casting of metals. In these applications, the refractory materials are made from mixtures of chromite and magnesite. The use is declining because of the environmental regulations due to the possibility of the formation of chromium (VI).

Catalysts

Several chromium compounds are used as catalysts for processing hydrocarbons. For example the Phillips catalysts for the production of polyethylene are mixtures of chromium and silicon dioxide or mixtures of chromium and titanium and aluminium oxide. Fe-Cr mixed oxides are employed as high-temperature catalysts for the water gas shift reaction. Copper chromite is a useful hydrogenation catalyst.

Other Use

- Chromium (IV) oxide (CrO_2) is a magnetic compound. Its ideal shape anisotropy, which imparts high coercivity and remnant magnetization, made it a compound superior to the γ-Fe_2O_3. Chromium (IV) oxide is used to manufacture magnetic tape used in high-performance audio tape and standard audio cassettes. Chromates can prevent corrosion of steel under wet conditions, and therefore, chromates are added to drilling muds.

- Chromium (III) oxide is a metal polish known as green rouge.
- Chromic acid is a powerful oxidizing agent and is a useful compound for cleaning laboratory glassware of any trace of organic compounds. It is prepared *in situ* by dissolving potassium dichromate in concentrated sulfuric acid, which is then used to wash the apparatus. Sodium dichromate is sometimes used because of its higher solubility (50 g/L versus 200 g/L respectively). The use of dichromate cleaning solutions is now phased out due to the high toxicity and environmental concerns. Modern cleaning solutions are highly effective and chromium free. Potassium dichromate is a chemical reagent, used as a titrating agent. It is also used as a mordant (*i.e.*, a fixing agent) for dyes in fabric.

Biological Role

Trivalent chromium (Cr (III) or Cr^{3+}) occurs in trace amounts in foods and waters, and appears to be benign. In contrast, hexavalent chromium or Cr^{6+}) is very toxic and mutagenic when inhaled. Cr (VI) has not been established as a carcinogen when in solution, although it may cause allergic contact dermatitis (ACD).

Chromium deficiency, involving a lack of Cr (III) in the body, or perhaps some complex of it, such as glucose tolerance factor is controversial, or is at least extremely rare. Chromium has no verified biological role and has been classified by some as not essential for mammals. However, other reviews have regarded it as an essential trace element in humans.

Chromium deficiency has been attributed to only three people on long-term parenteral nutrition, which is when a patient is fed a liquid diet through intravenous drips for long periods of time.

Although no biological role for chromium has ever been demonstrated, dietary supplements for chromium include chromium (III) picolinate, chromium (III) polynicotinate, and related materials. The benefit of those supplements is questioned by some studies. The use of chromium-containing dietary supplements is controversial, owing to the absence of any verified biological role, the expense of these supplements, and the complex effects of their use. The popular dietary supplement chromium picolinate complex generates chromosome damage in hamster cells (due to the picolinate ligand). In the United States the dietary guidelines for daily chromium uptake were lowered in 2001 from 50–200 μg for an adult to 35 μg (adult male) and to 25 μg (adult female).

No comprehensive, reliable database of chromium content of food currently exists. Data reported prior to 1980 is unreliable due to analytical error. Chromium content of food varies widely due to differences in soil mineral content, growing season, plant cultivar, and contamination during processing. In addition, large amounts of chromium (and nickel) leech into food cooked in stainless steel.

Precautions

Water insoluble chromium (III) compounds and chromium metal are not considered a health hazard, while the toxicity and carcinogenic properties of chromium (VI) have been known for a long time. Because of the specific transport mechanisms, only limited amounts of **chromium (III)** enter the cells. Several *in vitro* studies indicated that high concentrations of chromium (III) in the cell can lead to DNA damage. Acute oral toxicity ranges between 1.5 and 3.3 mg/kg. The proposed beneficial effects of chromium (III) and the use as dietary supplements yielded some controversial results, but recent reviews suggest that moderate uptake of chromium (III) through dietary supplements poses no risk.

Cr (VI)

The acute oral toxicity for chromium (VI) ranges between 50 and 150 µg/kg. In the body, chromium (VI) is reduced by several mechanisms to chromium (III) already in the blood before it enters the cells. The chromium(III) is excreted from the body, whereas the chromate ion is transferred into the cell by a transport mechanism, by which also sulfate and phosphate ions enter the cell. The acute toxicity of chromium(VI) is due to its strong oxidational properties. After it reaches the blood stream, it damages the kidneys, the liver and blood cells through oxidation reactions. Hemolysis, renal and liver failure are the results of these damages. Aggressive dialysis can improve the situation.

The carcinogenity of chromate dust is known for a long time, and in 1890 the first publication described the elevated cancer risk of workers in a chromate dye company. Three mechanisms have been proposed to describe the genotoxicity of chromium (VI). The first mechanism includes highly reactive hydroxyl radicals and other reactive radicals which are by products of the reduction of chromium (VI) to chromium (III). The second process includes the direct binding of chromium(V), produced by reduction in the cell, and chromium (IV) compounds to the DNA. The last mechanism attributed the genotoxicity to the binding to the DNA of the end product of the chromium (III) reduction.

Chromium salts (chromates) are also the cause of allergic reactions in some people. Chromates are often used to manufacture, amongst other things, leather products, paints, cement, mortar and anti-corrosives. Contact with products containing chromates can lead to allergic contact dermatitis and irritant dermatitis, resulting in ulceration of the skin, sometimes referred to as "chrome ulcers". This condition is often found in workers that have been exposed to strong chromate solutions in electroplating, tanning and chrome-producing manufacturers.

Environmental Issues

As chromium compounds were used in dyes and paints and the tanning of leather, these compounds are often found in soil and groundwater at abandoned industrial sites, now needing environmental cleanup and remediation per the treatment of brownfield land. Primer paint containing hexavalent chromium is still widely used for aerospace and automobile refinishing applications.

In 2010, the Environmental Working Group studied the drinking water in 35 American cities. The study was the first nationwide analysis measuring the presence of the chemical in U.S. water systems. The study found measurable hexavalent chromium in the tap water of 31 of the cities sampled, with Norman, Oklahoma, at the top of list; 25 cities had levels that exceeded California's proposed limit. Note: Concentrations of Cr (VI) in US municipal drinking water supplies reported by EWG are within likely, natural background levels for the areas tested and not necessarily indicative of industrial pollution (CalEPA Fact Sheet), as asserted by EWG. This factor was not taken into consideration in their report.

ORGANOCHROMIUM CHEMISTRY

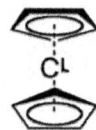

Organochromium chemistry is a branch of organometallic chemistry that deals with organic compounds containing a chromium to carbon bond and their reactions. The field is of some relevance to organic synthesis. The relevant oxidation states for chromium range from -2 to +6.

History

The first organochromium compound was described in 1919 by Franz Hein. He treated phenylmagnesium bromide with chromium (III) chloride to give a new product (after hydrolysis) which he incorrectly identified as pentaphenyl chromium bromide (Ph_5CrBr). Years later, in 1957 H.H. Zeiss *et. al.* repeated Hein's experiments and correctly arrived at a cationic bisarene chromium sandwich compound (Ar_2Cr^+). Bis (benzene) chromium itself was discovered around the same time in 1956 by Ernst Otto Fischer by reaction of chromium (III) chloride, benzene and aluminum chloride. The related compound chromocene was discovered a few years earlier in 1953 also by Fischer.

$$2\ \bigcirc\ +\ CrCl_3 \xrightarrow[\substack{140\ °C \\ \text{High pressure}}]{AlCl_3} Cr^+ \xrightarrow[\substack{\text{aq.} \\ NaOH}]{Na_2S_2O_4} Cr$$

In another development, Anet and Leblanc also in 1957 prepared a benzyl chromium solution from benzyl bromide and chromium (II) perchlorate. This reaction involves one-electron oxidative addition of the carbon-bromine bond, a process which was shown by Kochi to be a case of double single electron transfer, first to give the benzyl free radical and then to the benzyl anion.

G. Wilke *et. al.* introduced tris-(η-allyl) chromium in 1963 as an early Ziegler-Natta catalyst (but not successful in the long-run) Chromocene compounds were first employed in ethylene polymerization in 1972 by Union Carbide and continue to used today in the industrial production of high-density polyethylene.

The organochromium compound (phenylmethoxycarbene)pentacarbonyl-chromium, $Ph(OCH_3)C=Cr(CO)_5$ was the first carbene complex to be crystallo-graphically characterized by Fischer in 1967 (now called a Fischer carbene). The first ever carbyne, this one also containing chromium, made its debut in 1973.

carbene   carbyne

The first example of a proposed metal-metal quintuple bond is found in a compound of the type $[CrAr]_2$, where Ar is a bulky aryl ligand.

Applications in Organic Synthesis

Although organochromium chemistry is heavily employed in industrial catalysis, relatively few reagents have been developed for applications in organic synthesis. Two are the Nozaki-Hiyama-Kishi reaction (transmetallation with organonickel intermediate) and the Takai olefination (oxidation of Cr (II) to Cr (III) while replacing halogens). In a niche exploit, certain tricarbonylchromium complexes display benzylic activation.

Organochromium Compounds

Organochromium compounds can be divided into these broad compound classes :

- Sandwich compounds : chromocenes Cp_2Cr and Bis (benzene) chromium derivatives $(ArH)_2Cr$. More commonly studied are half-sandwich complexes like $(ArH)Cr(CO)_3$.
- Chromium carbenes $(R_1)(R_2)C::CrL_n$ and carbynes $(R:::CrL_n)$
- Chromium (III) complexes $RCrL_5$.

Ethylene Polymerization and Oligomerization

Chromium catalysts are important in ethylene polymerization. The **Phillips catalyst** are prepared by impregnating chromium (VI) oxide on silica followed activation in dry air at high temperatures. The bright yellow catalyst becomes reduced by the ethylene to afford a probable Cr (II) species that is catalytically active. A related catalytic systems developed by Union Carbide and DSM are also based on silica with chromocene and other chromium complexes. How these catalysts work is unclear. One model system describes it as co-ordination polymerization :

polyethylene
Mn = 10,000 - 20,000

With two THF ligands the catalyst is stable but in dichloromethane one ligand is lost to form a 13 electron chromium intermediate. This enables side-on addition of an ethylene unit and a polymer chain can grow by migratory insertion.

Chromium compounds also catalyse the trimerization of ethylene to produce the monomer 1-hexene.

Comparisons with Heavier Group 6 Organometallics

The heavier group 6 elements molybdenum and tungsten form organometallic compounds similar to those for chromium but also with differences. Whereas Cr (III) aquo alkyl compounds are well studied, the corresponding Mo (III) and W (III) compounds are not. Whereas chromocene is a stable compound, the related molybdenocene and tungstenocene are highly reactive. On the other hand, Mo and W readily form derivatives of the type Cp_2MX_2, whereas the smaller Cr does not form such clamshell compounds. Homoleptic alkyl and aryl complexes of the type R_4M are rare, and hexamethyl tungsten has no analogue in Cr chemistry.

Similar are the carbonyls such as molybdenum hexacarbonyl and tungsten hexacarbonyl and the related carbene and carbyne complexes. Compounds of the type [CpM(CO)3]2 are known for all three metals, *e.g.* Cyclopentadienylmolybdenum tricarbonyl dimer. The chromium compound is however prone to homolysis of the Cr-Cr bond owing to steric crowding.

In the Kauffmann olefination, molybdenum (III) chloride and methyllithium form an organometallic complex capable of carbonyl olefination.

ISOTOPES OF CHROMIUM

Naturally occurring **chromium (Cr)** is composed of four stable isotopes; ^{50}Cr, ^{52}Cr, ^{53}Cr, and ^{54}Cr with ^{52}Cr being the most abundant (83.789% natural abundance). ^{50}Cr is suspected of decaying by $\beta^+\beta^+$ to ^{50}Ti with a half-life of (more than) 1.8×10^{17} years. Twenty-two radioisotopes, all which are entirely synthetic, have been characterized with the most stable being ^{51}Cr with a half-life of 27.7 days. All of the remaining radioactive isotopes have half-lives that are less than 24 hours and the majority of these have half-lives that are less than 1 minute, the least stable being ^{66}Cr with a half-life of 10 milliseconds. This element also has 2 meta states, ^{45}Cr$_m$, the more stable one, and ^{59}Cr$_m$, the least stable isotope or isomer.

^{53}Cr is the radiogenic decay product of ^{53Mn}. Chromium isotopic contents are typically combined with manganese isotopic contents and have found application in isotope geology. Mn-Cr isotope ratios reinforce the evidence from ^{26Al} and ^{107Pd} for the early history of the solar system. Variations in ^{53}Cr/^{52}Cr and Mn/Cr ratios from several meteorites indicate an initial ^{53}Mn/^{55}Mn ratio that suggests Mn-Cr isotope systematics must result from in-situ decay of ^{53}Mn in differentiated planetary bodies. Hence ^{53}Cr provides additional evidence for nucleosynthetic processes immediately before coalescence of the solar system. The same isotope is preferentially involved in certain leaching reactions, thereby allowing its abundance in seawater sediments to be used as a proxy for atmospheric oxygen concentrations.

The isotopes of chromium range from ^{42}Cr to ^{67}Cr. The primary decay mode before the most abundant stable isotope, ^{52}Cr, is electron capture and the primary mode after is beta decay.

Standard atomic mass : 51.9961(6) u

Table

nuclide symbol	Z(p)	N(n)	isotopic mass (u) excitation energy	half-life	decay mode(s)	daughter isotope(s)	nuclear spin	representative isotopic composition (mole fraction)	range of natural variation (mole fraction)
^{42}Cr	24	18	42.00643(32)#	14(3) ms [13(+4-2) ms]	β^+ (>99.9%)	^{42}V	0+		
					2p (<.1%)	^{40}Ti			
^{43}Cr	24	19	42.99771(24)#	21.6(7) ms	β^+ (71%)	^{43}V	(3/2+)		
					β^+, p (23%)	^{42}Ti			
					β^+, 2p (6%)	^{41}Sc			
					β^+, α (<.1%)	^{39}Sc			

(Contd...)

(Contd...)

^{44}Cr	24	20	43.98555(5)#	54(4) ms [53(+4-3) ms]	β$^+$ (93%)	^{44}V	0+		
					β$^+$, p (7%)	^{43}Ti			
^{45}Cr	24	21	44.97964(54)	50(6) ms	β$^+$ (73%)	^{45}V	7/2-#		
					β$^+$, p (27%)	^{44}Ti			
^{45m}Cr			50(100)# keV β$^+$	1# ms ^{45}V	IT	^{45}Cr	3/2+#		
^{46}Cr	24	22	45.968359(21)	0.26(6) s	β$^+$	^{46}V	0+		
^{47}Cr	24	23	46.962900(15)	500(15) ms	β$^+$	^{47}V	3/2-		
^{48}Cr	24	24	47.954032(8)	21.56(3) h	β$^+$	^{48}V	0+		
^{49}Cr	24	25	48.9513357(26)	42.3(1) min	β$^+$	^{49}V	5/2-		
^{50}Cr	24	26	49.9460442(11)	**Observationally Stable**			0+	0.04345(13)	0.04294-0.04345
^{51}Cr	24	27	50.9447674(11)	27.7025(24) d	EC	^{51}V	7/2-		
^{52}Cr	24	28	51.9405075(8)	**Stable**			0+	0.83789(18)	0.83762-0.83790
^{53}Cr	24	29	52.9406494(8)	**Stable**			3/2-	0.09501(17)	0.09501-0.09553
^{54}Cr	24	30	53.9388804(8)	**Stable**			0+	0.02365(7)	0.02365-0.02391
^{55}Cr	24	31	54.9408397(8)	3.497(3) min	β$^-$	**^{55}Mn**	3/2-		
^{56}Cr	24	32	55.9406531(20)	5.94(10) min	β$^-$	^{56}Mn	0+		
^{57}Cr	24	33	56.943613(2)	21.1(10) s	β$^-$	^{37}Mn	(3/2-)		
^{58}Cr	24	34	57.94435(22)	7.0(3) s	β$^-$	^{58}Mn	0+		
^{59}Cr	24	35	58.94859(26)	460(50) ms	β$^-$	^{59}Mn	5/2-#		
^{59m}Cr			503.0(17) keV	96(20) µs			(9/2+)		
^{60}Cr	24	36	59.95008(23)	560(60) ms	β$^-$	^{60}Mn	0+		
^{61}Cr	24	37	60.95472(27)	261(15) ms	β$^-$ (>99.9%)	^{61}Mn	5/2-#		
					β$^-$, n (<.1%)	^{60}Mn			
^{62}Cr	24	38	61.95661(36)	199(9) ms	β$^-$ (>99.9%)	^{62}Mn	0+		
					β$^-$, n	^{61}Mn			
^{63}Cr	24	39	62.96186(32)#	129(2) ms	β$^-$	^{63}Mn	(1/2-)#		
					β$^-$, n	^{62}Mn			
^{64}Cr	24	40	63.96441(43)#	43(1) ms	β$^-$	^{64}Mn	0+		
^{65}Cr	24	41	64.97016(54)#	27(3) ms	β$^-$	^{65}Mn	(1/2-)#		
^{66}Cr	24	42	65.97338(64)#	10(6) ms	β$^-$	^{66}Mn	0+		
^{67}Cr	24	43	66.97955(75)#	10# ms [>300 ns]	β$^-$	^{67}Mn	1/2-#		

1. Abbreviations :

   EC : Electron capture

   IT : Isomeric transition

2   Bold for stable isotopes

3.   Suspected of decaying by $\beta^+\beta^+$ decay to ^{50}Ti with a half-life of no less than 1.3×10^{18} a

Notes

- Values marked # are not purely derived from experimental data, but at least partly from systematic trends. Spins with weak assignment arguments are enclosed in parentheses.

- Uncertainties are given in concise form in parentheses after the corresponding last digits. Uncertainty values denote one standard deviation, except isotopic composition and standard atomic mass from IUPAC which use expanded uncertainties.

- Nuclide masses are given by IUPAP Commission on Symbols, Units, Nomenclature, Atomic Masses and Fundamental Constants (SUNAMCO)

- Isotope abundances are given by IUPAC Commission on Isotopic Abundances and Atomic Weights.

Chapter 7

MANGANESE

Manganese is a chemical element, designated by the symbol Mn. It has the atomic number 25. It is found as a free element in nature (often in combination with iron), and in many minerals. Manganese is a metal with important industrial metal alloy uses, particularly in stainless steels.

Historically, manganese is named for various black minerals (such as pyrolusite) from the same region of Magnesia in Greece which gave names to similar-sounding magnesium, Mg, and magnetite, an ore of the element iron, Fe. By the mid-18th century, Swedish chemist Carl Wilhelm Scheele had used pyrolusite to produce chlorine. Scheele and others were aware that pyrolusite (now known to be manganese dioxide) contained a new element, but they were not able to isolate it. Johan Gottlieb Gahn was the first to isolate an impure sample of manganese metal in 1774, by reducing the dioxide with carbon.

Manganese phosphating is used as a treatment for rust and corrosion prevention on steel. Depending on their oxidation state, manganese ions have various colours and are used industrially as pigments. The permanganates of alkali and alkaline earth metals are powerful oxidizers. Manganese dioxide is used as the cathode (electron acceptor) material in zinc-carbon and alkaline batteries.

In biology, manganese (II) ions function as cofactors for a large variety of enzymes with many functions. Manganese enzymes are particularly essential in detoxification of superoxide free radicals in organisms that must deal with elemental oxygen. Manganese also functions in the oxygen-evolving complex of photosynthetic plants. The element is a required trace mineral for all known living organisms. In larger amounts, and apparently with far greater activity by inhalation, it can cause a poisoning syndrome in mammals, with neurological damage which is sometimes irreversible.

CHARACTERISTICS

Physical Properties

Manganese is a silvery-gray metal that resembles iron. It is hard and very brittle, difficult to fuse, but easy to oxidize. Manganese metal and its common ions are paramagnetic. Manganese tarnishes slowly in air and "rusts" like iron, in water containing dissolved oxygen.

HISTORY

The origin of the name manganese is complex. In ancient times, two black minerals from Magnesia in what is now modern Greece, were both called magnes from their place of origin, but were thought to differ in gender. The male magnes attracted iron, and was the iron ore we now know as lodestone or magnetite, and which probably gave us the term magnet. The female magnes ore did not attract iron, but was used to decolorize glass. This feminine magnes was later called magnesia, known now in modern times as pyrolusite or manganese dioxide. Neither this mineral nor manganese itself is magnetic. In the 16th century, manganese dioxide was called manganesum (note the two n's instead of one) by glass-makers, possibly as a corruption and concatenation of two words, since alchemists and glass-makers eventually had to differentiate a magnesia negra (the black ore) from magnesia alba (a white ore, also from Magnesia, also useful in glass-making). Michele Mercati called magnesia negra manganesa, and finally the metal isolated from it became known as manganese (German : Mangan). The name magnesia eventually was then used to refer only to the white magnesia alba (magnesium oxide), which provided the name magnesium for that free element, when it was eventually isolated, much later.

Several oxides of manganese, for example manganese dioxide, are abundant in nature, and owing to their colour, these oxides have been used as since the Stone Age. The cave paintings in Gargas contain manganese as pigments and these cave paintings are 30,000 to 24,000 years old.

Manganese compounds were used by Egyptian and Roman glass-makers, to either remove colour from glass or add colour to it. The use as "glass-makers soap" continued through the Middle Ages until modern times and is evident in 14th-century glass from Venice.

Because of the use in glass-making, manganese dioxide was available to alchemists, the first chemists, and was used for experiments. Ignatius Gottfried Kaim and Johann Glauber (17th century) discovered that manganese dioxide could be converted to permanganate, a useful laboratory reagent. By the mid-18th century, the Swedish chemist Carl Wilhelm Scheele used manganese dioxide to produce chlorine. First, hydrochloric acid, or a mixture of dilute sulfuric acid and sodium chloride was made to react with manganese dioxide, later hydrochloric acid from the Leblanc process was used and the manganese dioxide was recycled by the Weldon process. The production of chlorine and hypochlorite containing bleaching agents was a large consumer of manganese ores.

Scheele and other chemists were aware that manganese dioxide contained a new element, but they were not able to isolate it. Johan Gottlieb Gahn was the first to isolate an impure sample of manganese metal in 1774, by reducing the dioxide with carbon.

The manganese content of some iron ores used in Greece led to the speculations that the steel produced from that ore contains inadvertent amounts of manganese, making the Spartan steel exceptionally hard. Around the beginning of the 19th century, manganese was used in steel-making and several patents were granted. In 1816, it was noted that adding manganese to iron made it harder, without making it any more brittle. In 1837, British academic James Couper noted an association between heavy exposures to manganese in mines with a form of Parkinson's disease. In 1912, manganese phosphating electro-chemical conversion coatings for protecting firearms against rust and corrosion were patented in the United States, and have seen widespread use ever since.

The invention of the Leclanché cell in 1866 and the subsequent improvement of the batteries containing manganese dioxide as cathodic depolarizer increased the demand of manganese dioxide. Until the introduction of the nickel-cadmium battery and lithium-containing batteries, most batteries contained manganese. The zinc-carbon battery and the alkaline battery normally use industrially produced manganese dioxide, because natural occurring manganese dioxide contains impurities. In the 20th century, manganese dioxide has seen wide commercial use as the chief cathodic material for commercial disposable dry cells and dry batteries of both the standard (zinc-carbon) and alkaline types.

OCCURRENCE AND PRODUCTION

Manganese makes up about 1000 ppm (0.1%) of the Earth's crust, making it the 12th most abundant element there. Soil contains 7–9000 ppm of manganese with an average of 440 ppm. Seawater has only 10 ppm manganese and the atmosphere contains $0.01 \, \mu g/m^3$. Manganese occurs principally as pyrolusite (MnO_2), braunite, ($Mn^{2+}Mn^{3+}_6$), psilomelane ($Ba,H_2O)_2Mn_5O_{10}$, and to a lesser extent as rhodochrosite ($MnCO_3$).

The most important manganese ore is pyrolusite (MnO_2). Other economically important manganese ores usually show a close spatial relation to the iron ores. Land-based resources are large but irregularly distributed. About 80% of the known world manganese resources are found in South Africa; other important manganese deposits are in Ukraine, Australia, India, China, Gabon and Brazil. In 1978, 500 billion tons of manganese nodules were estimated to exist on the ocean floor. Attempts to find economically viable methods of harvesting manganese nodules were abandoned in the 1970s.

In South Africa most identified deposits are located near Hotazel in the Northern Cape Province with an estimated 15 billion tons in 2011. In 2011 South Africa was the world's largest producer of manganese producing 3.4 million tons.

Manganese is mined in South Africa, Australia, China, Brazil, Gabon, Ukraine, India and Ghana and Kazakhstan. US Import Sources (1998–2001) : Manganese ore : Gabon, 70%; South Africa, 10%; Australia, 9%; Mexico, 5%; and other, 6%. Ferro-manganese : South Africa, 47%; France, 22%; Mexico, 8%; Australia, 8%; and other, 15%. Manganese contained in all manganese imports : South Africa, 31%; Gabon, 21%; Australia, 13%; Mexico, 8%; and other, 27%.

For the production of ferro-manganese, the manganese ore is mixed with iron ore and carbon, and then reduced either in a blast furnace or in an electric arc furnace. The resulting ferro-manganese has a manganese content of 30 to 80%. Pure manganese used for the production of iron-free alloys is produced by leaching manganese ore with sulfuric acid and a subsequent electro-winning process.

In 1972 the CIA's Project Azorian, through billionaire Howard Hughes, commissioned the ship Hughes Glomar Explorer with the cover story of harvesting manganese nodules from the sea floor. That triggered a rush of activity to attempt to collect manganese nodules, which was not actually practical. The real mission of Hughes Glomar Explorer was to raise a sunken Soviet submarine, the K-129, with the goal of retrieving Soviet code books.

APPLICATIONS

Manganese has no satisfactory substitute in its major applications, which are related to metallurgical alloy use. In minor applications, (*e.g.*, manganese phosphating), zinc and sometimes vanadium are viable substitutes.

Steel

Manganese is essential to iron and steel production by virtue of its sulfur-fixing, deoxidizing, and alloying properties. Steel-making, including its iron-making component, has accounted for most manganese demand, presently in the range of 85% to 90% of the total demand. Among a variety of other uses, manganese is a key component of low-cost stainless steel formulations.

Small amounts of manganese improve the workability of steel at high temperatures, because it forms a high-melting sulfide and therefore prevents the formation of a liquid iron sulfide at the grain boundaries. If the manganese content reaches 4%, the embrittlement of the steel becomes a dominant feature. The embrittlement decreases at higher manganese concentrations and reaches an acceptable level at 8%. Steel containing 8 to 15% of manganese can have a high tensile strength of up to 863 MPa. Steel with 12% manganese was used for British steel helmets. This steel composition was discovered in 1882 by Robert Hadfield and is still known as Hadfield steel.

Aluminium Alloys

The second large application for manganese is as an alloying agent for aluminium. Aluminium with a manganese content of roughly 1.5% has an increased resistance against corrosion due to the formation of grains absorbing impurities which would lead to galvanic corrosion. The corrosion-resistant aluminium alloys 3004 and 3104 with a manganese content of 0.8 to 1.5% are the alloys used for most of the beverage cans. Before year 2000, more than 1.6 million tonnes have been used of those alloys; with a content of 1% manganese, this amount would need 16,000 tonnes of manganese.

Other Uses

Methylcyclopentadienyl manganese tricarbonyl is used as an additive in unleaded gasoline to boost octane rating and reduce engine knocking. The manganese in this unusual organometallic compound is in the +1 oxidation state.

Manganese (IV) oxide (manganese dioxide, MnO_2) is used as a reagent in organic chemistry for the oxidation of benzylic alcohols (*i.e.* adjacent to an aromatic ring). Manganese dioxide has been used since antiquity to oxidatively neutralize the greenish tinge in glass caused by trace amounts of iron contamination. MnO_2 is also used in the manufacture of oxygen and chlorine, and in drying black paints. In some preparations, it is a brown pigment that can be used to make paint and is a constituent of natural umber.

Manganese (IV) oxide was used in the original type of dry cell battery as an electron acceptor from zinc, and is the blackish material found when opening carbon–zinc type flashlight cells. The manganese dioxide is reduced to the manganese oxide-hydroxide MnO(OH) during discharging, preventing the formation of hydrogen at the anode of the battery.

$$MnO_2 + H_2O + -e \rightarrow MnO(OH) + OH-$$

The same material also functions in newer alkaline batteries (usually battery cells), which use the same basic reaction, but a different electrolyte mixture. In 2002, more than 230,000 tons of manganese dioxide was used for this purpose.

The metal is very occasionally used in coins; until 2000, the only United States coin to use manganese was the "wartime" nickel from 1942 to 1945. An alloy of 75% copper and 25% nickel was traditionally used for the production of nickel coins. However, because of shortage of nickel metal during the war, it was substituted by more available silver and manganese, thus resulting in an alloy of 56% copper, 35% silver and 9% manganese. Since 2000, dollar coins, for example the Sacagawea dollar and the Presidential $1 coins, are made from a brass containing 7% of manganese with a pure copper core. In both cases of nickel and dollar, the use of manganese in the coin was to duplicate the electro-magnetic properties of a previous identically sized and valued coin, for vending purposes. In the case of the later U.S. dollar coins, the manganese alloy was an attempt to duplicate properties of the copper/nickel alloy used in the previous Susan B. Anthony dollar.

Manganese compounds have been used as pigments and for the colouring of ceramics and glass. The brown colour of ceramic is sometimes based on manganese compounds. In the glass industry, manganese compounds are used for two effects. Manganese (III) reacts with iron (II) to induce a strong green colour in glass by forming less-colored iron (III) and slightly pink manganese (II), compensating for the residual colour of the iron (III). Larger amounts of manganese are used to produce pink colored glass.

BIOLOGICAL ROLE

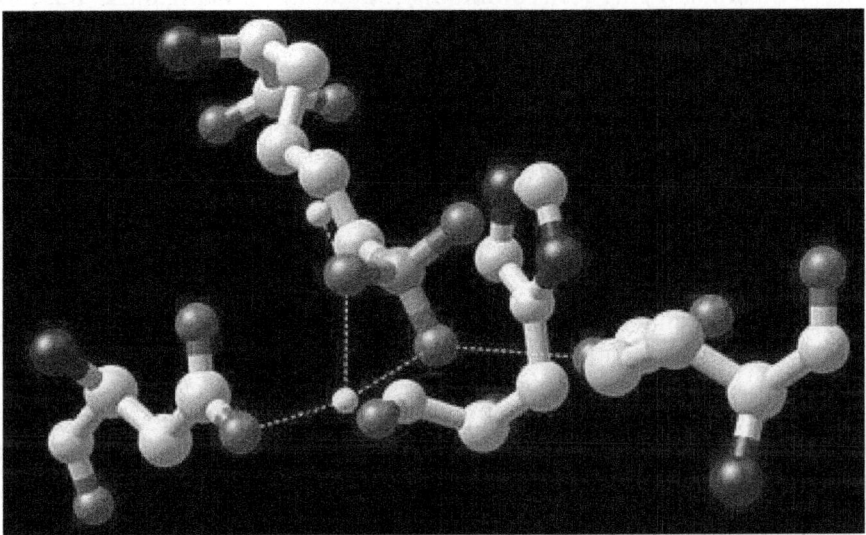

Fig. : Reactive center of arginase with boronic acid inhibitor–the manganese atoms are shown in yellow.

Manganese is an essential trace nutrient in all known forms of life. The classes of enzymes that have manganese cofactors are very broad, and include oxidoreductases, transferases, hydrolases, lyases, isomerases, ligases, lectins, and integrins. The reverse transcriptases of many retroviruses (though not lentiviruses such as HIV) contain manganese. The best-known manganese-containing polypeptides may be arginase, the diphtheria toxin, and Mn-containing superoxide dismutase (Mn-SOD).

Mn-SOD is the type of SOD present in eukaryotic mitochondria, and also in most bacteria (this fact is in keeping with the bacterial-origin theory of mitochondria). The Mn-SOD enzyme is probably one of the most ancient, for nearly all organisms living in the presence of oxygen use it to deal with the toxic effects of superoxide, formed from the 1-electron reduction of dioxygen. Exceptions include a few kinds of bacteria, such as Lactobacillus plantarum and related lactobacilli, which use a different non-enzymatic mechanism, involving manganese (Mn^{2+}) ions complexed with polyphosphate directly for this task, indicating how this function possibly evolved in aerobic life.

The human body contains about 12 mg of manganese, which is stored mainly in the bones; in the tissue, it is mostly concentrated in the liver and kidneys. In the human brain, the manganese is bound to manganese metalloproteins, most notably glutamine synthetase in astrocytes.

Manganese is also important in photosynthetic oxygen evolution in chloroplasts in plants. The oxygen-evolving complex (OEC) is a part of photo-system II contained in the thylakoid membranes of chloroplasts; it is responsible for the terminal photo-oxidation of water during the light reactions of photo-synthesis, and has a metalloenzyme core containing four atoms of manganese. For this reason, most broad-spectrum plant fertilizers contain manganese.

PRECAUTIONS

Manganese compounds are less toxic than those of other widespread metals, such as nickel and copper. However, exposure to manganese dusts and fumes should not exceed the ceiling value of 5 mg/m even for short periods because of its toxicity level. Manganese poisoning has been linked to impaired motor skills and cognitive disorders.

The permanganate exhibits a higher toxicity than the manganese (II) compounds. The fatal dose is about 10 g, and several fatal intoxications have occurred. The strong oxidative effect leads to necrosis of the mucous membrane. For example, the esophagus is affected if the permanganate is swallowed. Only a limited amount is absorbed by the intestines, but this small amount shows severe effects on the kidneys and on the liver.

In 2005, a study suggested a possible link between manganese inhalation and central nervous system toxicity in rats.

Manganese exposure in United States is regulated by Occupational Safety and Health Administration.

Generally, exposure to ambient Mn air concentrations in excess of 5 µg Mn/m^3 can lead to Mn-induced symptoms. Increased ferroportin protein expression in human embryonic kidney (HEK293) cells is associated with decreased intracellular Mn concentration and attenuated cytotoxicity, characterized by the reversal of Mn-reduced glutamate uptake and diminished lactate dehydrogenase leakage.

ENVIRONMENTAL HEALTH CONCERNS

Manganese in Drinking Water

Waterborne manganese has a greater bio-availability than dietary manganese. According to results from a 2010 study, higher levels of exposure to manganese in drinking water are associated with increased intellectual impairment and reduced intelligence quotients in school-age children. It is hypothesized that long-term exposure to the naturally occurring manganese in shower water puts up to 8.7 million Americans at risk.

Manganese in Gasoline

Methylcyclopentadienyl manganese tricarbonyl (MMT) is a gasoline additive used to replace lead compounds for unleaded gasolines, to improve the octane number in low octane number petrol distillates. It functions as an antiknock agent by the action of the carbonyl groups. Fuels containing manganese tend to form manganese carbides, which damage exhaust valves. The need to use lead or manganese compounds is merely historic, as the availability of reformation processes which create high-octane rating fuels increased. The use of such fuels directly or in mixture with non-reformed distillates is universal in developed countries (EU, Japan, etc.). In USA the imperative to provide the lowest possible price per volume on motor fuels (low fuel taxation rate) and lax legislation of fuel content caused refineries to use MMT. Compared to 1953, levels of manganese in air have dropped. Many racing competitions specifically ban manganese compounds in racing fuel (cart, minibike). MMT contains 24.4–25.2% manganese. There is strong correlation between elevated atmospheric manganese concentrations and automobile traffic density.

ROLE IN NEUROLOGICAL DISORDERS

Manganism

Manganese over-exposure is most frequently associated with manganism, a rare neurological disorder associated with excessive manganese ingestion or inhalation. Historically, persons employed in the production or processing of manganese alloys have been at risk for developing manganism; however, current health and safety regulations protect workers in developed nations. The disorder was first described in 1837 by British academic John Couper, who studied two patients who were manganese grinders.

Manganism is a biphasic disorder. In its early stages, an intoxicated person may experience depression, mood swings, compulsive behaviours, and psychosis. Early neurological symptoms give way to late-stage manganism, which resembles Parkinson's disease. Symptoms include weakness, monotone and slowed speech, an expressionless face, tremor, forward-leaning gait, inability to walk backwards without falling, rigidity, and general problems with dexterity, gait and balance. Unlike Parkinson's disease, manganism is not associated with loss of smell and patients are typically unresponsive to treatment with L-DOPA. Symptoms of late-stage manganism become more severe over time even if the source of exposure is removed and brain manganese levels return to normal.

Childhood Developmental Disorders

Several recent studies attempt to examine the effects of chronic low-dose manganese overexposure on development in children. The earliest study of this kind was conducted in the Chinese province of Shanxi. Drinking water there had been contaminated through improper sewage irrigation and contained 240–350 μg Mn/L. Although WMn concentrations at or below 300 μg Mn/L are considered safe by the US EPA and 400 μg Mn/L are considered safe by the World Health

Organization, the 92 children sampled (between 11 and 13 years of age) from this province displayed lower performance on tests of manual dexterity and rapidity, short-term memory, and visual identification when compared to children from an uncontaminated area. More recently, a study of 10-years-old children in Bangladesh showed a relationship between WMn concentration in well water and diminished IQ scores. A third study conducted in Quebec examined school children between the ages of 6 and 15 living in homes that received water from a well containing 610 µg Mn/L; controls lived in homes that received water from a 160 µg Mn/L well. Children in the experimental group showed increased hyperactive and oppositional behaviours.

Neurodegenerative Diseases

Chronic low-dose manganese intoxication is strongly implicated in a number of neurodegenerative disorders, including Alzheimer's disease, Parkinson's disease, and amyotrophic lateral sclerosis. It may also play a role in the development of multiple sclerosis, restless leg syndrome, and Huntington's disease. A protein called DMT1 is the major transporter involved in manganese absorption from the intestine, and may be the major transporter of manganese across the blood–brain barrier. DMT1 also transports inhaled manganese across the nasal epithelium. The putative mechanism of action is that manganese over-exposure and/or dysregulation lead to oxidative stress, mitochondrial dysfunction, glutamate-mediated excitoxicity, and aggregation of proteins.

Chemical Properties

The most common oxidation states of manganese are +2, +3, +4, +6 and +7, though oxidation states from −3 to +7 are observed. Mn^{2+} often competes with Mg^{2+} in biological systems. Manganese compounds where manganese is in oxidation state +7, which are restricted to the unstable oxide Mn_2O_7 and compounds of the intensely purple permanganate anion MnO_4^-, are powerful oxidizing agents. Compounds with oxidation states +5 (blue) and +6 are strong oxidizing agents and are vulnerable to disproportionation.

The most stable oxidation state for manganese is +2, which has a pale pink colour, and many manganese (II) compounds are known, such as manganese (II) sulfate and manganese (II) chloride. This oxidation state is also seen in the mineral rhodochrosite (manganese (II) carbonate). The +2 oxidation of Mn results from removal of the two 4s electrons, leaving a "high spin" ion in which all five of the 3d orbitals contain a single electron. Absorption of visible light by this ion is accomplished only by a spin-forbidden transition in which one of the d electrons must pair with another, to give the atom a change in spin of two units. The unlikeliness of such a transition is seen in the uniformly pale and almost colourless nature of Mn (II) compounds relative to other oxidation states of manganese.

Table : Oxidation states of manganese.

0	$Mn_2(CO)_{10}$
+1	$MnC_5H_4CH_3(CO)_3$
+2	$MnCl_2$
+3	MnF_3
+4	MnO_2
+5	K_3MnO_4
+6	K_2MnO_4
+7	$KMnO_4$

Common oxidation states are in bold.

The +3 oxidation state is known in compounds like manganese (III) acetate, but these are quite powerful oxidizing agents and also prone to disproportionation in solution to manganese (II) and manganese (IV). Solid compounds of manganese (III) are characterized by their preference for distorted octahedral co-ordination due to the Jahn-Teller effect and its strong purple-red colour. The oxidation state 5+ can be obtained if manganese dioxide is dissolved in molten sodium nitrite. Manganate (VI) salts can also be produced by dissolving Mn compounds, such as manganese dioxide, in molten alkali while exposed to air. Permanganate (+7 oxidation state) compounds are purple, and can give glass a violet colour. Potassium permanganate, sodium permanganate and barium permanganate are all potent oxidizers. Potassium permanganate, also called Condy's crystals, is a commonly used laboratory reagent because of its oxidizing properties and finds use as a topical medicine (for example, in the treatment of fish diseases). Solutions of potassium permanganate were among the first stains and fixatives to be used in the preparation of biological cells and tissues for electron microscopy.

Isotopes of Manganese

Naturally occurring manganese (Mn) is composed of 1 stable isotope, ^{55}Mn. 25 radioisotopes have been characterized with the most stable being ^{53}Mn with a half-life of 3.7 million years, ^{54}Mn with a half-life of 312.3 days, and ^{52}Mn with a half-life of 5.591 days. All of the remaining radioactive isotopes have half-lives that are less than 3 hours and the majority of these have half-lives that are less than 1 minute, but only ^{45}Mn has an unknown half-life. The least stable is ^{44}Mn with a half-life shorter than 105 nanoseconds. This element also has 3 meta states.

Manganese is part of the iron group of elements which are thought to be synthesized in large stars shortly before supernova explosion. ^{53}Mn decays to ^{53}Cr with a half-life of 3.7 million years. Because of its relatively short half-life, ^{53}Mn occurs only in tiny amounts due to the action of cosmic rays on iron in rocks. Manganese isotopic contents are typically combined with chromium isotopic contents and have found application in isotope geology and radiometric dating. Mn-Cr isotopic ratios reinforce the evidence from ^{26}Al and ^{107}Pd for the early history of the solar system. Variations in ^{53}Cr/^{52}Cr and Mn/Cr ratios from several meteorites indicate an initial ^{53}Mn/^{55}Mn ratio that suggests Mn-Cr isotopic systematics must result from in-situ decay of ^{53}Mn in differentiated planetary bodies. Hence ^{53}Mn provides additional evidence for nucleosynthetic processes immediately before coalescence of the solar system.

The isotopes of manganese range in atomic weight from 46 u (^{46}Mn) to 65 u (^{65}Mn). The primary decay mode before the most abundant stable isotope, ^{55}Mn, is electron capture and the primary mode after is beta decay.

Standard atomic mass : 54.938045(5) u

Table

nuclide symbol	Z(p)	N(n)	isotopic mass (u)	half-life	decay mode(s)	daughter isotope(s)	nuclear spin	representative isotopic composition (mole fraction)	range of natural variation (mole fraction)
	excitation energy								
^{44}Mn	25	19	44.00687(54)#	<105 ns	p	^{43}Cr	(2-)#		
^{45}Mn	25	20	44.99451(32)#	unknown	p	^{44}Cr	(7/2-)#		
^{46}Mn	25	21	45.98672(12)#	37(3) ms	β+ (78%)	^{46}Cr	(4+)		
					β+, p (22%)	^{45}V			
					β+, α (<1%)	^{42}Ti			
					β+, 2p (<1%)	^{44}Ti			
^{46m}Mn	150(100)# keV			1# ms	β+	^{46}Cr	1-#		
^{47}Mn	25	22	46.97610(17)#	100(50) ms	β+	^{47}Cr	5/2-#		
					β+, p (3.4%)	^{46}V			
^{48}Mn	25	23	47.96852(12)	158.1(22) ms	β+ (99.71%)	^{48}Cr	4+		
					β+, p	^{47}V			
					β+, α (6×10−4%)	^{44}Ti			
^{49}Mn	25	24	48.959618(26)	382(7) ms	β+	^{49}Cr	5/2-		
^{50}Mn	25	25	49.9542382(11)	283.29(8) ms	β+	^{50}Cr	0+		
^{50m}Mn	229(7) keV			1.75(3) min	β+	^{50}Cr	5+		
^{51}Mn	25	26	50.9482108(11)	46.2(1) min	β+	^{51}Cr	5/2-		

(Contd...)

(Contd...)

nuclide symbol	Z(p)	N(n)	isotopic mass (u)	half-life	decay mode(s)	daughter isotope(s)	nuclear spin	representative isotopic composition (mole fraction)	range of natural variation (mole fraction)
			excitation energy						
^{52}Mn	25	27	51.9455655(21)	5.591(3) d	β+	^{52}Cr	6+		
^{52m}Mn	377.749(5) keV IT			21.1(2) min 52Mn	β+ (98.25%)	^{52}Cr	2+		
^{53}Mn	25	28	52.9412901(9)	3.7(4)×106 a	EC	^{53}Cr	7/2-	trace	
^{54}Mn	25	29	53.9403589(14)	312.03(3) d	EC 99.99%	^{54}Cr	3+		
					β-(2.9×10−4%)	^{54}Fe			
					β+ (5.76×10−7%)	^{54}Cr			
^{55}Mn	25	30	54.9380451(7)	Stable			5/2-	1.0000	
^{56}Mn	25	31	55.9389049(7)	2.5789(1) h	β-	^{56}Fe	3+		
^{57}Mn	25	32	56.9382854(20)	85.4(18) s	β-	^{57}Fe	5/2-		
^{58}Mn	25	33	57.93998(3)	3.0(1) s	β-	^{58}Fe	1+		
^{58m}Mn	71.78(5) keV IT (<.1%)			65.2(5) s 58Mn	β-(>99.9%)	^{58}Fe	(4)+		
^{59}Mn	25	34	58.94044(3)	4.59(5) s	β-	^{59}Fe	(5/2)-		
^{60}Mn	25	35	59.94291(9)	51(6) s	β-	^{60}Fe	0+		
^{60m}Mn	271.90(10) keV IT			1.77(2) s 60Mn	β-	^{60}Fe	3+		
^{61}Mn	25	36	60.94465(24)	0.67(4) s	β-	^{61}Fe	(5/2)-		
^{62}Mn	25	37	61.94843(24)	671(5) ms	β-(>99.9%)	^{62}Fe	(3+)		
					β-, n (<.1%)	^{61}Fe			
^{62m}Mn	0(150)# keV			92(13) ms			(1+)		
^{63}Mn	25	38	62.95024(28)	275(4) ms	β-	^{63}Fe	5/2-#		
^{64}Mn	25	39	63.95425(29)	88.8(25) ms	β-(>99.9%)	^{64}Fe	(1+)		
					β-, n (<.1%)	^{63}Fe			
^{64m}Mn	135(3) keV			>100 μs					
^{65}Mn	25	40	64.95634(58)	92(1) ms	β-(>99.9%)	^{65}Fe	5/2-#		
					β-, n (<.1%)	^{64}Fe			
^{66}Mn	25	41	65.96108(43)#	64.4(18) ms	β-(>99.9%)	^{66}Fe			
					β-, n (<.1%)	^{65}Fe			
^{67}Mn	25	42	66.96414(54)#	45(3) ms	β-	^{67}Fe	5/2-#		
^{68}Mn	25	43	67.96930(64)#	28(4) ms					
^{69}Mn	25	44	68.97284(86)#	14(4) ms			5/2-#		

Abbreviations :

EC : Electron capture

IT : Isomeric transition

1. Bold for stable isotopes

Notes

- Values marked # are not purely derived from experimental data, but at least partly from systematic trends. Spins with weak assignment arguments are enclosed in parentheses.
- Uncertainties are given in concise form in parentheses after the corresponding last digits. Uncertainty values denote one standard deviation, except isotopic composition and standard atomic mass from IUPAC which use expanded uncertainties.
- Nuclide masses are given by IUPAP Commission on Symbols, Units, Nomenclature, Atomic Masses and Fundamental Constants (SUNAMCO)
- Isotope abundances are given by IUPAC Commission on Isotopic Abundances and Atomic Weights.

Chapter 8

PHOTOSYNTHETIC REACTION CENTRE

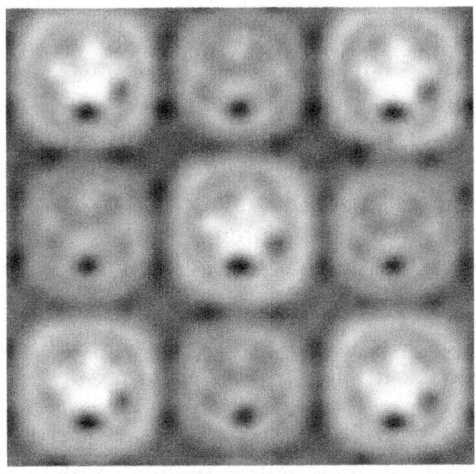

Fig. : Electron micrograph of 2D crystals of the LH1-Reaction centre photosynthetic unit.

A photosynthetic reaction centre (or photosynthetic reaction center) is a complex of several proteins, pigments and other co-factors assembled together to execute the primary energy conversion reactions of photosynthesis. Molecular excitations, either originating directly from sunlight or transferred as excitation energy *via* light-harvesting antenna systems, give rise to electron transfer reactions along a series of protein-bound co-factors. These co-factors are light-absorbing molecules (also named chromophores or pigments) such as chlorophyll and phaeophytin, as well as quinones. The energy of the photon is used to promote an electron to a higher molecular energy level of a pigment. The free energy created is then used to reduce a chain of nearby electron acceptors, which have subsequently higher redox-potentials. These electron transfer steps are the initial phase of a series of energy conversion reactions, ultimately resulting in the production of chemical energy during photosynthesis.

TRANSFORMING LIGHT ENERGY INTO CHARGE SEPARATION

Reaction centres are present in all green plants, algae, and many bacteria. Although these species are separated by billions of years of evolution, the reaction centres are homologous for all photosynthetic species. In contrast, a large variety in light-harvesting complexes exist between the photosynthetic species. Green plants and algae have two different types of reaction centres that are part of larger super-complexes known as photo-system I P700 and photo-system II P680. The structures of these super-complexes are large, involving multiple light-harvesting complexes. The reaction centre found in Rhodopseudomonas bacteria is currently best understood, since it was the first reaction centre of known structure and has fewer polypeptide chains than the examples in green plants.

A reaction centre is laid out in such a way that it captures the energy of a photon using pigment molecules and turns it into a usable form. Once the light energy has been absorbed directly by the pigment molecules, or passed to them by resonance transfer from a surrounding light-harvesting complex, they release two electrons into an electron transport chain.

Light is made up of small bundles of energy called photons. If a photon with the right amount of energy hits an electron, it will raise the electron to a higher energy level. Electrons are most stable at their lowest energy level, what is also called its ground state. In this state, the electron is in the orbit that has the least amount of energy. Electrons in higher energy levels can return to ground state in a manner analogous to a ball falling down a staircase. In doing so, the electrons release energy. This is the process that is exploited by a photosynthetic reaction centre.

When an electron rises to a higher energy level, decrease in the reduction potential of the molecule in which the electron resides occurs. This means that the molecule has a greater tendency to donate electrons, the key to the conversion of light energy to chemical energy. In green plants, the electron transport chain that follows has many electron acceptors including phaeophytin, quinone, plastoquinone, cytochrome bf, and ferredoxin, which result in the reduced molecule NADPH. The passage of the electron through the electron transport chain also results in the pumping of protons (hydrogen ions) from the chloroplast's stroma into the lumen, resulting in a proton gradient across the thylakoid membrane that can be used to synthesise ATP using ATP synthase. Both the ATP and NADPH are used in the Calvin cycle to fix carbon dioxide into triose sugars.

BACTERIA

Structure

The bacterial photosynthetic reaction centre has been an important model to understand the structure and chemistry of the biological process of capturing light energy. In the 1960s, Roderick Clayton was the first to purify the reaction centre complex from purple bacteria. However, the first crystal structure was determined in 1982 by Hartmut Michel, Johann Deisenhofer and Robert Huber for which they shared the Nobel Prize in 1988. This was also significant, since it was the first structure for any membrane protein complex.

Four different sub-units were found to be important for the function of the photosynthetic reaction centre. The L and M sub-units, shown in blue and purple in the image of the structure, both span the lipid bilayer of the plasma membrane. They are structurally similar to one another, both having 5 transmembrane alpha helices. Four bacteriochlorophyll b (BChl-b) molecules, two bacteriophaeophytin b molecules (BPh) molecules, two quinones (QA and QB), and a ferrous ion are associated with the L and M sub-units. The H sub-unit, shown in gold, lies on the cytoplasmic side of the plasma membrane. A cytochrome sub-unit, here not shown, contains four c-type haems and is located on the periplasmic surface of the membrane. The latter sub-unit is not a general structural motif in photosynthetic bacteria. The L and M sub-units bind the functional and light-interacting cofactors, shown here in green.

Reaction centres from different bacterial species may contain slightly altered bacterio-chlorophyll and bacterio-phaeophytin chromophores as functional cofactors. These alterations cause shifts in the colour of light that can be absorbed, thus creating specific niches for photosynthesis. The reaction centre contains two pigments that serve to collect and transfer the energy from photon absorption : BChl and Bph. BChl roughly resembles the chlorophyll molecule found in green plants, but, due to minor structural differences, its peak absorption wavelength is shifted into the infrared, with wavelengths as long as 1000 nm. Bph has the same structure as BChl, but the central magnesium ion is replaced by two protons. This alteration causes both an absorbance maximum shift and a lowered redox-potential.

Mechanism

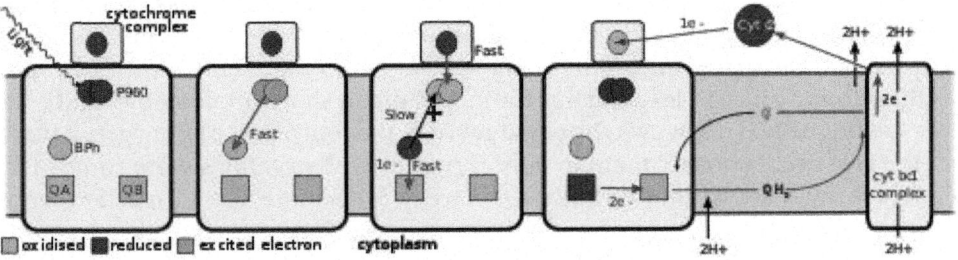

Fig. : The light reaction.

The process starts when light is absorbed by two BChl molecules (a dimer) that lie near the periplasmic side of the membrane. This pair of chlorophyll molecules, often called the "special pair", absorbs photons between 870 nm and 960 nm, depending on the species and, thus, is called P870 (for the species rhodobacter sphaeroides) or P960 (for rhodopseudomonas viridis), with P standing for "pigment"). Once P absorbs a photon, it ejects an electron, which is transferred through another molecule of Bchl to the BPh in the L sub-unit. This initial charge separation yields a positive charge on P and a negative charge on the BPh. This process takes place in 10 picoseconds (10−11 seconds).

The charges on the specialpair + and the BPh-could undergo charge recombination in this state. This would waste the high-energy electron and convert the absorbed light energy into heat. Several factors of the reaction centre structure serve to prevent this. First, the transfer of an electron from BPh-to P960+ is relatively slow compared to two other redox reactions in the reaction centre. The faster reactions involve the transfer of an electron from BPh-(BPh-is oxidised to BPh) to the electron acceptor quinone (QA), and the transfer of an electron to P960+ (P960+ is reduced to P960) from a heme in the cytochrome sub-unit above the reaction centre.

The high-energy electron that resides on the tightly bound quinone molecule Q_A is transferred to an exchangeable quinone molecule Q_B. This molecule is loosely associated with the protein and is fairly easy to detach. Two of the high-energy electrons are required to fully reduce Q_B to QH_2, taking up two protons from the cytoplasm in the process. The reduced quinone QH_2 diffuses through the membrane to another protein complex (cytochrome bc1-complex) where it is oxidised. In the process the reducing power of the QH_2 is used to pump protons across the membrane to the periplasmic space. The electrons from the cytochrome bc_1-complex are then transferred through a soluble cytochrome c intermediate, called cytochrome c_2, in the periplasm to the cytochrome sub-unit. Thus, the flow of electrons in this system is cyclical.

GREEN PLANTS

Oxygenic Photosynthesis

In 1772, the chemist Joseph Priestley carried out a series of experiments relating to the gases involved in respiration and combustion. In his first experiment, he lit a candle and placed it under an upturned jar. After a short period of time, the candle burned out. He carried out a similar experiment with a mouse in the confined space of the burning candle. He found that the mouse died a short time after the candle had been extinguished. However, he could revivify the foul air by placing green plants in the area and exposing them to light. Priestley's observations were some of the first experiments that demonstrated the activity of a photosynthetic reaction centre.

In 1779, Jan Ingenhousz carried out more than 500 experiments spread out over 4 months in an attempt to understand what was really going on. He wrote up his discoveries in a book entitled Experiments upon Vegetables. Ingenhousz took green plants and immersed them in water inside a transparent tank. He observed many bubbles rising from the surface of the leaves whenever the plants were exposed to light. Ingenhousz collected the gas that was given off by the plants and performed several different tests in attempt to determine what the gas was. The test that finally revealed the identity of the gas was placing a smouldering taper into the gas sample and having it relight. This test proved it was oxygen, or, as Joseph Priestley had called it, 'de-phlogisticated air'.

In 1932, Professor Robert Emerson and an undergraduate student, William Arnold, used a repetitive flash technique to precisely measure small quantities of oxygen evolved by chlorophyll in the algae Chlorella. Their experiment proved the existence of a photosynthetic unit. Gaffron and Wohl later interpreted the experiment and realized that the light absorbed by the photosynthetic unit was transferred. This reaction occurs at the reaction centre of photo-system II and takes place in cyanobacteria, algae and green plants.

Photo-system II

Photo-system II is the photo-system that generates the two electrons that will eventually reduce NADP+ in Ferredoxin-NADP-reductase. Photo-system II is present on the thylakoid membranes inside chloroplasts, the site of photosynthesis in green plants. The structure of Photo-system II is remarkably similar to the bacterial reaction centre, and it is theorized that they share a common ancestor.

The core of photo-system II consists of two sub-units referred to as D1 and D2. These two sub-units are similar to the L and M sub-units present in the bacterial reaction centre. Photo-system II differs from the bacterial reaction centre in that it has many additional sub-units that bind additional chlorophylls to increase efficiency. The overall reaction catalysed by photo-system II is :

$$2Q + 2H_2O \xrightarrow{\text{light}} O_2 + 2QH_2$$

Q represents plastoquinone, the oxidized form of Q. QH_2 represents plastoquinol, the reduced form of Q. This process of reducing quinone is comparable to that which takes place in the bacterial reaction centre. Photo-system II obtains electrons by oxidizing water in a process called photolysis. Molecular oxygen is a by-product of this process, and it is this reaction that supplies the atmosphere with oxygen. The fact that the oxygen from green plants originated from water was first deduced by the Canadian-born American biochemist Martin David Kamen. He used a natural, stable isotope of oxygen, O_{18} to trace the path of the oxygen, from water to gaseous molecular oxygen. This reaction is catalysed by a reactive centre in photo-system II containing four manganese ions.

The reaction begins with the excitation of a pair of chlorophyll molecules similar to those in the bacterial reaction centre. Due to the presence of chlorophyll a, as opposed to bacteriochlorophyll, photo-system II absorbs light at a shorter wavelength. The pair of chlorophyll molecules at the reaction centre are often referred to as P680. When the photon has been absorbed, the resulting high-energy electron is transferred to a nearby phaeophytin molecule. This is above and to the right of the pair on the diagram and is coloured grey. The electron travels from the phaeophytin molecule through two plastoquinone molecules, the first tightly bound, the second loosely bound. The tightly bound molecule is shown above the phaeophytin molecule and is coloured red. The loosely bound molecule is to the left of this and is also coloured red. This flow of electrons is similar to that of the bacterial reaction centre. Two electrons are required to fully reduce the loosely bound plastoquinone molecule to QH_2 as well as the uptake of two protons.

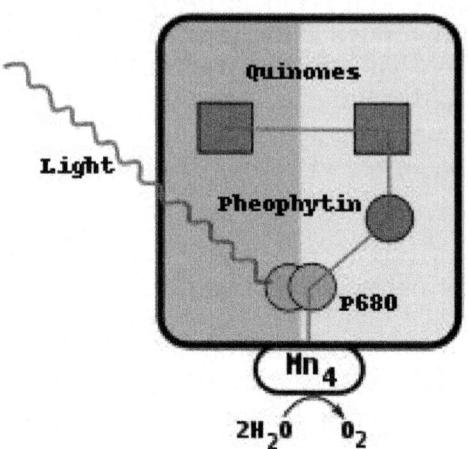

The difference between photo-system II and the bacterial reaction centre is the source of the electron that neutralizes the pair of chlorophyll *a* molecules. In the bacterial reaction centre, the electron is obtained from a reduced compound haem group in a cytochrome sub-unit or from a water-soluble cytochrome-c protein.

Once photoinduced charge separation has taken place, the P680 molecule carries a positive charge. P680 is a very strong oxidant and extracts electrons from two water molecules that are bound at the manganese centre directly below the pair. This centre, below and to the left of the pair in the diagram, contains four manganese ions, a calcium ion, a chloride ion, and a tyrosine residue. Manganese is efficient because it is capable of existing in four oxidation states : Mn^{2+}, Mn^{3+}, Mn^{4+} and Mn^{5+}. Manganese also forms strong bonds with oxygen-containing molecules such as water.

Every time the P680 absorbs a photon, it emits an electron, gaining a positive charge. This charge is neutralized by the extraction of an electron from the manganese centre, which sits directly below it. The process of oxidizing two molecules of water requires four electrons. The water molecules that are oxidized in the manganese centre are the source of the electrons that reduce the two molecules of Q to QH_2. To date, this water-splitting catalytic centre cannot be reproduced by any man-made catalyst.

Photo-system I

After the electron has left photo-system II it is transferred to a cytochrome b6f complex and then to plastocyanin, a blue copper protein and electron carrier. The plastocyanin complex carries the electron that will neutralize the pair in the next reaction centre, photo-system I.

As with photo-system II and the bacterial reaction centre, a pair of chlorophyll a molecules initiates photoinduced charge separation. This pair is referred to as P700. 700 Is a reference to the wavelength at which the chlorophyll molecules absorb light maximally. The P700 lies in the centre of the protein. Once photoinduced charge separation has been initiated, the electron travels down a pathway through a chlorophyll a molecule situated directly above the P700, through a quinone molecule situated directly above that, through three 4Fe-4S clusters, and finally to an interchangeable ferredoxin complex. Ferredoxin is a soluble protein containing a 2Fe-2S cluster co-ordinated by four cysteine residues. The positive charge left on the P700 is neutralized by the transfer of an electron from plastocyanin. Thus the overall reaction catalysed by photo-system I is :

$$Pc(Cu^+) + Fd_{ox} \xrightarrow{\text{light}} Pc(Cu^{2+}) + Fd_{red}$$

The co-operation between photo-systems I and II creates an electron flow from H_2O to $NADP^+$. This pathway is called the 'Z-scheme' because the redox diagram from P680 to P700 resembles the letter z.

CUBOID

In geometry, a cuboid is a convex polyhedron bounded by six quadrilateral faces, whose polyhedral graph is the same as that of a cube. While some mathematical literature refers to any such polyhedron as a cuboid, other sources use "cuboid" to refer to a shape of this type in which each of the faces is a rectangle (and so each pair of adjacent faces meets in a right angle); this more restrictive type of cuboid is also known as a rectangular cuboid, right cuboid, rectangular box, rectangular hexahedron, right rectangular prism, or rectangular parallelepiped.

General Cuboids

By Euler's formula the number of faces ('F'), vertices (V), and edges (E) of any convex polyhedron are related by the formula "F + V " = E + 2. In the case of a cuboid this gives 6 + 8 = 12 + 2; that is, like a cube, a cuboid has 6 faces, 8 vertices, and 12 edges.

Along with the rectangular cuboids, any parallelepiped is a cuboid of this type, as is a square frustum (the shape formed by truncation of the apex of a square pyramid).

Rectangular Cuboid

In a rectangular cuboid, all angles are right angles, and opposite faces of a cuboid are equal. It is also a right rectangular prism. The term "rectangular or oblong prism" is ambiguous. Also the term rectangular parallelepiped or orthogonal parallelepiped is used.

The square cuboid, square box, or right square prism (also ambiguously called square prism) is a special case of the cuboid in which at least two faces are squares. The cube is a special case of the square cuboid in which all six faces are squares.

Rectangular cuboid	
Type	Prism
Faces	6 rectangles
Edges	12
Vertices	8
Symmetry group	D_2h, [2,2], (*222), order 8
Schläfli symbol	{ }×{ }×{ } or { }³
Coxeter diagram	● ● ●
Dual polyhedron	Rectangular fusil
Properties	convex, zonohedron, isogonal

If the dimensions of a cuboid are a, b and c, then its volume is abc and its surface area is $2ab + 2ac + 2bc$.

The length of the space diagonal is :

$$d = \sqrt{a^2 + b^2 + c^2}.$$

Cuboid shapes are often used for boxes, cupboards, rooms, buildings, etc. Cuboids are among those solids that can tessellate 3-dimensional space. The shape is fairly versatile in being able to contain multiple smaller cuboids, *e.g.* sugar cubes in a box, boxes in a cupboard, cupboards in a room, and rooms in a building.

A cuboid with integer edges as well as integer face diagonals is called an Euler brick, for example with sides 44, 117 and 240. A perfect cuboid is an Euler brick whose space diagonal is also an integer. It is currently unknown whether a perfect cuboid actually exists.

BUTTERFLY THEOREM

Let M be the mid-point of a chord PQ of a circle, through which two other chords AB and CD are drawn; AD and BC intersect chord PQ at X and Y correspondingly. Then M is the midpoint of XY.

A formal proof of the theorem is as follows : Let the perpendiculars XX' and XX'' be dropped from the point X on the straight lines AM and DM respectively.

Similarly, let YY' and YY' be dropped from the point Y perpendicular to the straight lines BM and CM respectively.

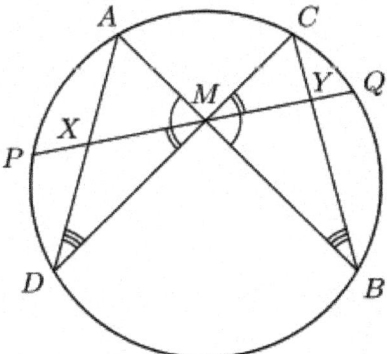

Fig. : The butterfly theorem is a classical result in Euclidean geometry, which can be stated as follows :

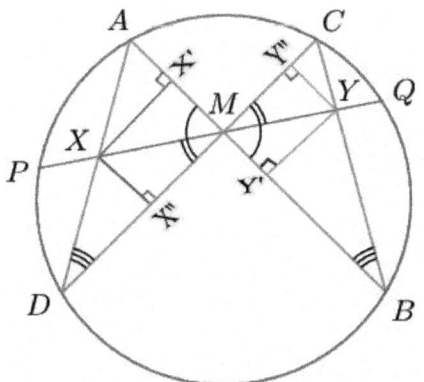

Fig. : Proof of Butterfly theorem.

Now, since

$\Delta MXX' \sim \Delta MYY'$,

$$\frac{MX}{MY} = \frac{XX'}{YY'},$$

$\Delta MXX'' \sim \Delta MYY''$

$$\frac{MX}{MY} = \frac{XX''}{YY''},$$

$\Delta AXX' \sim \Delta CYY''$,

$$\frac{XX'}{YY''} = \frac{AX}{CY},$$

$\Delta DXX'' \sim \Delta BYY'$,

$$\frac{XX''}{YY'} = \frac{DX}{BY},$$

From the preceding equations, it can be easily seen that :

$$\left(\frac{MX}{MY}\right)^2 = \frac{XX'}{YY'}\frac{XX''}{YY''},$$

$$= \frac{AX \cdot DX}{CY \cdot BY},$$

$$= \frac{PX \cdot QX}{PY \cdot QY},$$

$$= \frac{(PM - XM) \cdot (MQ + XM)}{(PM + MY) \cdot (QM - MY)},$$

$$= \frac{(PM)^2 - (MX)^2}{(PM)^2 - (MY)^2},$$

since $PM = MQ$

Now,

$$\frac{(MX)^2}{(MY)^2} = \frac{(PM)^2 - (MX)^2}{(PM)^2 - (MY)^2}.$$

So, it can be concluded that $MX = MY$, or M is the mid-point of XY.

CHAIN GEOMETRY

What is a Chain Geometry?

Let R be a ring with unity and let F be a field contained in R as a subring. The *chain geometry* $\Sigma(F,R)$ is an incidence structure with point set $\mathbb{P}(R)$, *i.e.* the *projective line over F*. The projective line over F can be considered as a subset of $\mathbb{P}(R)$. Its images under the group of projectivities are the *chains*.

The Classical Example

Here R is the field $\mathbb{C}$ of complex numbers and F is the field $\mathbb{R}$ of reals. The corresponding chain geometry is an algebraic model for the *Euclidean Möbius plane, i.e.,* the Euclidean plane with one extra "point at infinity". The $\mathbb{R}$-chains correspond exactly to the circles and lines of the plane (with the extra assignment that all lines contain the point at infinity). The term *chain* goes back to Karl G. C. von Staudt .

Another model for the projective line $\mathbb{P}(\mathbb{C})$ is a *sphere* in Euclidean 3-space; the $\mathbb{R}$-chains appear exactly as the circles on the sphere.

The link between these two geometric models is the *stereographic projection* from the "north pole" N : Each point of the sphere (other than N) is projected to a plane which is parallel to the plane of the "equator". The point N, however, is mapped to the point at infinity.

The second picture on the left hand side illustrates a theorem due to Λ. Miquel in the Euclidean Möbius plane. The underlying configuration consists of 8 points and 6 circles such that there are

- 4 points on each circle
- 3 circles through each point.

This theorem holds - mutatis mutandis - in many other chain geometries.

Fig. : Stereographic projection.   **Fig. : Miquel's theorem.**

Distant Graphs

A basic notion for the projective line $\mathbb{P}(R)$ over a ring R is the relation *distant*. Non-distant points are also called *parallel*. This terminology goes back to the projective line over the *real dual numbers*, where parallel points represent parallel spears of the Euclidean plane .

The points of $\mathbb{P}(R)$ are the vertices and the unordered pairs of distant points are the edges of the *distant graph*.

For example, if GF(2) is the field of *integers modulo* 2 then the projective line over GF(2) × GF(2), i. e the ring of *double numbers* over GF(2), has a distant graph which is depicted on the very left. It has 9 vertices, the shaded triangles serve only for better visualisation.

If R is the ring of *integers modulo* 4 or the ring of *dual numbers over* GF(2), then the distant graph of $\mathbb{P}(R)$ is isomorphic to a *regular octahedron*. So, non-isomorphic rings may have isomorphic distant graphs.

One of our main results is as follows: The distant graph of a projective line $\mathbb{P}(R)$ is *connected* if, and only if, the underlying ring R is a GE_2-*ring*. This means that the group $\mathrm{GL}_2(R)$ of invertible 2x2-matrices over R is generated by the set of all elementary and invertible diagonal matrices.

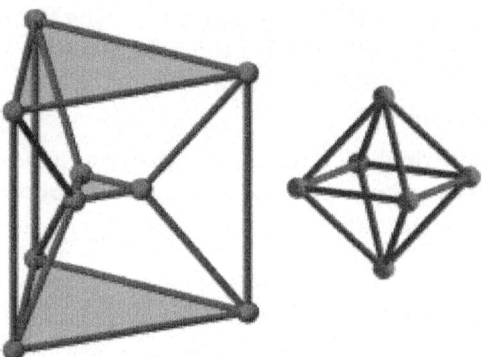

Fig. : Distant graphs of projective lines.

Another Famous Model: Blaschke's Cylinder

By a result due to Wilhelm Blaschke, the projective line over the ring $R = \mathbb{R}[\varepsilon]$, $\varepsilon^2=0$, of *real dual numbers* can be represented as a cylinder in affine 3-space. The $\mathbb{R}$-chains of $\mathbb{P}(\mathbb{R}[\varepsilon])$ correspond with the non-degenerate conics on this cylinder. Two points of the cylinder are non-distant (parallel) if, and only if, they are on a common generator.

Up to isomorphism the chain geometry based on $(F,R) = (\mathbb{R},\mathbb{R}[\varepsilon])$ coincides with the *Euclidean Laguerre geometry* (*spears* and *oriented circles* of the Euclidean plane) and the *3-dimensional space-time geometry* with signature (2,1).

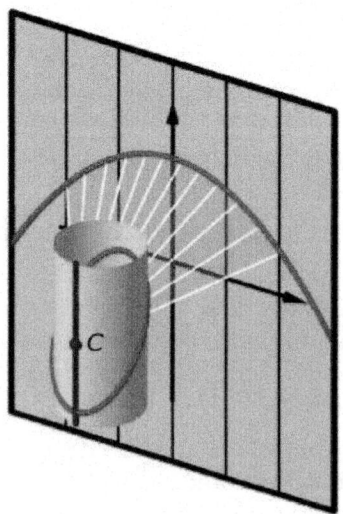

Fig. : Stereographic projection of Blaschke's cylinder.

Let us consider the *stereographic projection* of Blaschke's cylinder. All but one generators of the cylinder are mapped to a pencil of parallel lines. They turn the image plane into an isotropic plane. The exceptional blue generator contains the centre C of projection and it is mapped onto a *generator at infinity* which has to be added to the affine isotropic plane. (This is *not* the projective closure of an affine plane!)

The images of the conics on the cylinder fall into two classes: A conic that does not pass through C goes over to a *parabola* with isotropic diameters. A conic that runs through the centre C is mapped onto a non-isotropic line.

So the chain geometry based on $(\mathbb{R},\mathbb{R}[\varepsilon])$ is also an algebraic model for the *isotropic plane* with one extra *generator at infinity*.

Projective Models

The link between the projective line $\mathbb{P}(R)$ over a ring R and classical projective geometry is given by a *projective model*. Here the points of $\mathbb{P}(R)$ appear as certain *subspaces* of a projective space over a field F. Algebraically, such a model is based upon an injective homomorphism of R into the endomorphism ring of an F-vector space.

We present some low dimensional examples:

Matrix Rings. The projective line over the ring R of 2 by 2 matrices over a field F (viz. $R = F^{2\times2}$) is represented by the set of *all lines* of a 3-dimensional projective space over F.

If F is embedded into R via scalar matrices $\mathrm{diag}(x,x)$, then each F-chain appears as a regulus and, conversely, each regulus arises in this way. Here a *regulus* is the set of all lines that meet three mutually skew lines (called transversals).

The picture on the left hand side shows a regulus in the real projective space ($F=\mathbb{R}$) together with two of its transversals. (In our image the regulus is one family of generators of a *hyperbolic paraboloid*.)

Similar results hold if R is the endomorphism ring of an arbitrary vector space over F, but the definition of a regulus is more sophisticated, especially in the infinite dimensional case.

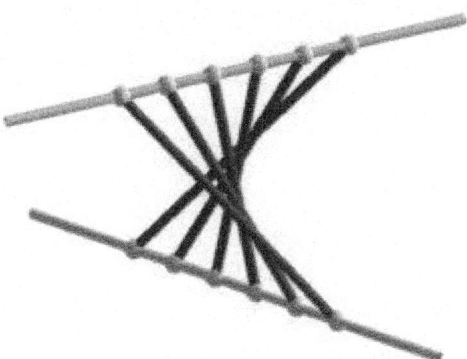

Fig. : Regulus with two transversal lines.

Complex Numbers. As has been sketched above, the projective line over $\mathbb{R}^{2\times2}$ corresponds to the set of all lines of the real projective 3-space. If the complex numbers $\mathbb{C}$ are embedded in $\mathbb{R}^{2\times2}$ in the usual way, then each $\mathbb{C}$-chain appears as *elliptic linear congruences* of lines (regular spread) and, conversely, each elliptic congruence arises in this way.

The picture shows one such congruence of lines. It has the remarkable property to send precisely one of its lines through each point of the space. For better visualisation we have partitioned the congruence into reguli that belong to the quadrics of a pencil of *hyperboloids of revolution*; also the common axis of revolution (not drawn !) and the line at infinity of the horizontal planes (not drawn either !) belongs to the congruence.

If we choose just one elliptic congruence of lines then it is a model of the complex projective line $\mathbb{P}(\mathbb{C})$.

The $\mathbb{R}$-sublines of $\mathbb{P}(\mathbb{C})$ appear as the reguli contained in the congruence. You can see some of those reguli on our picture.

In the Euclidean Möbius plane those reguli correspond to a pencil of concentric circles, the two extra-lines represent their common midpoint and the point at infinity.

These results remain true (with some minor modifications) if an arbitrary quadratic field extension L/F is taken instead of $\mathbb{C}/\mathbb{R}$.

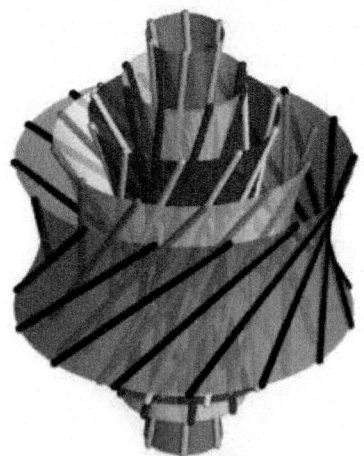

Fig. : Elliptic linear congruence (regular spread).

Double Numbers. Another example that admits an easy visualisation is the ring $F \times F$ of *double numbers* over a field F.

A projective model is given by a *hyperbolic linear congruence* of lines in a projective 3-space over F, i.e., the set of all lines that meet two skew axes.

As above, the F-chains appear as the reguli contained in the congruence. Here F is embedded in $F \times F$ by identifying each element x of F with the pair (x,x).

Similar results hold for the ring F^n: A projective model in a $(2n-1)$-dimensional projective space is given by all $(n-1)$-dimensional subspaces that meet n lines in general position.

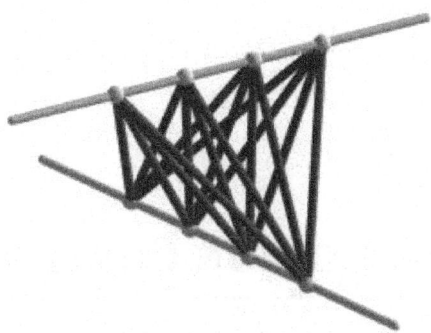

Fig. : Hyperbolic linear congruence.

Dual Numbers. The projective line over the ring $F[\varepsilon]$, $\varepsilon^2=0$, of *dual numbers* over a field F can be represented by a subset of a *parabolic linear congruence* of lines. Such a congruence has just one axis which belongs to the congruence. The axis is the only line of the congruence that represents no point of the dual projective line. As in the hyperbolic case, all lines of the congruence through a point on the axis form a pencil. The mapping that assigns to each point P of the axis the carrier plane of the pencil with centre P is a projectivity.

The F-chains appear exactly as the reguli contained in the congruence.

Fig. : Parabolic linear congruence.

Ternions. The ring of *ternions* over a field F is isomorphic to the ring of upper triangular 2x2 matrices over F. The projective line over this ring is represented by lines of a *special linear complex*, i.e., the set of all lines in a 3-dimensional projective space that meet a fixed axis; only this axis represents no point of the projective line over the ternions.

The F-chains appear as the reguli that have the axis of the complex among their transversal lines.

Fig. : Special linear complex.

Chapter 9

ORGANOIRON CHEMISTRY

Organoiron chemistry is the chemistry of organometallic compounds containing a carbon-to-iron chemical bond. Organoiron compounds are relevant in organic synthesis as reagents such as iron pentacarbonyl, diiron nonacarbonyl and disodium tetracarbonylferrate. Iron adopts oxidation states from Fe(-II) through to Fe (IV). Although iron is generally less active in many catalytic applications, it is less expensive and "greener" than other metals. Organoiron compounds feature a wide range of ligands that support the Fe-C bond; as with other organometals, these supporting ligands prominently include phosphines, carbon monoxide, and cyclopentadienyl, but hard ligands such as amines are employed as well.

IRON CARBONYLS

The Binary Carbonyls and Their Anions

Important iron carbonyls are the three neutral binary carbonyls, iron pentacarbonyl, diiron nonacarbonyl, and triiron dodecacarbonyl. One or more carbonyl ligands in these compounds can be replaced by a variety of other ligands (dienes, phosphines).

Iron carbonyls have been used in stoichiometric carbonylation reactions, *e.g.* for the conversion of alkyl bromides to aldehydes. Disodium tetracarbonylferrate, "Collman's Reagent," can be alkylated followed by carbonylation to give the acyl derivatives that undergo protonolysis to afford aldehydes :

$$LiFe(CO)_4 + H^+ \rightarrow RCHO \text{ (+ iron containing products)}$$

Similar iron acyls can be accessed by treating iron pentacarbonyl with organolithium compounds :

$$ArLi + Fe(CO)_5 \rightarrow LiFe(CO)_4C(O)R$$

In this case, the carbanion attacks a CO ligand. In a complementary reaction, Collman's reagent can be used to convert acyl chlorides to aldehydes. Similar reactions can be achieved with $[HFe(CO)_4]^-$ salts.

Fe(CO)$_3$ Derivatives

Iron diene complexes are usually prepared from Fe(CO)$_5$ or Fe$_2$(CO)$_9$. Derivatives are known for common dienes are cyclohexadiene, norbornadiene and cycloocta-diene, but even cyclobutadiene can be stabilized. In the complex with butadiene, the diene adopts a cis-conformation. Iron carbonyls are used as a protective group for dienes in hydrogenations and Diels-Alder reactions. Cyclobutadieneiron tri-carbonyl is prepared from 3,4-dichlorocyclobutene and Fe$_2$(CO)$_9$.

Cyclohexadienes, many derived from Birch reduction of aromatic compounds, form derivatives Fe(CO)$_3$. The affinity of the Fe(CO)$_3$ unit for conjugated dienes is manifested in the ability of iron carbonyls catalyse the isomerisations of 1,5-cy-clooctadiene to 1,3-cyclooctadiene. Cyclohexadiene complexes undergo hydride abstraction to give cyclohexadienyl cations, which add nucleophiles.

The enone complex (benzylideneacetone) iron tricarbonyl serves as a source of the Fe(CO)$_3$ sub-unit and is employed to prepare other derivatives. It is used complementarily to Fe$_2$(CO)$_9$.

Alkynes form many compounds with upon reaction with iron carbonyls. These include cyclobutadiene derivatives, ferroles" of the formula Fe$_2$(C$_4$R$_4$)(CO)$_6$, as well as cyclopentadienone and cyclobutadiene derivatives.

Sulfur and Phosphorus Derivative

Complexes of the type Fe$_2$(SR)$_2$(CO)$_6$ and Fe$_2$(PR$_2$)$_2$(CO)$_6$ form, usually by the re-action of thiols and secondary phosphines with iron carbonyls. The thiolates can also be obtained from the tetrahedrane Fe$_2$S$_2$(CO)$_6$.

CYCLOPENTADIENYL DERIVATIVES, INCLUDING FERROCENES

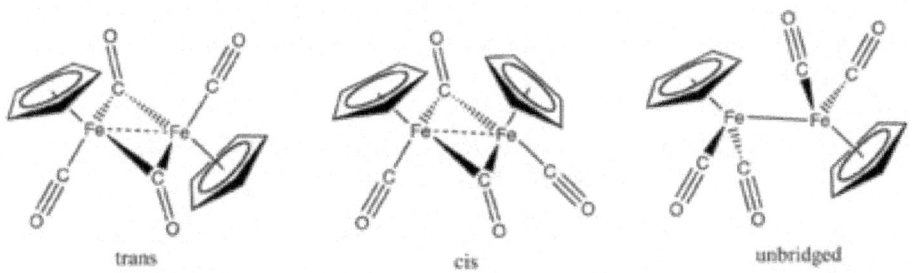

trans   cis   unbridged

Ferrocene and its Derivatives

The rapid growth of organometallic chemistry in the 20th century can be traced to the discovery of ferrocene, a very stable compound which foreshadowed the synthesis of many related sandwich compounds. Ferrocene is formed by reaction of sodium cyclopentadienide with iron (II) chloride :

$$2\,NaC_5H_5 + FeCl_2 \rightarrow Fe(C_5H_5)_2 + 2\,NaCl$$

Ferrocene displays diverse reactivity localized on the cyclopentadienyl ligands, including Friedel-Crafts reactions and lithation. Ferrocene is also a structurally unusual scaffold as illustrated by the popularity of ligands such as 1,1'-bis(diphenylphosphino)ferrocene, which are useful in catalysis. Treatment of ferrocene with aluminium trichloride and benzene gives the cation $[CpFe(C_6H_6)]^+$. Oxidation of ferrocene gives the blue 17e species ferrocenium. Derivatives of fullerene can also act as a highly substituted cyclopentadienyl ligand.

Fp$_2$ and its Derivatives

$Fe(CO)_5$ reacts with dicyclopentadiene to give the cyclopentadienyliron dicarbonyl dimer ($[FeCp(CO)_2]_2$). Reduction of this species with sodium gives "NaFp" (Fp = $[FeCp(CO)_2]^-$), a potent nucleophile and precursor to many derivatives of the type $CpFe(CO)_2R$. Fp can also be synthesized photochemically using UV-Visible light. The derivative $[FpCH_2S(CH_3)_2]$ has been used in cyclopropanations. Fp-acyl compounds are prochiral, and studies have exploited the chiral derivatives $CpFe(PPh_3)(CO)acyl$. Pyrolysis of Fp$_2$ gives the cuboidal cluster $[FeCp(CO)]_4$.

Polyhapto Organic Ligands

Stable complexes with iron with and without CO ligands are known for a wide variety of polyunsaturated hydrocarbons, *e.g.* cycloheptatriene, azulene, cyclooctatetraene (COT), and bullvalene. The compound $Fe(COT)_2$ is well known, $Fe_3(COT)_3$ was described in 2009 as the reaction product of $Fe(COT)_2$ with a catalytic amount of an persistent carbene. It can be regarded as an organic version of triiron dodecacarbonyl

PHOSPHINE- AND AMINE-FE (II) COMPLEXES

As for other organometallic compounds, organoiron (II) complexes in the absence of Cp ligands are commonly complemented by tertiary diphosphines and to a lesser extent amine/imine ligands.

ORGANOIRON COMPOUNDS IN ORGANIC SYNTHESIS AND HOMOGENEOUS CATALYSIS

Because of its low cost and low toxicity of its salts, iron is often employed as a stoichiometric reagent. Iron role as a catalyst in organic reactions is overshadowed by the related chemistries of cobalt and nickel. Some main categories are :

- Addition reactions for example the Aldol reaction and Michael reaction
- Substitution notably coupling reactions. Ferric chloride is a well known catalyst in the Friedel–Crafts reaction. Compounds of the type $[(\eta^3-allyl)Fe(CO)^{4+}X^-$ are allyl cation synthons in allylic substitution. Likewise the compound $Ph(CO_2)$ $Fe^+(\eta_2-vinyl(OEt))BF_4^-$ is a masked vinyl cation. Disodium tetracarbonylferrate can be regarded a CO dianion synthon.

- Cycloadditions, for example cyclopropanation using $CpFe(CO)_2CH_2S^+(CH_3)_2BF_4^-$. Also Ene reaction
- Hydrogenation and reduction, example catalyst Knölker complex.
- Isomerization reactions and rearrangement reactions
- Cross-coupling reactions. Iron compounds such as $Fe(acac)_3$ catalyze a wide range of cross-coupling reactions with one substrate an aryl or alkyl Grignard and the other substrate an aryl, alkenyl, or acyl organohalide. In the related Kumada coupling the catalysts are based on palladium and nickel.

BIOCHEMISTRY

In the area of bioorganometallic chemistry, organoiron species are found at the active sites of the three hydrogenase enzymes as well as carbon monoxide dehydrogenase.

Chapter 10

COBALT

Cobalt is a chemical element with symbol Co and atomic number 27. Like nickel, cobalt in the Earth's crust is found only in chemically combined form, save for small deposits found in alloys of natural meteoric iron. The free element, produced by reductive smelting, is a hard, lustrous, silver-gray metal.

Cobalt-based blue pigments (cobalt blue) have been used since ancient times for jewelry and paints, and to impart a distinctive blue tint to glass, but the colour was later thought by alchemists to be due to the known metal bismuth. Miners had long used the name kobold ore (German for goblin ore) for some of the blue-pigment producing minerals; they were so named because they were poor in known metals, and gave poisonous arsenic-containing fumes upon smelting. In 1735, such ores were found to be reducible to a new metal (the first discovered since ancient times), and this was ultimately named for the kobold.

Today, some cobalt is produced specifically from various metallic-lustered ores, for example cobaltite ($CoAsS$), but the main source of the element is as a by-product of copper and nickel mining. The copper belt in the Democratic Republic of the Congo and Zambia yields most of the cobalt mined worldwide.

Cobalt is primarily used as the metal, in the preparation of magnetic, wear-resistant and high-strength alloys. Its compounds cobalt silicate and cobalt (II) aluminate ($CoAl_2O_4$, cobalt blue) give a distinctive deep blue colour to glass, smalt, ceramics, inks, paints and varnishes. Cobalt occurs naturally as only one stable isotope, cobalt-59. Cobalt-60 is a commercially important radioisotope, used as a radioactive tracer and for the production of high intensity gamma rays.

Cobalt is the active center of coenzymes called cobalamins, the most common example of which is vitamin B_{12}. As such it is an essential trace dietary mineral for all animals. Cobalt in inorganic form is also an active nutrient for bacteria, algae and fungi.

CHARACTERISTICS

Cobalt is a ferromagnetic metal with a specific gravity of 8.9. The Curie temperature is 1115°C and the magnetic moment is 1.6–1.7 Bohr magnetons per atom. Cobalt has a relative permeability two thirds that of iron. Metallic cobalt occurs as two crystallographic structures : hcp and fcc. The ideal transition temperature between the hcp and fcc structures is 450°C, but in practice, the energy difference is so small that random intergrowth of the two is common.

Cobalt is a weakly reducing metal that is protected from oxidation by a passivating oxide film. It is attacked by halogens and sulfur. Heating in oxygen produces Co_3O_4 which loses oxygen at 900°C to give the monoxide CoO. The metal reacts with Fluorine gas (F_2) at 520 K to give CoF_3; with chlorine (Cl_2), bromine (Br_2) and iodine (I_2), the corresponding binary halides are formed. It does not react with hydrogen gas (H_2) or nitrogen gas (N_2) even when heated, but it does react with boron, carbon, phosphorus, arsenic and sulphur. At ordinary temperatures, it reacts slowly with mineral acids, and very slowly with moist, but not with dry, air.

COMPOUNDS

Common oxidation states of cobalt include +2 and +3, although compounds with oxidation states ranging from −3 to +4 are also known. A common oxidation state for simple compounds is +2 (cobalt (II)). These salts form the pink-coloured metal aquo complex $[Co(H_2O)_6]_2^+$ in water. Addition of chloride gives the intensely blue $[CoCl_4]_2^-$.

Oxygen and Chalcogen Compounds

Several oxides of cobalt are known. Green cobalt (II) oxide (CoO) has rocksalt structure. It is readily oxidized with water and oxygen to brown cobalt (III) hydroxide $_3$). At temperatures of 600–700°C, CoO oxidizes to the blue cobalt (II,III) oxide, which has a spinel structure. Black cobalt (III) oxide is also known. Cobalt oxides are anti-ferromagnetic at low temperature : CoO (Néel temperature 291 K) and Co_3O_4 (Néel temperature : 40 K), which is analogous to magnetite, with a mixture of +2 and +3 oxidation states.

The principal chalcogenides of cobalt include the black cobalt (II) sulfides, CoS_2, which adopts a pyrite-like structure, and cobalt (III) sulfide. Pentlandite is metal-rich.

Halides

Four dihalides of cobalt (II) are known : cobalt(II) fluoride (CoF2, pink), cobalt (II) chloride (CoCl$_2$, blue), cobalt (II) bromide (CoBr$_2$, green), cobalt (II) iodide (CoI$_2$, blue-black). These halides exist in anhydrous and hydrated forms. Whereas the anhydrous dichloride is blue, the hydrate is red.

The reduction potential for the reaction

Co_3^{++} e-→ Co_2^+ is +1.92 V, beyond that for chlorine to chloride, +1.36 V. As a consequence cobalt (III) and chloride would result in the cobalt (III) being reduced to cobalt (II). Because the reduction potential for fluorine to fluoride is so high, +2.87 V, cobalt (III) fluoride is one of the few simple stable cobalt (III) compounds. Cobalt (III) fluoride, which is used in some fluorination reactions, reacts vigorously with water.

Coordination Compounds

As for all metals, molecular compounds and polyatomic ions of cobalt are classified as coordination complexes, that is molecules or ions that contain cobalt linked to several ligands. The principles of electronegativity and hardness–softness of a series of ligands can be used to explain the usual oxidation state of the cobalt. For example Co^{+3} complexes tend to have ammine ligands. As phosphorus is softer than nitrogen, phosphine ligands tend to feature the softer Co^{2+} and Co^+, an example being tris(triphenylphosphine)cobalt(I) chloride ($(P(C_6H_5)_3)_3CoCl$). The more electronegative (and harder) oxide and fluoride can stabilize Co^{4+} and Co^{5+} derivatives, e.g. caesium hexafluorocobaltate (Cs_2CoF_6) and potassium percobaltate (K_3CoO_4).

Alfred Werner, a Nobel-prize winning pioneer in coordination chemistry, worked with compounds of empirical formula $CoCl_3(NH_3)_6$. One of the isomers determined was cobalt (III) hexammine chloride. This coordination complex, a "typical" Werner-type complex, consists of a central cobalt atom coordinated by six ammine ligands orthogonal to each other and three chloride counteranions. Using chelating ethylenediamine ligands in place of ammonia gives tris(ethylenediamine) cobalt (III) chloride ($[Co(en)_3]Cl_3$), which was one of the first co-ordination complexes that was resolved into optical isomers. The complex exists as both either right- or left-handed forms of a "three-bladed propeller". This complex was first isolated by Werner as yellow-gold needle-like crystals.

HISTORY

Cobalt compounds have been used for centuries to impart a rich blue colour to glass, glazes and ceramics. Cobalt has been detected in Egyptian sculpture and Persian jewelry from the third millennium BC, in the ruins of Pompeii (destroyed in 79 AD), and in China dating from the Tang dynasty (618–907 AD) and the Ming dynasty (1368–1644 AD).

Cobalt has been used to colour glass since the Bronze Age. The excavation of the Uluburun shipwreck yielded an ingot of blue glass, which was cast during the 14th century BC. Blue glass items from Egypt are coloured with copper, iron, or cobalt. The oldest cobalt-coloured glass was from the time of the Eighteenth dynasty in Egypt (1550–1292 BC). The location where the cobalt compounds were obtained is unknown.

The word cobalt is derived from the German kobalt, from kobold meaning "goblin", a superstitious term used for the ore of cobalt by miners. The first attempts at smelting these ores to produce metals such as copper or nickel failed, yielding simply powder (cobalt (II) oxide) instead. Also, because the primary ores of cobalt always contain arsenic, smelting the ore oxidized the arsenic content into the highly toxic and volatile arsenic oxide, which also decreased the reputation of the ore for the miners.

Swedish chemist Georg Brandt is credited with discovering cobalt circa 1735, showing it to be a new previously unknown element different from bismuth and other traditional metals, and calling it a new "semi-metal." He was able to show that compounds of cobalt metal were the source of the blue colour in glass, which previously had been attributed to the bismuth found with cobalt. Cobalt became the first metal to be discovered since the pre-historical period, during which all the known metals (iron, copper, silver, gold, zinc, mercury, tin, lead and bismuth) had no recorded discoverers.

During the 19th century, a significant part of the world's production of cobalt blue (a dye made with cobalt compounds and alumina) and smalt (cobalt glass powdered for use for pigment purposes in ceramics and painting) was carried out at the Norwegian Blaafarveværket. The first mines for the production of smalt in the 16th to 18th century were located in Norway, Sweden, Saxony and Hungary. With the discovery of cobalt ore in New Caledonia in 1864 the mining of cobalt in Europe declined. With the discovery of ore deposits in Ontario, Canada in 1904 and the discovery of even larger deposits in the Katanga Province in the Congo in 1914 the mining operations shifted again. With the Shaba conflict starting in 1978, the main source for cobalt, the copper mines of Katanga Province, nearly stopped their production. The impact on the world cobalt economy from this conflict was however smaller than expected. Cobalt being a rare metal and the pigment being highly toxic, the industry had already established effective ways for recycling cobalt materials and in some cases was able to change to cobalt-free alternatives.

In 1938, John Livingood and Glenn T. Seaborg discovered cobalt-60. This isotope was famously used at Columbia University in the 1950s to establish parity violation in radioactive beta decay.

After World War II, the US wanted to be sure it was never short of the ore needed for military cobalt uses (as the Germans had been during that war) and explored for cobalt within the U.S. border. A good supply of the ore needed was found in Idaho near Blackbird canyon in the side of a mountain. The firm Calera Mining Company got production started at the site.

OCCURRENCE

The stable form of cobalt is created in supernovas *via* the r-process. It comprises 0.0029% of the Earth's crust and is one of the first transition metals.

Free cobalt (the native metal) is not found in on Earth due to the amount of oxygen in the atmosphere and chlorine in the ocean. Oxygen and chlorine are abundant enough in the upper layers of the Earth's crust so as to make native metal cobalt formation extremely rare. Except as recently delivered in meteoric iron, pure cobalt in native metal form is unknown on Earth. Though the element is of medium abundance, natural compounds of cobalt are numerous. Small amounts of cobalt compounds are found in most rocks, soil, plants, and animals.

In nature, cobalt is frequently associated with nickel, and both are characteristic components of meteoric iron, though cobalt is much less abundant in iron meteorites than nickel. As with nickel, cobalt in meteoric iron alloys may have been well enough protected from oxygen and moisture to occur as the free metal, a state which otherwise is not seen with either element in the ancient terrestrial crust.

Cobalt in compound form occurs as a minor component of copper and nickel minerals. It is the major metallic component in combination with sulfur and arsenic in the sulfidic cobaltite, safflorite, glaucodot (AsS), and skutterudite minerals. The mineral cattierite is similar to pyrite and occurs together with vaesite in the copper deposits of the Katanga Province. Upon contact with the atmosphere, weathering occurs and the sulfide minerals oxidize to form pink erythrite ("cobalt glance" : $Co_3(AsO_4)_2 \cdot 8H_2O$) and spherocobaltite.

PRODUCTION

The main ores of cobalt are cobaltite, erythrite, glaucodot and skutterudite, but most cobalt is obtained not by active mining of cobalt ores, but rather by reducing cobalt compounds that occur as by-products of nickel and copper mining activities.

In 2005, the copper deposits in the Katanga Province (former Shaba province) of the Democratic Republic of the Congo were the top producer of cobalt with almost 40% world share, reports the British Geological Survey. The political situation in the Congo influences the price of cobalt significantly.

The Mukondo Mountain project, operated by the Central African Mining and Exploration Company in Katanga, may be the richest cobalt reserve in the world. It is estimated to be able to produce about one third of total global production of cobalt in 2008. In July 2009 CAMEC announced a long term agreement under which CAMEC would deliver its entire annual production of cobalt in concentrate from Mukondo Mountain to Zhejiang Galico Cobalt & Nickel Materials of China.

Several methods exist for the separation of cobalt from copper and nickel. They depend on the concentration of cobalt and the exact composition of the used ore. One separation step involves froth flotation, in which surfactants bind to different ore components, leading to an enrichment of cobalt ores. Subsequent roasting converts the ores to the cobalt sulfate, whereas the copper and the iron are oxidized to the oxide. The leaching with water extracts the sulfate together with the arsenates. The residues are further leached with sulfuric acid yielding a solution of copper sulfate. Cobalt can also be leached from the slag of the copper smelter.

The products of the above-mentioned processes are transformed into the cobalt oxide. This oxide is reduced to the metal by the aluminothermic reaction or reduction with carbon in a blast furnace.

APPLICATIONS

The main application of cobalt is as the free metal, in production of certain high performance alloys.

Alloys

Cobalt-based superalloys consume most of the produced cobalt. The temperature stability of these alloys makes them suitable for use in turbine blades for gas turbines and jet aircraft engines, though nickel-based single crystal alloys surpass them in this regard. Cobalt-based alloys are also corrosion and wear-resistant. This makes them useful in the medical field, where cobalt is often used (along with titanium) for orthopedic implants that do not wear down over time. The development of the wear-resistant cobalt alloys started in the first decade of the 19th century with the stellite alloys, which are cobalt-chromium alloys with varying tungsten and carbon content. The formation of chromium and tungsten carbides makes them very hard and wear resistant. Special cobalt-chromium-molybdenum alloys like Vitallium are used for prosthetic parts such as hip and knee replacements. Cobalt alloys are also used for dental prosthetics, where they are useful to avoid allergies to nickel. Some high speed steel drill bits also use cobalt to increase heat and wear-resistance. The special alloys of aluminium, nickel, cobalt and iron, known as Alnico, and of samarium and cobalt (samarium-cobalt magnet) are used in permanent magnets. It is also alloyed with 95% platinum for jewelry purposes, yielding an alloy that is suitable for fine detailed casting and is also slightly magnetic.

Batteries

Lithium cobalt oxide ($LiCoO_2$) is widely used in lithium ion battery cathodes. The material is composed of cobalt oxide layers in which the lithium is intercalated. During discharging the lithium intercalated between the layers is set free as lithium ion. Nickel-cadmium (NiCd) and nickel metal hydride (NiMH) batteries also contain significant amounts of cobalt; the cobalt improves the oxidation capabilities of nickel in the battery.

Catalysts

Several cobalt compounds are used in chemical reactions as oxidation catalysts. Cobalt acetate is used for the conversion of xylene to terephthalic acid, the precursor to the bulk polymer polyethylene terephthalate. Typical catalysts are the cobalt carboxylates (known as cobalt soaps). They are also used in paints, varnishes, and inks as "drying agents" through the oxidation of drying oils. The same carboxylates are used to improve the adhesion of the steel to rubber in steel-belted radial tires.

Cobalt-based catalysts are also important in reactions involving carbon monoxide. Steam reforming, useful in hydrogen production, uses cobalt oxide-base catalysts. Cobalt is also a catalyst in the Fischer–Tropsch process, used in the hydrogenation of carbon monoxide into liquid fuels. The hydroformylation of alkenes often rely on cobalt octacarbonyl as the catalyst, although such processes have been partially displaced by more efficient iridium-and rhodium-based catalysts, *e.g.* the Cativa process.

The hydrodesulfurization of petroleum uses a catalyst derived from cobalt and molybdenum. This process helps to rid petroleum of sulfur impurities that interfere with the refining of liquid fuels.

Pigments and Colouring

Before the 19th century, the predominant use of cobalt was as a pigment. Since the Middle Ages, it has been involved in the production of smalt, a blue coloured glass. Smalt is produced by melting a mixture of the roasted mineral smaltite, quartz and potassium carbonate, yielding a dark blue silicate glass which is ground after the production. Smalt was widely used for the colouration of glass and as pigment for paintings. In 1780, Sven Rinman discovered cobalt green and in 1802 Louis Jacques Thénard discovered cobalt blue. The two varieties of cobalt blue pigment, cobalt blue (cobalt aluminate) and cobalt green (a mixture of cobalt (II) oxide and zinc oxide), were used as pigments for paintings because of their superior stability.

Radioisotopes

Cobalt-60 (Co-60 or ^{60}Co) is useful as a gamma ray source because it can be produced in predictable quantity and high activity by bombarding cobalt with neutrons. It produces two gamma rays with energies of 1.17 and 1.33 MeV.

Its uses include external beam radiotherapy, sterilization of medical supplies and medical waste, radiation treatment of foods for sterilization (cold pasteurization), industrial radiography (*e.g.* weld integrity radiographs), density measurements (*e.g.* concrete density measurements), and tank fill height switches. The metal has the unfortunate habit of producing a fine dust, causing problems with radiation protection. Cobalt from radiotherapy machines has been a serious hazard when not disposed of properly, and one of the worst radiation contamination accidents in North America occurred in 1984, after a discarded radiotherapy unit containing cobalt-60 was mistakenly disassembled in a junkyard in Juarez, Mexico.

Cobalt-60 has a radioactive half-life of 5.27 years. This decrease in activity requires periodic replacement of the sources used in radiotherapy and is one reason why cobalt machines have been largely replaced by linear accelerators in modern radiation therapy.

Cobalt-57 (Co-57 or ^{57}Co) is a cobalt radioisotope most often used in medical tests, as a radiolabel for vitamin B$_{12}$ uptake, and for the Schilling test. Cobalt-57 is used as a source in Mössbauer spectroscopy and is one of several possible sources in X-ray fluorescence devices.

Nuclear weapon designs could intentionally incorporate ^{59}Co, some of which would be activated in a nuclear explosion to produce ^{60}Co. The ^{60}Co, dispersed as nuclear fallout, creates what is sometimes called a cobalt bomb.

Other Uses

Other uses of cobalt are in electroplating, owing to its attractive appearance, hardness and resistance to oxidation, and as ground coats for porcelain enamels.

BIOLOGICAL ROLE

Cobalt is essential to all animals. It is a key constituent of cobalamin, also known as vitamin B_{12}, which is the primary biological reservoir of cobalt as an "ultratrace" element. Bacteria in the guts of ruminant animals convert cobalt salts into vitamin B_{12}, a compound which can only be produced by bacteria or archaea. The minimum presence of cobalt in soils therefore, markedly improves the health of grazing animals, and an uptake of 0.20 mg/kg a day is recommended for them, as they can obtain vitamin B_{12} in no other way.

In the early 20th century during the development for farming of the North Island Volcanic Plateau of New Zealand, cattle suffered from what was termed "bush sickness". It was discovered that the volcanic soils lacked cobalt salts, which was necessary for cattle. The ailment was cured by adding small amounts of cobalt to fertilizers in the form of Superphosphate (at the time derived from Canadian sources).

In the 1930s "coast disease" of sheep in the Ninety Mile Desert of the Southeast of South Australia was found to be due to nutrient deficiencies of the trace elements cobalt and copper. The cobalt deficiency was overcome by the development of "cobalt bullets", dense pellets of cobalt oxide mixed with clay, which are orally inserted to lodge in the animal's rumen.

Non-ruminant herbivores produce vitamin B_{12} from bacteria in their colons which again make the vitamin from simple cobalt salts. However the vitamin cannot be absorbed from the colon, and thus non-ruminants must ingest feces to obtain the nutrient. Animals that do not follow these methods of getting vitamin B_{12} from their own gastrointestinal bacteria or that of other animals, must obtain the vitamin pre-made in other animal products in their diet, and they cannot benefit from ingesting simple cobalt salts.

The cobalamin-based proteins use corrin to hold the cobalt. Coenzyme B_{12} features a reactive C-Co bond, which participates in its reactions. In humans, B_{12} exists with two types of alkyl ligand : methyl and adenosyl. MeB_{12} promotes methyl ($-CH_3$) group transfers. The adenosyl version of B_{12} catalyzes re-arrangements in which a hydrogen atom is directly transferred between two adjacent atoms with concomitant exchange of the second substituent, X, which may be a carbon atom with substituents, an oxygen atom of an alcohol, or an amine. Methylmalonyl coenzyme A mutase (MUT) converts MMl-CoA to Su-CoA, an important step in the extraction of energy from proteins and fats.

Although far less common than other metalloproteins (*e.g.* those of zinc and iron), cobaltoproteins are known aside from B_{12}. These proteins include methionine aminopeptidase 2 an enzyme that occurs in humans and other mammals which does not use the corrin ring of B_{12}, but binds cobalt directly. Another non-corrin cobalt enzyme is nitrile hydratase, an enzyme in bacteria that are able to metabolize nitriles.

PRECAUTIONS

Cobalt is an essential element for life in minute amounts. The LD_{50} value for soluble cobalt salts has been estimated to be between 150 and 500 mg/kg. Thus, for a 100 kg person the LD_{50} for a single dose would be about 20 grams.

However, chronic cobalt ingestion has caused serious health problems at doses far less than the lethal dose. In 1966, the addition of cobalt compounds to stabilize beer foam in Canada led to a peculiar form of toxin-induced cardiomyopathy, which came to be known as beer drinker's cardiomyopathy.

After nickel and chromium, cobalt is a major cause of contact dermatitis.

REMEDIATION

Cobalt can be effectively absorbed by charred pigs bones; however this process is inhibited by copper and zinc; which have greater affinities to bone char.

Chapter 11

ORGANOCOBALT CHEMISTRY

Organocobalt chemistry is the chemistry of organometallic compounds containing a carbon to cobalt chemical bond. Organocobalt compounds are involved in several organic reactions and the important bio-molecule vitamin B_{12} has a cobalt-carbon bond. Many organocobalt compounds exhibit useful catalytic properties, the preeminent example being Dicobalt octacarbonyl. An early example of organocobalt chemistry is the carbonylation of azobenzene with dicobalt octacarbonyl as described by Murahashi & Horiie in 1956 :

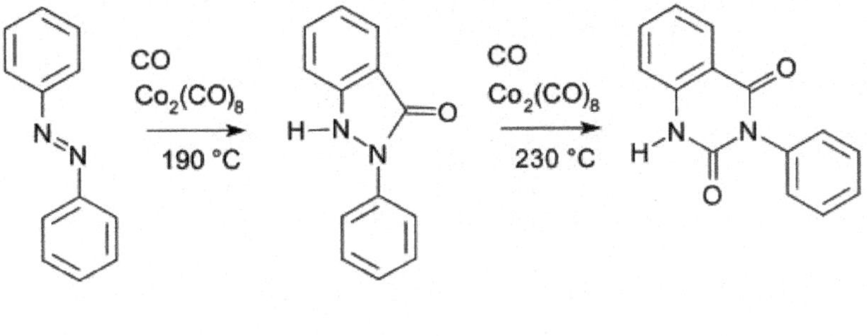

CARBONYL COMPLEXES

Dicobalt octacarbonyl reacts with hydrogen and alkenes to give aldehydes. This reaction is the basis of hydroformylation, the formation of aldehydes from an alkene, CO and hydrogen. A key intermediate is cobalt tetracarbonyl hydride ($HCo(CO)_4$). The original Ruhrchemie process produced propanal from ethene and syngas using cobalt carbonyl has been displaced by rhodium-based catalysts. Processes involving cobalt are practiced by BASF, EXXON, and Shell mainly for the production of C7-C14 alcohols used for the production of surfactants.

In hydrocarboxylations hydrogen is replaced by water or an alcohol and the reaction product is a carboxylic acid or an ester. An example of this reaction type is the conversion of butadiene to adipic acid.

Alkyne Derivatives of $Co_2(CO)_8$

Dicobalt octacarbonyl also reacts with alkynes to give "tetrahedranes" of the formula $Co_2(CO)_6(C_2R_2)$. Because the cobalt carbonyl centers can be removed later, it functions as a protective group for the alkyne. In the Nicholas reaction an alkyne group is also protected and at the same time the alpha-carbon position is activated for nucleophilic substitution.

Cyclization Reactions

Cobalt compounds react with dialkynes and dienes to cyclic intermediates in cyclometalation. Other alkynes, alkenes, nitriles or carbon monoxide can then insert themselves into the Co-C bond. Reaction types based on this concept are the Pauson–Khand reaction (CO insertion) and alkyne trimerization (notably with cyclopentadienylcobalt dicarbonyl).

CP, ALLYL, AND ALKENE COMPOUNDS

Sandwich Compounds

Organocobalt compounds are known with alkene, allyl, diene, and Cp ligands. A famous sandwich compound is cobaltocene, a 19-electron metallocene that is used as a reducing agent and a source of CpCo. Other sandwich compounds are CoCp and Co_2, with 20 electrons and 21 electrons, respectively. Reduction of cobalt (II) compounds in the presence of cyclooctadiene gives Co(cyclooctadiene)(cyclooctenyl), a synthetically versatile reagent

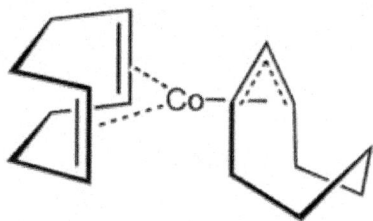

Fig. : Co(1,5-cyclooctadiene)(cyclooctenyl).

CpCo(CO)$_2$ and Derivatives

The half-sandwich compound cyclopentadienylcobalt dicarbonyl (CpCo(CO)$_2$) is a particularly versatile reagent because the CO ligands can be replaced and olefin and alkyne derivatives undergo reactions. A well studied reaction is alkyne trimerisation, which has been applied to the synthesis of a variety of complex structures.

VITAMIN B$_{12}$-TYPE COMPOUNDS

Cobalt is found in vitamin B$_{12}$ and related enzymes. These cofactors catalyze unusual reactions involving the intermediacy of Co-C bonds. In these reactions, the oxidation state of cobalt can vary from Co (III) to Co (I). In methylcobalamin the ligand is a methyl group, which is electrophilic. in vitamin B$_{12}$, the alkyl ligand is an adenosyl group. Related to vitamin B$_{12}$ are cobalt porphyrins, dimethylglyoximates, and related complexes of Schiff base ligands. These synthetic compounds also form alkyl derivatives that undergo diverse reactions reminiscent of the biological processes.

R = 5'-deoxyadenosyl, Me, OH, CN

Cobalt-mediated Radical Polymerization

The weak cobalt (III)-carbon bond in vitamin B_{12} analogues can be exploited in a type of controlled radical polymerization of acrylic and vinyl esters (*e.g.* vinyl acetate), acrylic acid and acrylonitrile. The reaction temperature is typically between 0 and 60°C. A Co-C bond containing radical initiator breaks up (by heat or by light) in a carbon free radical and a cobalt (II) radical species. The carbon radical starts polymer chain formation with monomer for instance an alkene as in any ordinary radical polymerization. Cobalt reversibly reforms the bond with the carbon radical terminus of the growing chain, which minimizes the concentration of radicals and suppresses undesirable termination reactions by recombination of two carbon radicals. The cobalt trapping reagent is called a persistent radical and the cobalt-capped polymer chain is said to be dormant. CMRP can be regarded as a series of carbometalation reactions of vinyl monomers. When the monomer possesses protons that can be easily abstracted by the cobalt radical, (catalytic) chain transfer may occur. The concept was introduced independently by two groups in 1994.

Cobalt mediated radical polymerization can proceed by catalytic chain transfer or by degenerative transfer :

P = polymeryl (slow 2nd order radical-radical termination)
R = small radical initiator (fast 2nd order radical-radical termination)

FISCHER-TROPSCH CATALYSIS

Cobalt catalysts (together with iron) are relevant in the Fischer-Tropsch process in which synthesis gas is converted to hydrocarbons. In this process, it is assumed that organocobalt intermediates form. An idealized reaction sequence is depicted below :

$M + CO \rightarrow M\text{-}CO$ ($M = Co, Fe$)

$M\text{-}CO + H_2 \rightarrow M\text{-}CH_3$

$M\text{-}CH_3 + CO \rightarrow OC\text{-}M\text{-}CH_3$

$OC\text{-}M\text{-}CH_3 \rightarrow M\text{-}(CO)\text{-}CH_3$

$M\text{-}(CO)\text{-}CH_3 + H_2 \rightarrow M\text{-}CH_2CH_3$

ISOTOPES OF COBALT

Naturally occurring **cobalt (Co)** is composed of 1 stable isotope, ^{59}Co. 28 radio-isotopes have been characterized with the most stable being ^{60}Co with a half-life of 5.2714 years, ^{57}Co with a half-life of 271.8 days, ^{56}Co with a half-life of 77.27 days, and ^{58}Co with a half-life of 70.86 days. All of the remaining radioactive isotopes have half-lives that are less than 18 hours and the majority of these have half-lives that are less than 1 second. This element also has 11 meta states, all of which have half-lives less than 15 minutes.

The isotopes of cobalt range in atomic weight from ^{47}Co to ^{75}Co. The primary decay mode for isotopes with atomic mass unit values less than that of the most abundant stable isotope, ^{59}Co, is electron capture and the primary mode of decay for those of greater than 59 atomic mass units is beta decay. The primary decay products before ^{59}Co are iron isotopes and the primary products after are nickel isotopes.

Radioactive isotopes can be produced by various nuclear reactions. For example, the isotope ^{57}Co is produced by cyclotron irradiation of iron. The principal reaction involved is the (d,n) reaction $^{56}Fe + {}^2H \rightarrow n + {}^{57}Co$.

Standard atomic mass : 58.933195(5) u

Use of Cobalt Radioisotopes in Medicine

Cobalt-60 (Co-60 or ^{60}Co) is a radioactive metal that is used in radiotherapy. It produces two gamma rays with energies of 1.17 MeV and 1.33 MeV. The ^{60}Co source is about 2 cm in diameter and as a result produces a geometric penumbra, making the edge of the radiation field fuzzy. The metal has the unfortunate habit of producing a fine dust, causing problems with radiation protection. The ^{60}Co source is useful for about 5 years but even after this point is still very radioactive, and so cobalt machines have fallen from favour in the Western world where linacs are common.

Cobalt-57 (Co-57 or ^{57}Co) is a radioactive metal that is used in medical tests; it is used as a radiolabel for vitamin B$_{12}$ uptake. It is useful for the Schilling test.

Industrial Uses for Radioactive Isotopes

Cobalt-60 (Co-60 or ^{60}Co) is useful as a gamma ray source because it can be produced — in predictable quantity, and high activity — by simply exposing natural cobalt to neutrons in a reactor for a given time. It is used for :

- Sterilization of medical supplies, and medical waste;
- Radiation treatment of foods for sterilization (cold pasteurization);
- Industrial radiography (*e.g.*, weld integrity radiographs);
- Density measurements (*e.g.*, concrete density measurements); and
- Tank fill height switches.

Cobalt-57 is used as a source in Mossbauer spectroscopy of iron-containing samples. The electron capture decay of the ^{57}Co forms an excited state of the ^{57}Fe nucleus, which in turn decays to the ground state with emission of a gamma ray. Measurement of the gamma ray spectrum provides information about the chemical state of the iron atom in the sample.

Table

nuclide symbol	Z(p)	N(n)	isotopic mass (u) / excitation energy	half-life	decay mode(s)	daughter	nu-clear spin	repre-sentative isotopic composi-tion (mole fraction)	range of natural varia-tion (mole frac-tion)
^{47}Co	27	20	47.01149(54)#				7/2-#		
^{48}Co	27	21	48.00176(43)#		p	^{47}Fe	6+#		
^{49}Co	27	22	48.98972(28)#	<35 ns	p (>99.9%)	^{48}Fe	7/2-#		
					β$^+$ (<.1%)	^{49}Fe			
^{50}Co	27	23	49.98154(18)#	44(4) ms	β$^+$, p (54%)	^{49}Mn	(6+)		
					β$^+$ (46%)	^{50}Fe			
^{51}Co	27	24	50.97072(16)#	60# ms [>200 ns]	β$^+$	^{51}Fe	7/2-#		
^{52}Co	27	25	51.96359(7)#	115(23) ms	β$^+$	^{52}Fe	(6+)		
^{52m}Co			380(100)# keV IT	104(11)# ms	β$^+$	^{52}Fe	2+#		
						^{52}Co			
^{53}Co	27	26	52.954219(19)	242(8) ms	β$^+$	^{53}Fe	7/2-#		
^{53m}Co			3197(29) keV p (1.5%)	247(12) ms	β$^+$	^{53}Fe			
						^{52}Fe			
^{54}Co	27	27	53.9484596(8)	193.28(7) ms	β$^+$	^{54}**Fe**	0+		
^{54m}Co			197.4(5) keV	1.48(2) min	β$^+$	^{54}**Fe**	(7)+		

(Contd...)

(*Contd...*)

^{55}Co	27	28	54.9419990(8)	17.53(3) h	β^+	^{55}Fe	7/2-		
^{56}Co	27	29	55.9398393(23)	77.233(27) d	β^+	^{56}**Fe**	4+		
^{57}Co	27	30	56.9362914(8)	271.74(6) d	EC	^{57}**Fe**	7/2-		
^{58}Co	27	31	57.9357528(13)	70.86(6) d	β^+	^{58}Fe	2+		
^{58m1}Co		24.95(6) keV		9.04(11) h	IT	^{58}Co	5+		
^{58m2}Co		53.15(7) keV		10.4(3) μs			4+		
^{59}Co	27	32	58.9331950(7)		**Stable**		7/2-	1.0000	
^{60}Co	27	33	59.9338171(7)	5.2713(8) a	β^-	^{60}**Ni**	5+		
^{60m}Co		58.59(1) keV β^- (.24%)		10.467(6) min ^{60}**Ni**	IT (99.76%)	^{60}Co	2+		
^{61}Co	27	34	60.9324758(10)	1.650(5) h	β^-	^{61}**Ni**	7/2-		
^{62}Co	27	35	61.934051(21)	1.50(4) min	β^-	^{62}**Ni**	2+		
^{62m}Co		22(5) keV IT (1%)		13.91(5) min ^{62}Co	β^- (99%)	^{62}**Ni**	5+		
^{63}Co	27	36	62.933612(21)	26.9(4) s	β^-	^{63}Ni	7/2		
^{64}Co	27	37	63.935810(21)	0.30(3) s	β^-	^{64}**Ni**	1+		
^{65}Co	27	38	64.936478(14)	1.20(6) s	β^-	^{65}Ni	(7/2)-		
^{66}Co	27	39	65.93976(27)	0.18(1) s	β^-	^{66}Ni	(3+)		
^{66m1}Co		175(3) keV		1.21(1) μs			(5+)		
^{66m2}Co		642(5) keV		>100 μs			(8-)		
^{67}Co	27	40	66.94089(34)	0.425(20) s	β^-	^{67}Ni	(7/2-)#		
^{68}Co	27	41	67.94487(34)	0.199(21) s	β^-	^{68}Ni	(7-)		
^{68m}Co		150(150)# keV		1.6(3) s			(3+)		
^{69}Co	27	42	68.94632(36)	227(13) ms	β^- (>99.9%)	^{69}Ni	7/2-#		
					β^-, n (<.1%)	^{68}Ni			
^{70}Co	27	43	69.9510(9)	119(6) ms	β^- (>99.9%)	^{70}Ni	(6-)		
					β^-, n (<.1%)	^{69}Ni			
^{70m}Co		200(200)# keV		500(180) ms			(3+)		
^{71}Co	27	44	70.9529(9)	97(2) ms	β^- (>99.9%)	^{71}Ni	7/2-#		
					β^-, n (<.1%)	^{70}Ni			
^{72}Co	27	45	71.95781(64)#	62(3) ms	β^- (>99.9%)	^{72}Ni			
					β^-, n (<.1%)	^{71}Ni			
^{73}Co	27	46	72.96024(75)#	41(4) ms			7/2-#		
^{74}Co	27	47	73.96538(86)#	50# ms [>300 ns]			0+		
^{75}Co	27	48	74.96833(86)#	40# ms [>300 ns]			7/2-#		

Abbreviations :

EC : Electron capture

IT : Isomeric transition

1. Bold for stable isotopes.

Notes

- Values marked # are not purely derived from experimental data, but at least partly from systematic trends. Spins with weak assignment arguments are enclosed in parentheses.
- Uncertainties are given in concise form in parentheses after the corresponding last digits. Uncertainty values denote one standard deviation, except isotopic composition and standard atomic mass from IUPAC which use expanded uncertainties.
- Nuclide masses are given by IUPAP Commission on Symbols, Units, Nomenclature, Atomic Masses and Fundamental Constants (SUNAMCO).
- Isotope abundances are given by IUPAC Commission on Isotopic Abundances and Atomic Weights.

Chapter 12

NICKEL

Nickel is a chemical element with the chemical symbol Ni and atomic number 28. It is a silvery-white lustrous metal with a slight golden tinge. Nickel belongs to the transition metals and is hard and ductile. Pure nickel shows a significant chemical activity that can be observed when nickel is powdered to maximize the exposed surface area on which reactions can occur, but larger pieces of the metal are slow to react with air at ambient conditions due to the formation of a protective oxide surface. Even then, nickel is reactive enough with oxygen that native nickel is rarely found on Earth's surface, being mostly confined to the interiors of larger nickel–iron meteorites that were protected from oxidation during their time in space. On Earth, such native nickel is always found in combination with iron, a reflection of those elements' origin as major end products of supernova nucleosynthesis. An iron–nickel mixture is thought to compose Earth's inner core.

The use of nickel (as a natural meteoric nickel–iron alloy) has been traced as far back as 3500 BC. Nickel was first isolated and classified as a chemical element in 1751 by Axel Fredrik Cronstedt, who initially mistook its ore for a copper mineral. The element's name comes from a mischievous sprite of German miner mythology, Nickel (similar to Old Nick), that personified the fact that copper-nickel ores resisted refinement into copper. An economically important source of nickel is the iron ore limonite, which often contains 1–2% nickel. Nickel's other important ore minerals include garnierite, and pentlandite. Major production sites include the Sudbury region in Canada (which is thought to be of meteoric origin), New Caledonia in the Pacific, and Norilsk in Russia.

Because of nickel's slow rate of oxidation at room temperature, it is considered corrosion-resistant. Historically, this has led to its use for plating metals such as iron and brass, in chemical apparatus, and in certain alloys that retain a high silvery polish, such as German silver. About 6% of world nickel production is still used for corrosion-resistant pure-nickel plating. Nickel was once a common component of coins, but has largely been replaced by cheaper iron for this purpose, especially since the metal is a skin allergen for some people. It was reintroduced into UK coins in 2012 despite objections from dermatologists.

Nickel is one of four elements that are ferromagnetic around room temperature. Alnico permanent magnets based partly on nickel are of intermediate strength between iron-based permanent magnets and rare-earth magnets. The metal is chiefly valuable in the modern world for the alloys it forms; about 60% of world production is used in nickel-steels (particularly stainless steel). Other common alloys, as well as some new superalloys, make up most of the remainder of world nickel use, with chemical uses for nickel compounds consuming less than 3% of production. As a compound, nickel has a number of niche chemical manufacturing uses, such as a catalyst for hydrogenation. Enzymes of some microorganisms and plants contain nickel as an active site, which makes the metal an essential nutrient for them.

CHARACTERISTICS

Atomic and Physical Properties

Nickel is a silvery-white metal with a slight golden tinge that takes a high polish. It is one of only four elements that are magnetic at or near room temperature, the others being iron, cobalt and gadolinium. Its Curie temperature is 355°C, meaning that bulk nickel is non-magnetic above this temperature. The unit cell of nickel is a face centered cube with the lattice parameter of 0.352 nm giving an atomic radius of 0.124 nm. Nickel belongs to the transition metals and is hard and ductile.

Electron Configuration Dispute

The nickel atom has two electron configurations, [Ar] $4s^2 3d^8$ and [Ar] $4s^1 3d^9$, which are very close in energy-the symbol [Ar] refers to the argon-like core structure. There is some disagreement as to which should be considered the lowest energy configuration. Chemistry textbooks quote the electron configuration of nickel as [Ar] $4s^2 3d^8$, or equivalently as [Ar] $3d^8 4s^2$. This configuration agrees with the Madelung energy ordering rule, which predicts that 4s is filled before 3d. It is supported by the experimental fact that the lowest energy state of the nickel atom is a $4s^2 3d^8$ energy level, specifically the $3d^8(3F) 4s^2 {}^3F$, J = 4 level.

However, each of these configurations in fact gives rise to a set of states of different energies. The two sets of energies overlap, and the average energy of states having configuration [Ar] $4s^1 3d^9$ is in fact lower than the average energy of states having configuration [Ar] $4s^2 3d^8$. For this reason, the research literature on atomic calculations quotes the ground state configuration of nickel as $4s^1 3d^9$.

Occurrence

On Earth, nickel occurs most often in combination with sulfur and iron in pentlandite, with sulfur in millerite, with arsenic in the mineral nickeline, and with arsenic and sulfur in nickel galena. Nickel is commonly found in iron meteorites as the alloys kamacite and taenite.

The bulk of the nickel mined comes from two types of ore deposits. The first are laterites, where the principal ore minerals are nickeliferous limonite : (Fe, Ni)O(OH) and garnierite (a hydrous nickel silicate) : $(Ni, Mg)_3Si_2O_5(OH)_4$. The second are magmatic sulfide deposits, where the principal ore mineral is pentlandite : $(Ni, Fe)_9S_8$.

Australia and New Caledonia have the biggest estimate reserves (45% all together).

In terms of World Resources, identified land-based resources averaging 1% nickel or greater contain at least 130 million tons of nickel (about the double of known reserves). About 60% is in laterites and 40% is in sulfide deposits.

Based on geophysical evidence, most of the nickel on Earth is postulated to be concentrated in the Earth's outer and inner cores. Kamacite and taenite are naturally occurring alloys of iron and nickel. For kamacite, the alloy is usually in the proportion of 90 : 10 to 95 : 5, although impurities (such as cobalt or carbon) may be present, while for taenite the nickel content is between 20% and 65%. Kamacite and taenite occur in nickel iron meteorites.

COMPOUNDS

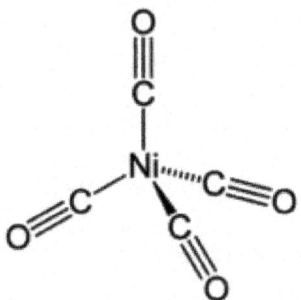

Fig. : Tetracarbonyl nickel.

The most common oxidation state of nickel is +2, but compounds of Ni^0, Ni^+, and Ni^{3+} are well known, as well as exotic oxidation states Ni^{2-}, Ni^{1-}, and Ni^{4+}.

Nickel(0)

Tetracarbonylnickel 4), discovered by Ludwig Mond, is a volatile, highly toxic liquid at room temperature. On heating, the complex decomposes back to nickel and carbon monoxide :

$$Ni(CO)^4 \rightleftharpoons Ni + 4\,CO$$

This behaviour is exploited in the Mond process for purifying nickel, as described above. The related nickel(0) complex bis(cyclooctadiene) nickel(0) is a useful catalyst in organonickel chemistry due to the easily displaced cod ligands.

Nickel (I)

Nickel (I) complexes are uncommon, one example being the tetrahedral complex $NiBr(PPh_3)_3$. Many feature Ni-Ni bonding, such as the dark red diamagnetic $K_4[Ni_2(CN)_6]$ prepared by reduction of $K_2[Ni_2(CN)_6]$ with sodium amalgam. This compound is oxidised in water, liberating H_2.

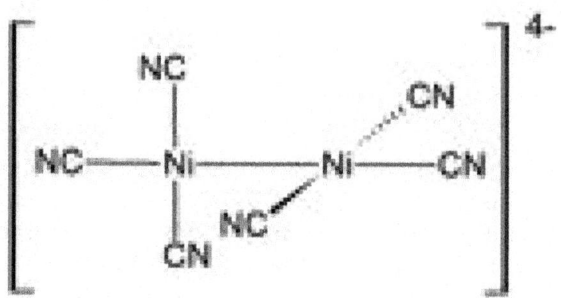

Fig. : Structure of $[Ni_2(CN)_6]^{2-}$ion.

Nickel (II)

Nickel (II) forms compounds with all common anions, *i.e.* the sulfide, sulfate, carbonate, hydroxide, carboxylates, and halides. Nickel (II) sulfate is produced in large quantities by dissolving nickel metal or oxides in sulfuric acid. It exists as both a hexa-and heptahydrates. This compound is useful for electroplating nickel. Common salts of nickel, such as the chloride, nitrate, and sulfate, dissolve in water to give green solutions containing the metal aquo complex $[Ni(H_2O)_6]^{2+}$.

The four halides form nickel compounds. The structures of these solids feature octahedral Ni centres. Nickel (II) chloride is most common, and its behaviour is illustrative of the other halides. Nickel (II) chloride is produced by dissolving nickel or its oxide in hydrochloric acid. It is usually encountered as the green hexahydrate, the formula of which is usually written $NiCl_2.(H_2O)_6$. When dissolved in water, this salt form the metal aquo complex $[Ni(H_2O)_6]^{2+}$. Dehydration of $NiCl_2$. $(H_2O)_6$ gives the yellow anhydrous $NiCl_2$.

Fig. : Nickel (III) antimonide.

Some tetra co-ordinate nickel (II) complexes, *e.g.* bis(triphenylphosphine) nickel chloride, exist both in tetrahedral and square planar geometries. The tetrahedral complexes are paramagnetic whereas the square planar complexes are diamagnetic. This equilibrium as well as the formation of octahedral complexes contrasts with the behaviour of the divalent complexes of the heavier group 10 metals, palladium (II) and platinum (II), which tend to adopt only square-planar geometry.

Nickelocene is known; it has an electron count of 20, making it relatively unstable.

Nickel(III) and (IV)

For simple compounds, nickel (III) and nickel (IV) only occurs with fluoride and oxides, with the exception of $KNiIO_6$, which can be considered as a formal salt of the $[IO_6]^{5-}$ ion. Ni(IV) is present in the mixed oxide $BaNiO_3$, while Ni (III) is present in nickel(III) oxide, which is used as the cathode in many rechargeable batteries, including nickel-cadmium, nickel-iron, nickel hydrogen, and nickel-metal hydride, and used by certain manufacturers in Li-ion batteries. Nickel (III) can be stabilized by σ-donor ligands such as thiols and phosphines.

HISTORY

Because the ores of nickel are easily mistaken for ores of silver, understanding of this metal and its use dates to relatively recent times. However, the unintentional use of nickel is ancient, and can be traced back as far as 3500 BC. Bronzes from what is now Syria had contained up to 2% nickel. Further, there are Chinese manuscripts suggesting that "white copper" (cupronickel, known as baitong) was used there between 1700 and 1400 BC. This Paktong white copper was exported to Britain as early as the 17th century, but the nickel content of this alloy was not discovered until 1822.

In medieval Germany, a red mineral was found in the Erzgebirge (Ore Mountains) that resembled copper ore. However, when miners were unable to extract any copper from it, they blamed a mischievous sprite of German mythology, Nickel (similar to Old Nick), for besetting the copper. They called this ore Kupfernickel from the German Kupfer for copper. This ore is now known to be nickeline or niccolite, a nickel arsenide. In 1751, Baron Axel Fredrik Cronstedt was trying to extract copper from kupfernickel—and instead produced a white metal that he named after the spirit that had given its name to the mineral, nickel. In modern German, Kupfernickel or Kupfer-Nickel designates the alloy cupronickel.

After its discovery, the only source for nickel was the rare Kupfernickel but, from 1824 on, the nickel was obtained as a by-product of cobalt blue production. The first large-scale producer of nickel was Norway, which exploited nickel-rich pyrrhotite from 1848 on. The introduction of nickel in steel production in 1889 increased the demand for nickel, and the nickel deposits of New Caledonia, which were discovered in 1865, provided most of the world's supply between 1875 and

1915. The discovery of the large deposits in the Sudbury Basin, Canada in 1883, in Norilsk-Talnakh, Russia in 1920, and in the Merensky Reef, South Africa in 1924 made large-scale production of nickel possible.

Fig. : Dutch coins made of pure nickel.

Nickel has been a component of coins since the mid-19th century. In the United States, the term "nickel" or "nick" originally applied to the copper-nickel Flying Eagle cent, which replaced copper with 12% nickel 1857–58, then the Indian Head cent of the same alloy from 1859–1864. Still later, in 1865, the term designated the three-cent nickel, with nickel increased to 25%. In 1866, the five-cent shield nickel (25% nickel, 75% copper) appropriated the designation. Along with the alloy proportion, this term has been used to the present in the United States. Coins of nearly pure nickel were first used in 1881 in Switzerland, and more notably 99.9% nickel five-cent coins were struck in Canada (the world's largest nickel producer at the time) during non-war years from 1922–1981, and their metal content made these coins magnetic. During the wartime period 1942–45, more or all nickel was removed from Canadian and U.S. coins, due to nickel's war-critical use in armor. Canada switched alloys again to plated steel during the Korean war, but was forced to stop making pure nickel "nickels" in 1981, reserving the pure 99.9% nickel alloy after 1968 only to its higher-value coins. Finally, in the 21st century, with rising nickel prices, most countries that formerly used nickel in their coins have abandoned the metal for cost reasons, and the U.S. five cents remains one of the few coins that still uses the metal for anything other than exterior plating.

WORLD PRODUCTION

Philippines, Indonesia, Russia, Canada and Australia are the world's largest producers of nickel, as reported by the US Geological Survey. The largest deposits of nickel in non-Russian Europe are located in Finland and Greece. Identified land-based resources averaging 1% nickel or greater contain at least 130 million tons of nickel. About 60% is in laterites and 40% is in sulfide deposits. In addition, extensive deep-sea resources of nickel are in manganese crusts and nodules covering large areas of the ocean floor, particularly in the Pacific Ocean.

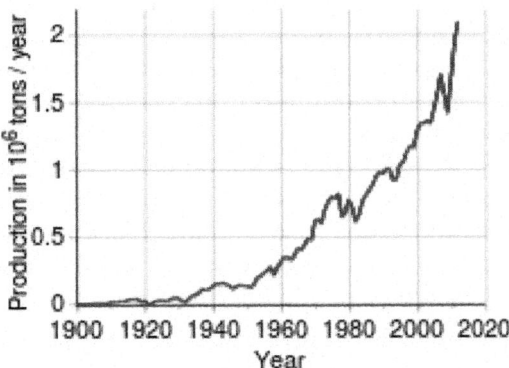

Fig. : Time trend of nickel production.

The one locality in the United States where nickel was commercially mined is Riddle, Oregon, where several square miles of nickel-bearing garnierite surface deposits are located. The mine closed in 1987. The Eagle mine project is a proposed new nickel mine in Michigan's upper peninsula.

Mine production and reserves	2012	2011	Reserves
Australia	230,000	215,000	20,000,000
Botswana	26,000	26,000	490,000
Brazil	140,000	209,000	7,500,000
Canada	220,000	220,000	3,300,000
China	91,000	89,800	3,000,000
Colombia	80,000	76,000	1,100,000
Cuba	72,000	71,000	5,500,000
Dominican Republic	24,000	21,700	970,000
Indonesia	320,000	290,000	3,900,000
Madagascar	22,000	5,900	1,600,000
New Caledonia	140,000	131,000	12,000,000
Philippines	330,000	270,000	1,100,000
Russia	270,000	267,000	6,100,000
South Africa	42,000	44,000	3,700,000
Other countries	120,000	103,000	4,600,000
World total (metric tons, rounded)	2,100,000	1,940,000	75,000,000

EXTRACTION AND PURIFICATION

Nickel is recovered through extractive metallurgy. Nickel is extracted from its ores by conventional roasting and reduction processes that yield a metal of greater than 75% purity. In many stainless steel applications, 75% pure nickel can be used without further purification, depending on the composition of the impurities.

Most sulfide ores have traditionally been processed using pyrometallurgical techniques to produce a matte for further refining. Recent advances in hydrometallurgy have resulted in significant nickel purification using these processes. Most sulfide deposits have traditionally been processed by concentration through a froth flotation process followed by pyrometallurgical extraction. In hydrometallurgical processes, nickel sulfide ores undergo flotation (differential flotation if Ni/Fe ratio is too low) and then smelted. After producing the nickel matte, further processing is done *via* the Sherritt-Gordon process. First, copper is removed by adding hydrogen sulfide, leaving a concentrate of only cobalt and nickel. Then, solvent extraction is used to separate the cobalt and nickel, with the final nickel concentration greater than 99%.

Electro-refining

A second common form of further refining involves the leaching of the metal matte into a nickel salt solution, followed by the electro-winning of the nickel from solution by plating it onto a cathode as electrolytic nickel.

Mond Process

Purification of nickel oxides to obtain the purest metal is performed *via* the Mond process, which increases the nickel concentrate to greater than 99.99% purity. This process was patented by L. Mond and has been in industrial use since before the beginning of the 20th century. In the process, nickel is reacted with carbon monoxide at around 40–80°C to form nickel carbonyl in the presence of a sulfur catalyst. Iron gives iron pentacarbonyl too, but this reaction is slow. If necessary, it may be separated by distillation. Dicobalt octacarbonyl is also formed in this process, but it decomposes to tetracobalt dodecacarbonyl at the reaction temperature to give a non-volatile solid.

Nickel is re-obtained from the nickel carbonyl by one of two processes. It may be passed through a large chamber at high temperatures in which tens of thousands of nickel spheres, called pellets, are constantly stirred. It then decomposes depositing pure nickel onto the nickel spheres. Alternatively, the nickel carbonyl may be decomposed in a smaller chamber at 230°C to create fine nickel powder. The resultant carbon monoxide is re-circulated and reused through the process. The highly pure nickel produced by this process is known as "carbonyl nickel".

Metal Value

The market price of nickel surged throughout 2006 and the early months of 2007; as of April 5, 2007, the metal was trading at 52,300 USD/tonne or 1.47 USD/oz. The price subsequently fell dramatically from these peaks, and as of 19 September, 2013 the metal was trading at 13,778 USD/tonne, or 0.39 USD/oz.

The US nickel coin contains 0.04 oz (1.25 g) of nickel, which at the April 2007 price was worth 6.5 cents, along with 3.75 grams of copper worth about 3 cents, making the metal value over 9 cents. Since the face value of a nickel is 5 cents, this made it an attractive target for melting by people wanting to sell the metals at a profit. However, the United States Mint, in anticipation of this practice, implemented new interim rules on December 14, 2006, subject to public comment for 30 days, which criminalize the melting and export of cents and nickels. Violators can be punished with a fine of up to $10,000 and/or imprisoned for a maximum of five years.

As of September 19, 2013, the melt value of a U.S. nickel (copper and nickel included) is $0.0450258, which is 90% of its face value.

APPLICATIONS

The fraction of global nickel production presently used for various applications is as follows : 46% for making nickel steels; 34% in non-ferrous alloys and superalloys; 14% electroplating, and 6% into other uses.

Nickel is used in many specific and recognizable industrial and consumer products, including stainless steel, alnico magnets, coinage, rechargeable batteries, electric guitar strings, microphone capsules, and special alloys. It is also used for plating and as a green tint in glass. Nickel is preeminently an alloy metal, and its chief use is in the nickel steels and nickel cast irons, of which there are many varieties. It is also widely used in many other alloys, such as nickel brasses and bronzes, and alloys with copper, chromium, aluminium, lead, cobalt, silver, and gold (Inconel, Incoloy, Monel, Nimonic).

A "horseshoe magnet" made of alnico nickel alloy. The composition of alnico alloys is typically 8–12% Al, 15–26% Ni, 5–24% Co, up to 6% Cu, up to 1% Ti, and the balance is Fe. The development of alnico began in 1931 when it was discovered that an alloy of iron, nickel, and aluminum had a coercivity double that of the best magnet steels of the time. Alnico magnets are now being replaced by rare earth magnets in many applications

Because of its resistance to corrosion, nickel has been occasionally used historically as a substitute for decorative silver. Nickel was also occasionally used in some countries after 1859 as a cheap coinage metal but in the later years of the 20th century was largely replaced by cheaper stainless steel (*i.e.*, iron) alloys, except notably in the United States.

Nickel is an excellent alloying agent for certain other precious metals, and so used in the so-called fire assay, as a collector of platinum group elements (PGE). As such, nickel is capable of full collection of all 6 PGE elements from ores, in addition to partial collection of gold. High-throughput nickel mines may also engage in PGE recovery (primarily platinum and palladium); examples are Norilsk in Russia and the Sudbury Basin in Canada.

Nickel foam or nickel mesh is used in gas diffusion electrodes for alkaline fuel cells.

Nickel and its alloys are frequently used as catalysts for hydrogenation reactions. Raney nickel, a finely divided nickel-aluminium alloy, is one common form, however related catalysts are also often used, including related 'Raney-type' catalysts.

Nickel is a naturally magnetostrictive material, meaning that, in the presence of a magnetic field, the material undergoes a small change in length. In the case of nickel, this change in length is negative (contraction of the material), which is known as negative magnetostriction and is on the order of 50 ppm.

Nickel is used as a binder in the cemented tungsten carbide or hardmetal industry and used in proportions of six to 12% by weight. Nickel can make the tungsten carbide magnetic and adds corrosion-resistant properties to the cemented tungsten carbide parts, although the hardness is lower than those of parts made with cobalt binder.

BIOLOGICAL ROLE

Although not recognized until the 1970s, nickel plays important roles in the biology of micro-organisms and plants. The plant enzyme urease (an enzyme that assists in the hydrolysis of urea) contains nickel. The NiFe-hydrogenases contain nickel in addition to iron-sulfur clusters. Such [NiFe]-hydrogenases characteristically oxidise H_2. A nickel-tetrapyrrole coenzyme, Cofactor F430, is present in the methyl coenzyme M reductase, which powers methanogenic archaea. One of the carbon monoxide dehydrogenase enzymes consists of an Fe-Ni-S cluster. Other nickel-containing enzymes include a rare bacterial class of superoxide dismutase and glyoxalase I enzymes in bacteria and several parasitic eukaryotic trypanosomal parasites (this enzyme in higher organisms, including yeast and mammals, uses divalent zinc, Zn^{2+}).

TOXICITY

In the US, the minimal risk level of nickel and its compounds is set to 0.2 $\mu g/m^3$ for inhalation during 15–364 days. Nickel sulfide fume and dust are believed carcinogenic, and various other nickel compounds may be as well. Nickel carbonyl, $[Ni(CO)_4]$, is an extremely toxic gas. The toxicity of metal carbonyls is a function of both the toxicity of the metal as well as the carbonyl's ability to give off highly toxic carbon monoxide gas, and this one is no exception; nickel carbonyl is also explosive in air.

In the US, the Tolerable Upper Limit of dietary nickel is 1000 µg/day, while estimated average ingestion is 69-162 µg/day. Large amounts of nickel (and chromium)--comparable to the estimated average ingestion above–leach into food cooked in stainless steel. For example the amount of nickel leached after 10 cooking cycle into one serving of tomato sauce averages to be 88 µg.

Sensitized individuals may show an allergy to nickel, affecting their skin, also known as dermatitis. Sensitivity to nickel may also be present in patients with pompholyx. Nickel is an important cause of contact allergy, partly due to its use in jewellery intended for pierced ears. Nickel allergies affecting pierced ears are often marked by itchy, red skin. Many earrings are now made nickel-free due to this problem. The amount of nickel allowed in products that come into contact with human skin is regulated by the European Union. In 2002, researchers found amounts of nickel being emitted by 1 and 2 Euro coins far in excess of those standards. This is believed due to a galvanic reaction. Nickel was voted Allergen of the Year in 2008 by the American Contact Dermatitis Society.

Reports also showed that both the nickel-induced activation of hypoxia-inducible factor and the up-regulation of hypoxia-inducible genes are due to depleted intra-cellular ascorbate levels. The addition of ascorbate to the culture medium increased the intra-cellular ascorbate level and reversed both the metal-induced stabilization of HIF-1-and HIF-1α-dependent gene expression.

Most toxic compound containing nickel is Cyclopentadienyl nickel nitrosyl, $(C_5H_5)NiNO$. It is a blood-red colour liquid.

ISOTOPES OF NICKEL

Naturally occurring nickel (Ni) is composed of five stable isotopes; 58Ni, 60Ni, 61Ni, 62Ni and 64Ni with 58Ni being the most abundant (68.077% natural abundance). 58Ni may decay by double beta-plus decay to 58Fe. 26 radioisotopes have been characterised with the most stable being 59Ni with a half-life of 76,000 years, 63Ni with a half-life of 100.1 years, and 56Ni with a half-life of 6.077 days. All of the remaining radioactive isotopes have half-lives that are less than 60 hours and the majority of these have half-lives that are less than 30 seconds. This element also has 1 meta state.

The isotopes of nickel range in atomic weight from 48Ni to 78Ni.

Nickel-48, discovered in 1999, is the most neutron-poor nickel isotope known. With 28 protons and 20 neutrons 48Ni is "doubly magic" (like 208Pb) and therefore unusually stable.

Nickel-56 is produced in large quantities in type Ia supernovae and the shape of the light curve of these supernovae corresponds to the decay of nickel-56 to cobalt-56 and then to iron-56.

Nickel-58 is the most abundant isotope of nickel, making up 68.077% of the natural abundance. Possible sources include electron capture from copper-58 and EC + p from zinc-59.

Nickel-59 is a long-lived cosmogenic radionuclide with a half-life of 76,000 years. 59Ni has found many applications in isotope geology. 59Ni has been used to date the terrestrial age of meteorites and to determine abundances of extraterrestrial dust in ice and sediment.

Nickel-60 is the daughter product of the extinct radionuclide 60Fe (half-life = 2.6 Ma). Because 60Fe had such a long half-life, its persistence in materials in the solar system at high enough concentrations may have generated observable variations in the isotopic composition of 60Ni. Therefore, the abundance of 60Ni present in extraterrestrial material may provide insight into the origin of the solar system and its early history/very early history. Unfortunately, nickel isotopes appear to have been heterogeneously distributed in the early solar system. Therefore, so far, no actual age information has been attained from 60Ni excesses. Other sources may also include beta decay from cobalt-60 and electron capture from copper-60.

Nickel-61 is the only stable isotope of nickel with a nuclear spin (I = 3/2), which makes it useful for studies by EPR spectroscopy.

Nickel-62 has the highest binding energy per nucleon of any isotope for any element, when including the electron shell in the calculation. More energy is released forming this isotope than any other, although fusion can form heavier isotopes. For instance, two 40Ca atoms can fuse to form 80Kr plus 4 electrons, liberating 77 keV per nucleon, but reactions leading to the iron/nickel region are more probable as they release more energy per baryon.

Nickel-64 is another isotope of nickel. Possible sources include beta decay from cobalt-64, and electron capture from copper-64

Nickel-78 is the element's heaviest isotope and is believed to have an important involvement in supernova nucleosynthesis of elements heavier than iron.

Standard atomic mass : 58.6934(2) u

Table

nuclide symbol	Z(p)	N(n)	isotopic mass (u) excitation energy	half-life	decay mode(s)	daughter isotope(s)	nuclear spin	representative isotopic composition (mole fraction)	range of natural variation (mole fraction)
48Ni	28	20	48.01975(54)#	10# ms [>500 ns]			0+		
49Ni	28	21	49.00966(43)#	13(4) ms [12(+5-3) ms]			7/2-#		
50Ni	28	22	49.99593(28)#	9.1(18) ms	β+	50Co	0+		

(Contd...)

(Contd...)

nuclide symbol	Z(p)	N(n)	isotopic mass (u)	half-life	decay mode(s)	daughter isotope(s)	nuclear spin	representative isotopic composition (mole fraction)	range of natural variation (mole fraction)
			excitation energy						
51Ni	28	23	50.98772(28)#	30# ms [>200 ns]	β+	51Co	7/2-#		
52Ni	28	24	51.97568(9)#	38(5) ms	β+ (83%)	52Co	0+		
					β+, p (17%)	51Fe			
53Ni	28	25	52.96847(17)#	45(15) ms	β+ (55%)	53Co	(7/2-)#		
					β+, p (45%)	52Fe			
54Ni	28	26	53.95791(5)	104(7) ms	β+	54Co	0+		
55Ni	28	27	54.951330(12)	204.7(17) ms	β+	55Co	7/2-		
56Ni	28	28	55.942132(12)	6.075(10) d	β+	56Co	0+		
57Ni	28	29	56.9397935(19)	35.60(6) h	β+	57Co	3/2-		
58Ni	28	30	57.9353429(7)	Observationally Stable			0+	0.680769(89)	
59Ni	28	31	58.9343467(7)	7.6(5)×104 a	EC (99%)	59Co	3/2-		
					β+ (1.5x10-5%)				
60Ni	28	32	59.9307864(7)	Stable			0+	0.262231(77)	
61Ni	28	33	60.9310560(7)	Stable			3/2-	0.011399(6)	
62Ni	28	34	61.9283451(6)	Stable			0+	0.036345(17)	
63Ni	28	35	62.9296694(6)	100.1(20) a	β-	63Cu	1/2-		
63mNi	87.15(11) keV			1.67(3) μs			5/2-		
64Ni	28	36	63.9279660(7)	Stable			0+	0.009256(9)	
65Ni	28	37	64.9300843(7)	2.5172(3) h	β-	65Cu	5/2-		
65mNi	63.37(5) keV			69(3) μs			1/2-		
66Ni	28	38	65.9291393(15)	54.6(3) h	β-	66Cu	0+		
67Ni	28	39	66.931569(3)	21(1) s	β-	67Cu	1/2-		
67mNi	1007(3) keV IT			13.3(2) μs 67Ni	β-	67Cu	9/2+		
68Ni	28	40	67.931869(3)	29(2) s	β-	68Cu	0+		
68m1Ni	1770.0(10) keV			276(65) ns			0+		
68m2Ni	2849.1(3) keV			860(50) μs			5-		
69Ni	28	41	68.935610(4)	11.5(3) s	β-	69Cu	9/2+		
69m1Ni	321(2) keV IT			3.5(4) s 69Ni	β-	69Cu	(1/2-)		
69m2Ni	2701(10) keV			439(3) ns					
70Ni	28	42	69.93650(37)	6.0(3) s	β-	70Cu	0+		
70mNi	2860(2) keV			232(1) ns			8+		
71Ni	28	43	70.94074(40)	2.56(3) s	β-	71Cu	1/2-#		
72Ni	28	44	71.94209(47)	1.57(5) s	β-(>99.9%)	72Cu	0+		
					β-, n (<.1%)	71Cu			

(Contd...)

(Contd...)

nuclide symbol	Z(p)	N(n)	isotopic mass (u)	half-life	decay mode(s)	daughter isotope(s)	nuclear spin	representative isotopic composition (mole fraction)	range of natural variation (mole fraction)
			excitation energy						
73Ni	28	45	72.94647(32)#	0.84(3) s	β-(>99.9%)	73Cu	(9/2+)		
					β-, n (<.1%)	72Cu			
74Ni	28	46	73.94807(43)#	0.68(18) s	β-(>99.9%)	74Cu	0+		
					β-, n (<.1%)	73Cu			
75Ni	28	47	74.95287(43)#	0.6(2) s	β-	75Cu	(7/2+)#		
					β-, n (1.6%)	74Cu			
76Ni	28	48	75.95533(97)#	470(390) ms [0.24(+55-24) s]	β-(>99.9%)	76Cu	0+		
					β-, n (<.1%)	75Cu			
77Ni	28	49	76.96055(54)#	300# ms [>300 ns]	β-	77Cu	9/2+#		
78Ni	28	50	77.96318(118)#	120# ms [>300 ns]	β-	78Cu	0+		

Abbreviations :

IT : Isomeric transition.

1.   Bold for stable isotopes.
2.   Believed to decay by β+β+ to ^{58}Fe with a half-life over 7×10^{20} years.
3.   Highest binding energy per nucleon of all nuclides.

Notes

•   Values marked # are not purely derived from experimental data, but at least partly from systematic trends. Spins with weak assignment arguments are enclosed in parentheses.

•   Uncertainties are given in concise form in parentheses after the corresponding last digits. Uncertainty values denote one standard deviation, except isotopic composition and standard atomic mass from IUPAC which use expanded uncertainties.

•   Nuclide masses are given by IUPAP Commission on Symbols, Units, Nomenclature, Atomic Masses and Fundamental Constants (SUNAMCO).

•   Isotope abundances are given by IUPAC Commission on Isotopic Abundances and Atomic Weights.

Chapter 13

THE CHEMISTRY OF COPPER

POSITION OF COPPER IN PERIODIC TABLE

Copper occupies the same family of the periodic table as silver and gold, since they each have one s-orbital electron on top of a filled electron shell which forms metallic bonds. This similarity in electron structure makes them similar in many characteristics. All have very high thermal and electrical conductivity, and all are malleable metals. Among pure metals at room temperature, copper has the second highest electrical and thermal conductivity, after silver.

OCCURRENCE AND EXTRACTION OF COPPER FORM COPPER PYRITES

Copper occurs both in combined state and free state. It also contains many ores. The important ores of copper are copper pyrites ($CuFeS_2$), cuprite and copper glance. The copper ores are mostly found in the north of India.

EXTRACTION

The extraction of copper also involves many steps. The ore used for extraction is copper pyrites, which is crushed, concentrated and then heated in the presence of air. During heating the moisture gets expelled and the copper pyrites gets converted to ferrous sulfide and cuprous sulfide.

$$2CuFeS_2 + O2 \rightarrow Cu2S + 2FeS + SO_2$$

Blast furnace is used to heat the mixture of roasted ore, powdered coke and sand. In the blast furnace oxidation reactions takes place. Ferrous sulfide forms ferrous oxide which combines with silica and forms the slag ($FeSiO_2$.

$$2FeS + SO_2 \rightarrow 2FeO + 2SO_2$$

$$FeO + SiO_2 \rightarrow FeSiO_3$$

Cuprous sulfide forms cuprous oxide which is partially converted to cuprous sulfide.

$$Cu_2S+3O_2 \rightarrow 2Cu_2O+2SO_2$$

$$Cu_2O+ FeS \rightarrow Cu_2S+FeO$$

This cuprous sulfide contains some amount of ferrous sulfide and this is called matte. Matte is removed from the base outlet of blast furnace. The removed matte is shifted into Bessemer converter which is lined inside with magnesium oxide. This converter has pipes through which hot air and SiO_2 are sent in. In this converter Cu_2S converts to Cu_2O and FeS converts to FeO. The ferrous oxide forms slag with SiO_2. The cuprous oxide formed reacts with Cu_2S and forms copper.

$$2Cu_2O+ Cu_2S \rightarrow 6Cu+SO_2$$

The copper thus formed is to be purified by electrolysis.

REACTIONS OF COPPER (II) IONS IN SOLUTION

The simplest ion that copper forms in solution is the typical blue hexaaquacopper (II) ion-$[Cu(H_2O)_6]^{2+}$.

REACTIONS OF HEXAAQUACOPPER (II) IONS WITH HYDROXIDE IONS

Hydroxide ions (from, say, sodium hydroxide solution) remove hydrogen ions from the water ligands attached to the copper ion. Once a hydrogen ion has been removed from two of the water molecules, you are left with a complex with no charge-a neutral complex. This is insoluble in water and a precipitate is formed.

$$[Cu(H_2O)_6]^{2+} + 2OH^- \rightarrow [Cu(H_2O)_4(OH)_2] + 2H_2O$$

In the test-tube, the colour change is :

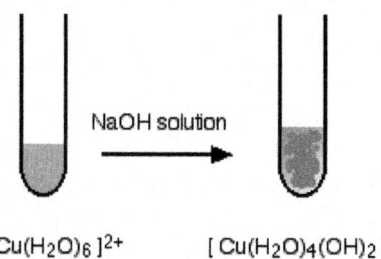

NaOH solution

$[Cu(H_2O)_6]^{2+}$ $[Cu(H_2O)_4(OH)_2]$

REACTIONS OF HEXAAQUACOPPER (II) IONS WITH AMMONIA SOLUTION

The ammonia acts as both a base and a ligand. With a small amount of ammonia, hydrogen ions are pulled off the hexaaqua ion exactly as in the hydroxide ion case to give the same neutral complex.

$$Cu(H_2)_6]^{2+}+2NH_3 \rightarrow [Cu(H_2O)_4(OH)_2]+2NH_{+4}$$

That precipitate dissolves if you add an excess of ammonia. The ammonia replaces water as a ligand to give tetraamminediaquacopper (II) ions. Notice that only 4 of the 6 water molecules are replaced.

$$Cu(H_2)^6]_{2+} + 4NH_3 \rightarrow [Cu(NH_3)_4(H_2O)_2]_{2+} + H_2O$$

The colour changes are :

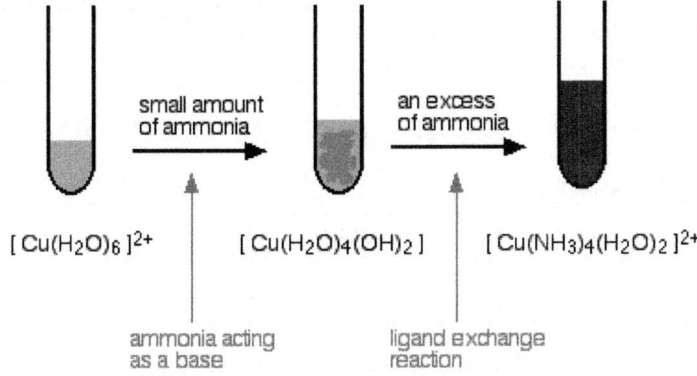

[Cu(H2O)6]2+   [Cu(H2O)4(OH)2]   [Cu(NH3)4(H2O)2]2+

small amount   an excess
of ammonia   of ammonia

ammonia acting   ligand exchange
as a base   reaction

THE REACTION OF HEXAAQUACOPPER (II) IONS WITH CARBONATE IONS

You simply get a precipitate of what you can think of as copper(II) carbonate.

$$Cu_2^+ + CO_{2-3} \rightarrow CuCO_3(s)$$

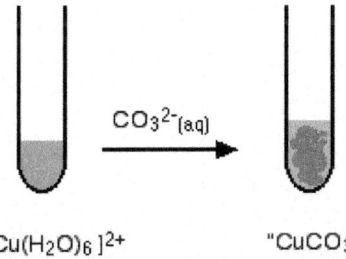

[Cu(H2O)6]2+   "CuCO3"

CO3²⁻(aq)

A LIGAND EXCHANGE REACTION INVOLVING CHLORIDE IONS

If you add concentrated hydrochloric acid to a solution containing hexaaquacopper (II) ions, the six water molecules are replaced by four chloride ions.

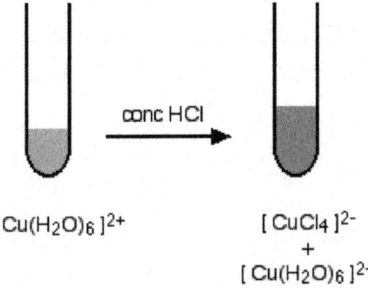

conc HCl

[Cu(H2O)6]2+   [CuCl4]2-
                              +
                       [Cu(H2O)6]2+

Fig. : The reaction taking place is reversible.

$$Cu(H_2O)_6]_{2+}+4Cl-\rightleftharpoons[CuCl_4]_{2-}+6H_2O$$

Because the reaction is reversible, you get a mixture of colours due to both of the complex ions. The colour of the tetrachlorocuprate (II) ion may also be described as olive-green or yellow. If you add water to the green solution, it returns to the blue colour.

THE REACTION OF HEXAAQUACOPPER(II) IONS WITH IODIDE IONS

The Simple Reaction

Copper (II) ions oxidise iodide ions to iodine, and in the process are themselves reduced to copper (I) iodide.

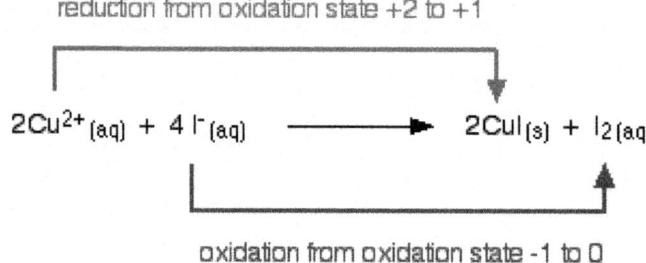

The initial mucky brown mixture separates into an off-white precipitate of copper (I) iodide under an iodine solution.

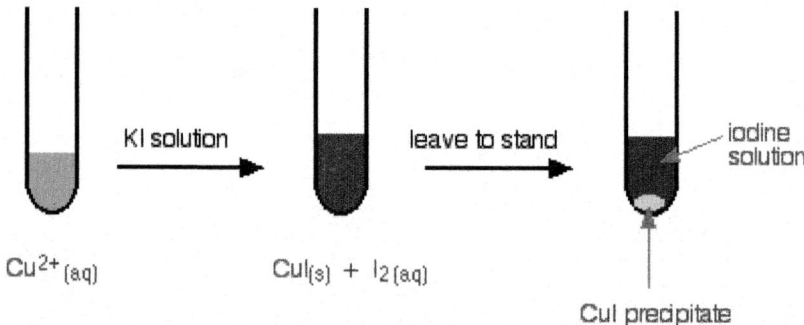

USING THIS REACTION TO FIND THE CONCENTRATION OF COPPER (II) IONS IN SOLUTION

If you pipette a known volume of a solution containing copper (II) ions into a flask, and then add an excess of potassium iodide solution, you get the reaction we have just described.

$$2Cu_{2+}+4I-\longrightarrow2CuI(s)+I_2(aq)$$

You can find the amount of iodine liberated by titration with sodium thiosulphate solution.

$$2S_2O_{2-3}(aq)+I_2(aq)\rightarrow S_4O_{2-6}(aq)+2I^-(aq)$$

As the sodium thiosulphate solution is run in from a burette, the colour of the iodine fades. When it is almost all gone, you add some starch solution. This reacts reversibly with iodine to give a deep blue starch-iodine complex which is much easier to see.

You add the last few drops of the sodium thiosulphate solution slowly until the blue colour disappears. If you trace the reacting proportions through the two equations, you will find that for every 2 moles of copper (II) ions you had to start with, you need 2 moles of sodium thiosulphate solution. If you know the concentration of the sodium thiosulphate solution, it is easy to calculate the concentration of the copper (II) ions.

SOME ESSENTIAL COPPER(I) CHEMISTRY

The Disproportionation of Copper (I) Ions in Solution

Copper (I) chemistry is limited by a reaction which occurs involving simple copper (I) ions in solution. This is a good example of disproportionation-a reaction in which something oxidises and reduces itself. Copper (I) ions in solution disproportionate to give copper (II) ions and a precipitate of copper. The reaction is :

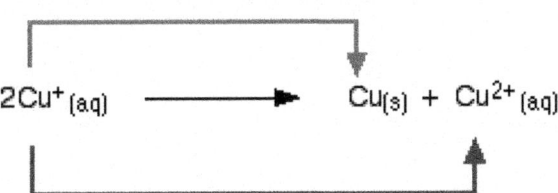

Any attempt to produce a simple copper (I) compound in solution results in this happening. For example, if you react copper (I) oxide with hot dilute sulfuric acid, you might expect to get a solution of copper (I) sulfate and water produced. In fact you get a brown precipitate of copper and a blue solution of copper(II) sulfate because of the disproportionation reaction.

$$Cu_2O+H_2SO_4\rightarrow Cu+CuSO_4+H_2O$$

STABALIZING THE COPPER(I) OXIDATION STATE

Insoluble Copper (I) Compounds

We've already seen that copper (I) iodide is produced as an off-white precipitate if you add potassium iodide solution to a solution containing copper (II) ions. The copper (I) iodide is virtually insoluble in water, and so the disproportionation reaction doesn't happen.

Similarly copper (I) chloride can be produced as a white precipitate (reaction described below). Provided this is separated from the solution and dried as quickly as possible, it remains white. In contact with water, though, it slowly turns blue as copper (II) ions are formed. The disproportionation reaction only occurs with simple copper (I) ions in solution.

Copper (I) Complexes

Forming copper (I) complexes (other than the one with water as a ligand) also stabilizes the copper (I) oxidation state. For example, both $[Cu(NH_3)_2]+$ and $[CuCl_2]$-are copper (I) complexes which don't disproportionate.

The chlorine-containing complex is formed if copper(I) oxide is dissolved in concentrated hydrochloric acid. You can think of this happening in two stages. First, you get copper (I) chloride formed :

$$Cu_2O(s)+2HCl(aq){\rightarrow}2CuCl(s)+H_2O(l)$$

But in the presence of excess chloride ions from the HCl, this reacts to give a stable, soluble copper(I) complex.

$$CuCl(s)+Cl-(aq){\rightarrow}[CuCl_2]-(aq)$$

You can get the white precipitate of copper (I) chloride (mentioned above) by adding water to this solution. This reverses the last reaction by stripping off the extra chloride ion.

Chapter 14

ZINC

Zinc, in commerce also spelter, is a metallic chemical element; it has the symbol Zn and atomic number 30. It is the first element of group 12 of the periodic table. Zinc is, in some respects, chemically similar to magnesium, because its ion is of similar size and its only common oxidation state is +2. Zinc is the 24th most abundant element in the Earth's crust and has five stable isotopes. The most common zinc ore is sphalerite (zinc blende), a zinc sulfide mineral. The largest mineable amounts are found in Australia, Asia, and the United States. Zinc production includes froth flotation of the ore, roasting, and final extraction using electricity (electro-winning).

Brass, which is an alloy of copper and zinc, has been used since at least the 10th century BC in Judea[1] and by the 7th century BC in Ancient Greece. Zinc metal was not produced in large scale until the 12th century in India, while the metal was unknown to Europe until the end of the 16th century. The mines of Rajasthan have given definite evidence of zinc production going back to 6th Century BC. To date the oldest evidence of pure zinc comes from Zawar, Rajasthan as early as 9th century AD, when distillation process was employed to make pure zinc. Alchemists burned zinc in air to form what they called "philosopher's wool" or "white snow."

The element was probably named by the alchemist Paracelsus after the German word Zinke. German chemist Andreas Sigismund Marggraf is normally given credit for discovering pure metallic zinc in 1746. Work by Luigi Galvani and Alessandro Volta uncovered the electro-chemical properties of zinc by 1800. Corrosion-resistant zinc plating of iron (hot-dip galvanizing) is the major application for zinc. Other applications are in batteries, small non-structural castings, and alloys, such as brass. A variety of zinc compounds are commonly used, such as zinc carbonate and zinc gluconate (as dietary supplements), zinc chloride (in deodorants), zinc pyrithione (anti-dandruff shampoos), zinc sulfide (in luminescent paints), and zinc methyl or zinc diethyl in the organic laboratory.

Zinc is an essential mineral of "exceptional biologic and public health importance". Zinc deficiency affects about two billion people in the developing world and is associated with many diseases. In children it causes growth retardation, delayed sexual maturation, infection susceptibility, and diarrhea, contributing to the death of about 800,000 children worldwide per year. Enzymes with a zinc atom in the reactive center are widespread in bio-chemistry, such as alcohol dehydrogenase in humans. Consumption of excess zinc can cause ataxia, lethargy and copper deficiency.

CHARACTERISTICS

Physical Properties

Zinc, also referred to in non-scientific contexts as spelter, is a bluish-white, lustrous, diamagnetic metal, though most common commercial grades of the metal have a dull finish. It is somewhat less dense than iron and has a hexagonal crystal structure.

The metal is hard and brittle at most temperatures but becomes malleable between 100 and 150°C. Above 210°C, the metal becomes brittle again and can be pulverized by beating. Zinc is a fair conductor of electricity. For a metal, zinc has relatively low melting (419.5°C, 787.1 F) and boiling points (907°C). Its melting point is the lowest of all the transition metals aside from mercury and cadmium.

Many alloys contain zinc, including brass, an alloy of copper and zinc. Other metals long known to form binary alloys with zinc are aluminium, antimony, bismuth, gold, iron, lead, mercury, silver, tin, magnesium, cobalt, nickel, tellurium and sodium. While neither zinc nor zirconium are ferromagnetic, their alloy $ZrZn$

2 exhibits ferromagnetism below 35 K.

Occurrence

Zinc makes up about 75 ppm (0.0075%) of the Earth's crust, making it the 24th most abundant element. Soil contains 5–770 ppm of zinc with an average of 64 ppm. Seawater has only 30 ppb zinc and the atmosphere contains 0.1–4 $\mu g/m^3$.

The element is normally found in association with other base metals such as copper and lead in ores. Zinc is a chalcophile, meaning the element has a low affinity for oxides and prefers to bond with sulfides. Chalcophiles formed as the crust solidified under the reducing conditions of the early Earth's atmosphere. Sphalerite, which is a form of zinc sulfide, is the most heavily mined zinc-containing ore because its concentrate contains 60–62% zinc.

Other minerals from which zinc is extracted include smithsonite (zinc carbonate), hemimorphite (zinc silicate), wurtzite (another zinc sulfide), and sometimes hydrozincite (basic zinc carbonate). With the exception of wurtzite, all these other minerals were formed as a result of weathering processes on the primordial zinc sulfides.

Identified world zinc resources total about 1.9 billion tonnes. Large deposits are in Australia, Canada and the United States with the largest reserves in Iran. At the current rate of consumption, these reserves are estimated to be depleted sometime between 2027 and 2055. About 346 million tonnes have been extracted throughout history to 2002, and one estimate found that about 109 million tonnes of that remains in use.

HISTORY

Ancient Use

Various isolated examples of the use of impure zinc in ancient times have been discovered. Zinc ores were used to make the zinc–copper alloy brass many centuries prior to the discovery of zinc as a separate element. Judean brass from the 14th to 10th centuries BC contains 23% zinc. Knowledge of how to produce brass spread to Ancient Greece by the 7th century BC but few varieties were made. Ornaments made of alloys that contain 80–90% zinc with lead, iron, antimony, and other metals making up the remainder, have been found that are 2500 years old. A possibly prehistoric statuette containing 87.5% zinc was found in a Dacian archaeological site.

The oldest known pills were made of the zinc carbonates hydrozincite and smithsonite. The pills were used for sore eyes, and were found aboard the roman ship Relitto del Pozzino, which wrecked in 140 BC.

The manufacture of brass was known to the Romans by about 30 BC. They made brass by heating powdered calamine (zinc silicate or carbonate), charcoal and copper together in a crucible. The resulting calamine brass was then either cast or hammered into shape and was used in weaponry. Some coins struck by Romans in the Christian era are made of what is probably calamine brass.

Strabo, in a passage taken from an earlier writer of the 4th century BC, mentions "drops of false silver", which when mixed with copper make brass. This may refer to small quantities of zinc by-product of smelting sulfide ores. Zinc in such remnants in melting ovens was usually discarded, as it was thought to be worthless. The Berne zinc tablet is a votive plaque dating to Roman Gaul made of an alloy that is mostly zinc.

The *Charaka Samhita*, thought to have been written in 500 BC or before, mentions a metal which, when oxidized, produces pushpanjan, thought to be zinc oxide. Zinc mines at *Zawar*, near Udaipur in India, have been active since the *Mauryan period*. The smelting of metallic zinc here however appears to have begun around the 12th century AD. One estimate is that this location produced an estimated million tonnes of metallic zinc and zinc oxide from the 12th to 16th centuries. Another estimate gives a total production of 60,000 tonnes of metallic zinc over this period. The Rasaratna Samuccaya, written in approximately the 13th century AD, mentions two types of zinc-containing ores; one used for metal extraction and another used for medicinal purposes.

Early Studies and Naming

Zinc was distinctly recognized as a metal under the designation of *Yasada* or *Jasada* in the medical Lexicon ascribed to the Hindu king *Madanapala* and written about the year 1374. Smelting and extraction of impure zinc by reducing calamine with wool and other organic substances was accomplished in the 13th century in India. The Chinese did not learn of the technique until the 17th century.

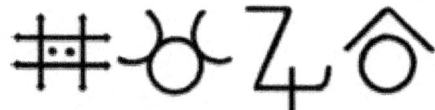

Fig. : Various alchemical symbols attributed to the element zinc.

Alchemists burned zinc metal in air and collected the resulting zinc oxide on a condenser. Some alchemists called this zinc oxide *lana philosophica*, Latin for "philosopher's wool", because it collected in wooly tufts while others thought it looked like white snow and named it nix album.

The name of the metal was probably first documented by Paracelsus, a Swiss-born German alchemist, who referred to the metal as "zincum" or "zinken" in his book Liber Mineralium II, in the 16th century. The word is probably derived from the German zinke, and supposedly meant "tooth-like, pointed or jagged" (metallic zinc crystals have a needle-like appearance). Zink could also imply "tin-like" because of its relation to German zinn meaning tin. Yet another possibility is that the word is derived from the Persian word seng meaning stone. The metal was also called Indian tin, tutanego, calamine, and spinter.

German metallurgist *Andreas Libavius* received a quantity of what he called "calay" of Malabar from a cargo ship captured from the Portuguese in 1596. Libavius described the properties of the sample, which may have been zinc. Zinc was regularly imported to Europe from the Orient in the 17th and early 18th centuries, but was at times very expensive.

Isolation

The isolation of metallic zinc in the West may have been achieved independently by several people. Postlewayt's Universal Dictionary, a contemporary source giving technological information in Europe, did not mention zinc before 1751 but the element was studied before then.

Flemish metallurgist P.M. de Respour reported that he extracted metallic zinc from zinc oxide in 1668. By the start of the 18th century, Étienne François Geoffroy described how zinc oxide condenses as yellow crystals on bars of iron placed above zinc ore being smelted. In Britain, John Lane is said to have carried out experiments to smelt zinc, probably at Landore, prior to his bankruptcy in 1726.

In 1738, *William Champion* patented in Great Britain a process to extract zinc from calamine in a vertical retort style smelter. His technology was somewhat similar to that used at Zawar zinc mines in Rajasthan but there is no evidence that he visited the Orient. Champion's process was used through 1851.

German chemist *Andreas Marggraf* normally gets credit for discovering pure metallic zinc even though Swedish chemist Anton von Swab distilled zinc from calamine four years before. In his 1746 experiment, Marggraf heated a mixture of calamine and charcoal in a closed vessel without copper to obtain a metal. This procedure became commercially practical by 1752.

Later Work

William Champion's brother, John, patented a process in 1758 for calcining zinc sulfide into an oxide usable in the retort process. Prior to this only calamine could be used to produce zinc. In 1798, Johann Christian Ruberg improved on the smelting process by building the first horizontal retort smelter. Jean-Jacques Daniel Dony built a different kind of horizontal zinc smelter in Belgium, which processed even more zinc. Italian doctor Luigi Galvani discovered in 1780 that connecting the spinal cord of a freshly dissected frog to an iron rail attached by a brass hook caused the frog's leg to twitch. He incorrectly thought he had discovered an ability of nerves and muscles to create electricity and called the effect "animal electricity". The galvanic cell and the process of galvanization were both named for Luigi Galvani and these discoveries paved the way for electrical batteries, galvanization and cathodic protection.

Galvani's friend, Alessandro Volta, continued researching this effect and invented the Voltaic pile in 1800. The basic unit of Volta's pile was a simplified galvanic cell, which is made of a plate of copper and a plate of zinc connected to each other externally and separated by an electrolyte. These were stacked in series to make the Voltaic cell, which in turn produced electricity by directing electrons from the zinc to the copper and allowing the zinc to corrode.

The non-magnetic character of zinc and its lack of colour in solution delayed discovery of its importance to bio-chemistry and nutrition. This changed in 1940 when carbonic anhydrase, an enzyme that scrubs carbon dioxide from blood, was shown to have zinc in its active site. The digestive enzyme carboxypeptidase became the second known zinc-containing enzyme in 1955.

PRODUCTION

Mining and Processing

Zinc is the fourth most common metal in use, trailing only iron, aluminium, and copper with an annual production of about 12 million tonnes. The world's largest zinc producer is Nyrstar, a merger of the Australian *OZ Minerals* and the Belgian *Umicore*. About 70% of the world's zinc originates from mining, while the remaining 30% comes from recycling secondary zinc. Commercially pure zinc is known as Special High Grade, often abbreviated SHG, and is 99.995% pure.

Worldwide, 95% of the zinc is mined from sulfidic ore deposits, in which sphalerite ZnS is nearly always mixed with the sulfides of copper, lead and iron. There are zinc mines throughout the world, with the main mining areas being China, Australia and Peru. China produced 29% of the global zinc output in 2010.

Zinc metal is produced using extractive metallurgy. After grinding the ore, froth flotation, which selectively separates minerals from gangue by taking advantage of differences in their hydrophobicity, is used to get an ore concentrate. A final concentration of zinc of about 50% is reached by this process with the remainder of the concentrate being sulfur (32%), iron (13%), and SiO_2 (5%).

Roasting converts the zinc sulfide concentrate produced during processing to zinc oxide :

$$2\,ZnS + 3\,O_2 \rightarrow 2\,ZnO + 2\,SO_2$$

The sulfur dioxide is used for the production of sulfuric acid, which is necessary for the leaching process. If deposits of zinc carbonate, zinc silicate or zinc spinel, like the Skorpion Deposit in Namibia are used for zinc production the roasting can be omitted.

For further processing two basic methods are used : pyrometallurgy or electrowinning. Pyrometallurgy processing reduces zinc oxide with carbon or carbon monoxide at 950°C (1,740°F) into the metal, which is distilled as zinc vapour. The zinc vapour is collected in a condenser. The below set of equations demonstrate this process :

$$2\,ZnO + C \rightarrow 2\,Zn + CO_2$$

$$ZnO + CO \rightarrow Zn + CO_2$$

In electro-winning, zinc is leached from the ore concentrate by sulfuric acid :

$$ZnO + H_2SO_4 \rightarrow ZnSO_4 + H_2O$$

Finally, the zinc is reduced by electrolysis.

$$2\,ZnSO_4 + 2\,H_2O \rightarrow 2\,Zn + 2\,H_2SO_4 + O_2$$

The sulfuric acid regenerated is recycled to the leaching step.

Environmental Impact

The production for sulfidic zinc ores produces large amounts of sulfur dioxide and cadmium vapour. Smelter slag and other residues of process also contain significant amounts of heavy metals. About 1.1 million tonnes of metallic zinc and 130 thousand tonnes of lead were mined and smelted in the Belgian towns of La Calamine and Plombières between 1806 and 1882. The dumps of the past mining operations leach significant amounts of zinc and cadmium, and, as a result, the sediments of the Geul River contain significant amounts of heavy metals. About two thousand years ago emissions of zinc from mining and smelting totaled 10 thousand tonnes a year. After increasing 10-fold from 1850, zinc emissions peaked at 3.4 million tonnes per year in the 1980s and declined to 2.7 million tonnes in the 1990s, although a 2005 study of the Arctic troposphere found that the concentrations there did not reflect the decline. Anthropogenic and natural emissions occur at a ratio of 20 to 1.

Levels of zinc in rivers flowing through industrial or mining areas can be as high as 20 ppm. Effective sewage treatment greatly reduces this; treatment along the Rhine, for example, has decreased zinc levels to 50 ppb. Concentrations of zinc as low as 2 ppm adversely affects the amount of oxygen that fish can carry in their blood.

Soils contaminated with zinc through the mining of zinc-containing ores, refining, or where zinc-containing sludge is used as fertilizer, can contain several grams of zinc per kilogram of dry soil. Levels of zinc in excess of 500 ppm in soil interfere with the ability of plants to absorb other essential metals, such as iron and manganese. Zinc levels of 2000 ppm to 180,000 ppm (18%) have been recorded in some soil samples.

APPLICATIONS

Major applications of zinc include (numbers are given for the US) :

1. Galvanizing (55%)
2. Alloys (21%)
3. Brass and bronze (16%)
4. Miscellaneous (8%).

Anti-corrosion and Batteries

The metal is most commonly used as an anti-corrosion agent. Galvanization, which is the coating of iron or steel to protect the metals against corrosion, is the most familiar form of using zinc in this way. In 2009 in the United States, 55% or 893 thousand tonnes of the zinc metal was used for galvanization.

Zinc is more reactive than iron or steel and thus will attract almost all local oxidation until it completely corrodes away. A protective surface layer of oxide and carbonate ($Zn_5(OH)_6$ forms as the zinc corrodes. This protection lasts even after the zinc layer is scratched but degrades through time as the zinc corrodes away. The zinc is applied electro-chemically or as molten zinc by hot-dip galvanizing or spraying. Galvanization is used on chain-link fencing, guard rails, suspension bridges, lightposts, metal roofs, heat exchangers, and car bodies.

The relative reactivity of zinc and its ability to attract oxidation to itself makes it an efficient sacrificial anode in cathodic protection (CP). For example, cathodic protection of a buried pipeline can be achieved by connecting anodes made from zinc to the pipe. Zinc acts as the anode (negative terminus) by slowly corroding away as it passes electric current to the steel pipeline. Zinc is also used to cathodically protect metals that are exposed to sea water from corrosion. A zinc disc attached to a ship's iron rudder will slowly corrode while the rudder stays unattacked. Other similar uses include a plug of zinc attached to a propeller or the metal protective guard for the keel of the ship.

With a standard electrode potential (SEP) of −0.76 volts, zinc is used as an anode material for batteries. (More reactive lithium (SEP −3.04 V) is used for anodes in lithium batteries). Powdered zinc is used in this way in alkaline batteries and sheets of zinc metal form the cases for and act as anodes in zinc–carbon batteries. Zinc is used as the anode or fuel of the zinc-air battery/fuel cell.

Alloys

A widely used alloy which contains zinc is brass, in which copper is alloyed with anywhere from 3% to 45% zinc, depending upon the type of brass. Brass is generally more ductile and stronger than copper and has superior corrosion resistance. These properties make it useful in communication equipment, hardware, musical instruments, and water valves.

Other widely used alloys that contain zinc include nickel silver, typewriter metal, soft and aluminium solder, and commercial bronze. Zinc is also used in contemporary pipe organs as a substitute for the traditional lead/tin alloy in pipes. Alloys of 85–88% zinc, 4–10% copper, and 2–8% aluminium find limited use in certain types of machine bearings. Zinc is the primary metal used in making American one cent coins since 1982. The zinc core is coated with a thin layer of copper to give the impression of a copper coin. In 1994, 33,200 tonnes (36,600 short tons) of zinc were used to produce 13.6 billion pennies in the United States.

Alloys of primarily zinc with small amounts of copper, aluminium, and magnesium are useful in die casting as well as spin casting, especially in the automotive, electrical, and hardware industries. These alloys are marketed under the name Zamak. An example of this is zinc aluminium. The low melting point together with the low viscosity of the alloy makes the production of small and intricate shapes possible. The low working temperature leads to rapid cooling of the cast products and therefore fast assembly is possible. Another alloy, marketed under the brand name Prestal, contains 78% zinc and 22% aluminium and is reported to be nearly as strong as steel but as malleable as plastic. This superplasticity of the alloy allows it to be molded using die casts made of ceramics and cement.

Similar alloys with the addition of a small amount of lead can be cold-rolled into sheets. An alloy of 96% zinc and 4% aluminium is used to make stamping dies for low production run applications for which ferrous metal dies would be too expensive. In building facades, roofs or other applications in which zinc is used as sheet metal and for methods such as deep drawing, roll forming or bending, zinc alloys with titanium and copper are used. Unalloyed zinc is too brittle for these kinds of manufacturing processes.

As a dense, inexpensive, easily worked material, zinc is used as a lead replacement. In the wake of lead concerns, zinc appears in weights for various applications ranging from fishing to tire balances and flywheels.

Cadmium zinc telluride (CZT) is a semi-conductive alloy that can be divided into an array of small sensing devices. These devices are similar to an integrated circuit and can detect the energy of incoming gamma ray photons. When placed behind an absorbing mask, the CZT sensor array can also be used to determine the direction of the rays.

Other Industrial Uses

Roughly one quarter of all zinc output in the United States, is consumed in the form of zinc compounds; a variety of which are used industrially. Zinc oxide is widely used as a white pigment in paints, and as a catalyst in the manufacture of rubber. It is also used as a heat disperser for the rubber and acts to protect its polymers from ultraviolet radiation (the same UV protection is conferred to plastics containing zinc oxide). The semi-conductor properties of zinc oxide make it useful in varistors and photocopying products. The zinc zinc-oxide cycle is a two step thermo-chemical process based on zinc and zinc oxide for hydrogen production.

Zinc chloride is often added to lumber as a fire retardant and can be used as a wood preservative. It is also used to make other chemicals. Zinc methyl (Zn is used in a number of organic syntheses. Zinc sulfide (ZnS) is used in luminescent pigments such as on the hands of clocks, X-ray and television screens, and luminous paints. Crystals of ZnS are used in lasers that operate in the mid-infrared part of the spectrum. Zinc sulfate is a chemical in dyes and pigments. Zinc pyrithione is used in antifouling paints.

Zinc powder is sometimes used as a propellant in model rockets. When a compressed mixture of 70% zinc and 30% sulfur powder is ignited there is a violent chemical reaction. This produces zinc sulfide, together with large amounts of hot gas, heat, and light. Zinc sheet metal is used to make zinc bars.

$_{64}$Zn, the most abundant isotope of zinc, is very susceptible to neutron activation, being transmuted into the highly radioactive $_{65}$Zn, which has a half-life of 244 days and produces intense gamma radiation. Because of this, Zinc Oxide used in nuclear reactors as an anti-corrosion agent is depleted of $_{64}$Zn before use, this is called depleted zinc oxide. For the same reason, zinc has been proposed as a salting material for nuclear weapons (cobalt is another, better-known salting material). A jacket of isotopically enriched $_{64}$Zn would be irradiated by the intense high-energy neutron flux from an exploding thermonuclear weapon, forming a large amount of $_{65}$Zn significantly increasing the radioactivity of the weapon's fallout. Such a weapon is not known to have ever been built, tested, or used. $_{65}$Zn is also used as a tracer to study how alloys that contain zinc wear out, or the path and the role of zinc in organisms.

Zinc dithiocarbamate complexes are used as agricultural fungicides; these include Zineb, Metiram, Propineb and Ziram. Zinc naphthenate is used as wood preservative. Zinc, in the form of ZDDP, is also used as an anti-wear additive for metal parts in engine oil.

Dietary Supplement

Zinc is included in most single tablet over-the-counter daily vitamin and mineral supplements. Preparations include zinc oxide, zinc acetate, and zinc gluconate. It is believed to possess anti-oxidant properties, which may protect against accelerated aging of the skin and muscles of the body; studies differ as to its effectiveness. Zinc also helps speed up the healing process after an injury. It is also suspected of being beneficial to the body's immune system. Indeed, zinc deficiency may have effects on virtually all parts of the human immune system.

Zinc deficiency has been associated with major depressive disorder (MDD), and zinc supplements may be an effective treatment.

Zinc serves as a simple, inexpensive, and critical tool for treating diarrheal episodes among children in the developing world. Zinc becomes depleted in the body during diarrhea, but recent studies suggest that replenishing zinc with a 10- to 14-days course of treatment can reduce the duration and severity of diarrheal episodes and may also prevent future episodes for up to three months.

Fig. : Zinc gluconate is one compound used for the delivery of zinc as a dietary supplement.

The Age-Related Eye Disease Study determined that zinc can be part of an effective treatment for age-related macular degeneration. Zinc supplementation is an effective treatment for acrodermatitis enteropathica, a genetic disorder affecting zinc absorption that was previously fatal to babies born with it.

Gastroenteritis is strongly attenuated by ingestion of zinc, and this effect could be due to direct anti-microbial action of the zinc ions in the gastrointestinal tract, or to the absorption of the zinc and re-release from immune cells (all granulocytes secrete zinc), or both. In 2011, researchers at John Jay College of Criminal Justice reported that dietary zinc supplements can mask the presence of drugs in urine. Similar claims have been made in web forums on that topic.

Although not yet tested as a therapy in humans, a growing body of evidence indicates that zinc may preferentially kill prostate cancer cells. Because zinc naturally homes to the prostate and because the prostate is accessible with relatively non-invasive procedures, its potential as a chemotherapeutic agent in this type of cancer has shown promise. However, other studies have demonstrated that chronic use of zinc supplements in excess of the recommended dosage may actually increase the chance of developing prostate cancer, also likely due to the natural buildup of this heavy metal in the prostate.

Zinc Lozenges and the Common Cold

There is strong evidence that zinc lozenges shorten the duration of colds. The most positive results have been found in studies in which zinc acetate was used, apparently because acetate does not bind zinc ions. Three high dose trials which used zinc acetate found an average 42% reduction in the duration of colds.

There is no concern of zinc toxicity in the dosages that were used in the zinc acetate trials with 80–100 mg/day of elemental zinc. The effect of zinc lozenges seems to take place locally in the oropharynx so that it is not a systemic effect, *i.e.*, the effect is not a dietary supplement effect.

Topical Use

Topical administration of zinc preparations include ones used on the skin, often in the form of zinc oxide. Zinc preparations can protect against sun-burn in the summer and wind-burn in the winter.[51] Applied thinly to a baby's diaper area (perineum) with each diaper change, it can protect against diaper rash.

Chelated zinc is used in toothpastes and mouthwashes to prevent bad breath. Zinc pyrithione is widely applied in shampoos because of its anti-dandruff function. Zinc ions are effective antimicrobial agents even at low concentrations.

Organic Chemistry

There are many important organozinc compounds. Organozinc chemistry is the science of organozinc compounds describing their physical properties, synthesis and reactions. Among important applications is the Frankland-Duppa Reaction in which an oxalate ester (ROCOCOOR) reacts with an alkyl halide R'X, zinc and hydrochloric acid to the α-hydroxycarboxylic esters RR'COHCOOR, the Reformatskii reaction which converts α-halo-esters and aldehydes to β-hydroxy-esters, the Simmons–Smith reaction in which the carbenoid (iodomethyl) zinc iodide reacts with alkene(or alkyne) and converts them to cyclopropane, the Addition reaction of organozinc compounds to carbonyl compounds. The Barbier reaction which is the zinc equivalent of the magnesium Grignard reaction and is better of the two. In presence of just about any water the formation of the organomagnesium halide will fail whereas the Barbier reaction can even take place in water. On the downside organozincs are much less nucleophilic than Grignards, are expensive and difficult to handle. Commercially available diorganozinc compounds are dimethylzinc, diethylzinc and diphenylzinc. In one study the active organozinc compound is obtained from much cheaper organobromine precursors :

The Negishi coupling is also an important reaction for the formation of new carbon carbon bonds between unsaturated carbon atoms in alkenes, arenes and alkynes. The catalysts are nickel and palladium. A key step in the catalytic cycle is a transmetalation in which a zinc halide exchanges its organic substituent for another halogen with the palladium (nickel) metal center. The Fukuyama coupling is another coupling reaction but this one with a thioester as reactant forming a ketone.

BIOLOGICAL ROLE

Zinc is an essential trace element, necessary for plants, animals, and micro-organisms. Zinc is found in nearly 100 specific enzymes (other sources say 300), serves as structural ions in transcription factors and is stored and transferred in metallothioneins. It is "typically the second most abundant transition metal in organisms" after iron and it is the only metal which appears in all enzyme classes.

In proteins, Zn ions are often co-ordinated to the amino acid side chains of aspartic acid, glutamic acid, cysteine and histidine. The theoretical and computational description of this zinc binding in proteins (as well as that of other transition metals) is difficult.

There are 2-4 grams of zinc distributed throughout the human body. Most zinc is in the brain, muscle, bones, kidney, and liver, with the highest concentrations in the prostate and parts of the eye. Semen is particularly rich in zinc, which is a key factor in prostate gland function and reproductive organ growth.

In humans, zinc plays "ubiquitous biological roles". It interacts with "a wide range of organic ligands", and has roles in the metabolism of RNA and DNA, signal transduction, and gene expression. It also regulates apoptosis. A 2006 study estimated that about 10% of human proteins potentially bind zinc, in addition to hundreds which transport and traffic zinc; a similar in silico study in the plant Arabidopsis thaliana found 2367 zinc-related proteins.

In the brain, zinc is stored in specific synaptic vesicles by glutamatergic neurons and can "modulate brain excitability". It plays a key role in synaptic plasticity and so in learning. However it has been called "the brain's dark horse" since it also can be a neurotoxin, suggesting zinc homeostasis plays a critical role in normal functioning of the brain and central nervous system.

Enzymes

Zinc is an efficient Lewis acid, making it a useful catalytic agent in hydroxylation and other enzymatic reactions. The metal also has a flexible co-ordination geometry, which allows proteins using it to rapidly shift conformations to perform biological reactions. Two examples of zinc-containing enzymes are carbonic anhydrase and carboxypeptidase, which are vital to the processes of carbon dioxide (CO_2) regulation and digestion of proteins, respectively.

In vertebrate blood, carbonic anhydrase converts CO_2 into bicarbonate and the same enzyme transforms the bicarbonate back into CO_2 for exhalation through the lungs. Without this enzyme, this conversion would occur about one million times slower at the normal blood pH of 7 or would require a pH of 10 or more. The non-related β-carbonic anhydrase is required in plants for leaf formation, the synthesis of indole acetic acid and alcoholic fermentation.

Carboxypeptidase cleaves peptide linkages during digestion of proteins. A co-ordinate covalent bond is formed between the terminal peptide and a C=O group attached to zinc, which gives the carbon a positive charge. This helps to create a hydrophobic pocket on the enzyme near the zinc, which attracts the non-polar part of the protein being digested.

Other Proteins

Zinc serves a purely structural role in zinc fingers, twists and clusters. Zinc fingers form parts of some transcription factors, which are proteins that recognize DNA base sequences during the replication and transcription of DNA. Each of the nine or ten $Zn2+$ions in a zinc finger helps maintain the finger's structure by co-ordinately binding to four amino acids in the transcription factor. The transcription factor wraps around the DNA helix and uses its fingers to accurately bind to the DNA sequence.

In blood plasma, zinc is bound to and transported by albumin (60%, low-affinity) and transferrin (10%). Since transferrin also transports iron, excessive iron reduces zinc absorption, and vice-versa. A similar reaction occurs with copper. The concentration of zinc in blood plasma stays relatively constant regardless of zinc intake. Cells in the salivary gland, prostate, immune system and intestine use zinc signaling as one way to communicate with other cells.

Zinc may be held in metallothionein reserves within micro-organisms or in the intestines or liver of animals. Metallothionein in intestinal cells is capable of adjusting absorption of zinc by 15–40%. However, inadequate or excessive zinc intake can be harmful; excess zinc particularly impairs copper absorption because metallothionein absorbs both metals.

Dietary Intake

In the U.S., the Recommended Dietary Allowance (RDA) is 8 mg/day for women and 11 mg/day for men. Median intake in the U.S. around 2000 was 9 mg/day for women and 14 mg/day in men. Oysters, lobster and red meats, especially beef, lamb and liver have some of the highest concentrations of zinc in food.

Zinc supplements should only be ingested when there is zinc deficiency or increased zinc necessity (*e.g.* after surgeries, traumata or burns). Persistent intake of high doses of zinc can cause copper deficiency.

The concentration of zinc in plants varies based on levels of the element in soil. When there is adequate zinc in the soil, the food plants that contain the most zinc are wheat (germ and bran) and various seeds (sesame, poppy, alfalfa, celery, mustard). Zinc is also found in beans, nuts, almonds, whole grains, pumpkin seeds, sunflower seeds and blackcurrant.

Other sources include fortified food and dietary supplements, which come in various forms. A 1998 review concluded that zinc oxide, one of the most common supplements in the United States, and zinc carbonate are nearly insoluble and poorly absorbed in the body. This review cited studies which found low plasma zinc concentrations after zinc oxide and zinc carbonate were consumed compared with those seen after consumption of zinc acetate and sulfate salts.

However, harmful excessive supplementation is a problem among the relatively affluent, and should probably not exceed 20 mg/day in healthy people, although the U.S. National Research Council set a Tolerable Upper Intake of 40 mg/day.

For fortification, however, a 2003 review recommended zinc oxide in cereals as cheap, stable, and as easily absorbed as more expensive forms. A 2005 study found that various compounds of zinc, including oxide and sulfate, did not show statistically significant differences in absorption when added as fortificants to maize tortillas. A 1987 study found that zinc picolinate was better absorbed than zinc gluconate or zinc citrate. However, a study published in 2008 determined that zinc glycinate is the best absorbed of the four dietary supplement types available.

Deficiency

Zinc deficiency is usually due to insufficient dietary intake, but can be associated with malabsorption, acrodermatitis enteropathica, chronic liver disease, chronic renal disease, sickle cell disease, diabetes, malignancy, and other chronic illnesses. Symptoms of mild zinc deficiency are diverse. Clinical outcomes include depressed growth, diarrhea, impotence and delayed sexual maturation, alopecia, eye and skin lesions, impaired appetite, altered cognition, impaired host defense properties, defects in carbohydrate utilization, and reproductive teratogenesis. Mild zinc deficiency depresses immunity, although excessive zinc does also. Animals with a diet deficient in zinc require twice as much food in order to attain the same weight gain as animals given sufficient zinc.

Groups at risk for zinc deficiency include the elderly, children in developing countries, and those with renal insufficiency. The zinc chelator phytate, found in seeds and cereal bran, can contribute to zinc malabsorption.

Despite some concerns, western vegetarians and vegans have not been found to suffer from overt zinc deficiencies any more than meat-eaters. Major plant sources of zinc include cooked dried beans, sea vegetables, fortified cereals, soyfoods, nuts, peas, and seeds. However, phytates in many whole-grains and fiber in many foods may interfere with zinc absorption and marginal zinc intake has poorly understood effects. There is some evidence to suggest that more than the US RDA of zinc daily may be needed in those whose diet is high in phytates, such as some vegetarians. These considerations must be balanced against the fact that there is a paucity of adequate zinc biomarkers, and the most widely used indicator, plasma zinc, has poor sensitivity and specificity. Diagnosing zinc deficiency is a persistent challenge.

Nearly two billion people in the developing world are deficient in zinc. In children it causes an increase in infection and diarrhea, contributing to the death of about 800,000 children worldwide per year. The World Health Organization advocates zinc supplementation for severe malnutrition and diarrhea. Zinc supplements help prevent disease and reduce mortality, especially among children with low birth weight or stunted growth. However, zinc supplements should not be administered alone, since many in the developing world have several deficiencies, and zinc interacts with other micro-nutrients.

SOIL REMEDIATION

The Ericoid Mycorrhizal Fungi Calluna, Erica and Vaccinium can grow in zinc metalliferous soils.

Agriculture

Zinc deficiency is crop plants' most common micro-nutrient deficiency; it is particularly common in high-pH soils. Zinc-deficient soil is cultivated in the cropland of about half of Turkey and India, a third of China, and most of Western Australia, and substantial responses to zinc fertilization have been reported in these areas. Plants that grow in soils that are zinc-deficient are more susceptible to disease. Zinc is primarily added to the soil through the weathering of rocks, but humans have added zinc through fossil fuel combustion, mine waste, phosphate fertilizers, limestone, manure, sewage sludge, and particles from galvanized surfaces. Excess zinc is toxic to plants, although zinc toxicity is far less widespread.

PRECAUTIONS

Toxicity

Although zinc is an essential requirement for good health, excess zinc can be harmful. Excessive absorption of zinc suppresses copper and iron absorption. The free zinc ion in solution is highly toxic to plants, invertebrates, and even vertebrate fish. The Free Ion Activity Model is well-established in the literature, and shows that just micromolar amounts of the free ion kills some organisms. A recent example showed 6 micromolar killing 93% of all Daphnia in water.

The free zinc ion is a powerful Lewis acid up to the point of being corrosive. Stomach acid contains hydrochloric acid, in which metallic zinc dissolves readily to give corrosive zinc chloride. Swallowing a post-1982 American one cent piece (97.5% zinc) can cause damage to the stomach lining due to the high solubility of the zinc ion in the acidic stomach.

There is evidence of induced copper deficiency in those taking 100–300 mg of zinc daily. A 2007 trial observed that elderly men taking 80 mg daily were hospitalized for urinary complications more often than those taking a placebo. The USDA RDA is 11 and 8 mg Zn/day for men and women, respectively. Even lower levels, closer to the RDA, may interfere with the utilization of copper and iron or adversely affect cholesterol. Levels of zinc in excess of 500 ppm in soil interfere with the ability of plants to absorb other essential metals, such as iron and manganese. There is also a condition called the zinc shakes or "zinc chills" that can be induced by the inhalation of freshly formed zinc oxide formed during the welding of galvanized materials. Zinc is a common ingredient of denture cream which may contain between 17 and 38 mg of zinc per gram. There have been cases of disability or even death due to excessive use of these products.

The U.S. Food and Drug Administration (FDA) has stated that zinc damages nerve receptors in the nose, which can cause anosmia. Reports of anosmia were also observed in the 1930s when zinc preparations were used in a failed attempt to prevent polio infections. On June 16, 2009, the FDA said that consumers should stop using zinc-based intranasal cold products and ordered their removal from store shelves. The FDA said the loss of smell can be life-threatening because people with impaired smell cannot detect leaking gas or smoke and cannot tell if food has spoiled before they eat it. Recent research suggests that the topical anti-microbial zinc pyrithione is a potent heat shock response inducer that may impair genomic integrity with induction of PARP-dependent energy crisis in cultured human keratinocytes and melanocytes.

Poisoning

In 1982, the United States Mint began minting pennies coated in copper but made primarily of zinc. With the new zinc pennies, there is the potential for zinc toxicosis, which can be fatal. One reported case of chronic ingestion of 425 pennies (over 1 kg of zinc) resulted in death due to gastrointestinal bacterial and fungal sepsis, while another patient, who ingested 12 grams of zinc, only showed lethargy and ataxia (gross lack of co-ordination of muscle movements). Several other cases have been reported of humans suffering zinc intoxication by the ingestion of zinc coins.

Pennies and other small coins are sometimes ingested by dogs, resulting in the need for medical treatment to remove the foreign body. The zinc content of some coins can cause zinc toxicity, which is commonly fatal in dogs, where it causes a severe hemolytic anemia, and also liver or kidney damage; vomiting and diarrhea are possible symptoms. Zinc is highly toxic in parrots and poisoning can often be fatal. The consumption of fruit juices stored in galvanized cans has resulted in mass parrot poisonings with zinc.

COMPOUNDS OF ZINC

Compounds of zinc are chemical compounds containing the element zinc which is a member of the group 12 of the periodic table. The oxidation state of most compounds is the group oxidation state of +2. Zinc may be classified as a post-transition main group element with zinc (II) having much chemical behaviour in common with copper (II). Many salts of zinc (II) are isomorphous with salts of magnesium (II) due to the ionic radii of the cations being almost the same. Zinc forms many complexes; metallo-proteins containing zinc are widespread in biological systems.

General Characteristics

Zinc atoms have an electronic configuration of $[Ar]3d^{10}4s^2$. When compounds in the +2 oxidation state are formed the s electrons are lost, so the bare zinc ion has the electronic configuration $[Ar]3d^{10}$. Examples of these zinc compounds include the oxide, ZnO and sulfide, ZnS, (zinc blende) in which the oxide and sulfide ions are tetrahedrally bound to four zinc ions. Many complexes, such as $ZnCl_4^{2-}$, are tetrahedral. Tetrahedrally coordinated zinc is found in metallo-enzymes such as carbonic anhydrase. However 6-co-ordinate octahedral complexes can also be found, such as the ion $[Zn(H_2O)_6]^{2+}$, which is present when a zinc salt is dissolved in water.

Many zinc (II) salts are isomorphous (have the same type of crystal structure) with the corresponding salts of magnesium (II) which results from the fact that Zn^{2+} and Mg^{2+} have almost identical ionic radii. This comes about because of the d-block contraction. Whilst calcium is somewhat larger than magnesium, there is a steady decrease in size as atomic number increases from calcium to zinc. By chance it is the ionic radius of zinc that is almost equal to that of magnesium. In most other respects the chemistry of zinc (II) most closely resembles the chemistry of copper (II), its neighbour in the periodic table, in which there is one less electron. However, whereas Cu^{2+} is classed as a transition metal ion by virtue of its electronic configuration, $[Ar]3d^9$, in which there is an incomplete d-shell, Zn^{2+} is best considered to be an ion of a post-transition main group element. The IUPAC periodic table places zinc in the d-block.

Some compounds with zinc in the oxidation state +1 are known. The compounds have the formula RZn_2R and they contain a $Zn - Zn$ bond analogous to the metal-metal bond in mercury(I) ion, $Hg2^{2+}$. In this respect zinc is similar to magnesium where low valent compounds containing a $Mg - Mg$ bond have been characterised.

No compounds of zinc in oxidation states other than +1 or +2 are known. Calculations indicate that a zinc compound with the oxidation state of +4 is unlikely to exist. Although higher oxidation states are more stable with the heavier elements of a group, the compound $HgF4$ was only characterized at 4 K in a neon/argon matrix.

Colour and Magnetism

Zinc compounds, like those of main group elements, are mostly colourless. Exceptions occur when the compound contains a coloured anion or ligand. Zinc selenide, $ZnSe$, however, is yellow, due to charge-transfer transitions and zinc telluride, $ZnTe$ is brown for the same reason. Zinc oxide turns yellow when heated due to the loss of some oxygen atoms and formation of a defect structure.

Compounds containing zinc and no other metal are all diamagnetic.

Reactivity of the Metal

Zinc is a strong reducing agent with a standard redox potential of -0.76 V. Pure zinc tarnishes rapidly in air, eventually forming a passive layer of basic zinc carbonate, $Zn_5(OH)_6CO_3$. The reaction of zinc with water is prevented by the passive layer. When this layer is penetrated by acids such as hydrochloric acid and sulfuric acid the reaction proceeds with the evolution of hydrogen gas.

$$Zn(s) + 2H+ (aq) \rightarrow Zn^{2+} (aq) + H_2 \uparrow$$

The hydrogen ion is reduced by accepting an electron from the reducing agent. The zinc metal is oxidised. Amalgamation with mercury, as in the Jones reductor also destroys the passive layer. Zinc reacts with alkalis as with acids. It reacts directly with oxidising non-metals such as chalcogens and halogens to form binary compounds.

Binary Compounds

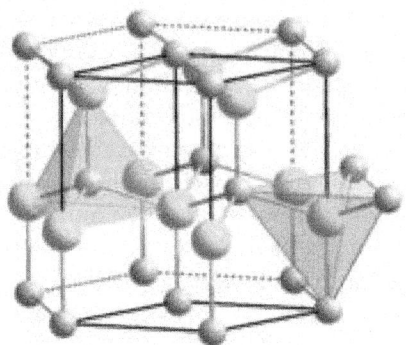

Fig. : The Wurtzite structure, showing the tetrahedral environment of both Zn and S atoms.

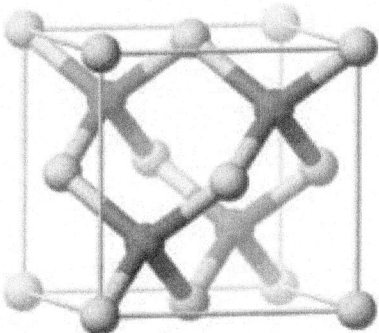

Fig. : A unit cell of zinc blende.

Zinc oxide, ZnO, is the most important manufactured compound of zinc, with a wide variety of uses. It crystallizes with the Wurtzite structure. It is amphoteric, dissolving in acids to give the aqueous Zn^{2+} ion and in alkali to give the tetrahedral hydroxo complex, $[Zn(OH)_4]^{2-}$. Zinc hydroxide, $Zn(OH)_2$ is also amphoteric.

Zinc sulfide, ZnS, crystallizes in two closely related structures, the Zinc blende structure and the Wurtzite structure which are common structures of compounds with the formula MA. Both Zn and S are tetrahedrally co-ordinated by the other ion. A useful property of ZnS is its phosphorescence. The other chalcogenides, ZnSe, and ZnTe, have applications in electronics and optics.

Of the four halides ZnF_2 has the most ionic character, whereas the others, $ZnCl_2$, $ZnBr_2$, and ZnI_2, have relatively low melting points and are considered to have more covalent character. The pnictogenides Zn_3N_2 (notable for its high melting point), Zn_3P_2, Zn_3As_2 and Zn_3Sb_2, have various applications. Other binary compounds of zinc include the peroxide ZnO_2, the hydride ZnH_2, and the carbide ZnC_2.

Salts

The nitrate $Zn(NO_3)_2$ (used as oxidizing agent), the chlorate $Zn(ClO_3)_2$, the sulfate $ZnSO_4$ (known as "white vitriol"), the phosphate $Zn_3(PO_4)_2$ (used as primer pigment), the molybdate $ZnMoO_4$ (used as white pigment), the chromate $ZnCrO_4$ (one of the few coloured zinc compounds), the arsenite $Zn(AsO_2)_2$ (colourless powder) and the arsenate octahydrate $Zn(AsO_4)_2 \bullet 8H_2O$ (white powder, also referred to as koettigite) are a few examples of other common inorganic compounds of zinc. The latter two compounds are both used in insecticides and wood preservatives. One of the simplest examples of an organic compound of zinc is the acetate $Zn(O_2CCH_3)_2$, which has several medicinal applications. Zinc salts are usually fully dissociated in aqueous solution. Exceptions occur when the anion can form a complex, such as in the case of zinc sulfate, where the complex $[Zn(H_2O)n(SO_4]$ may be formed, $(\log K = ca. 2.5)$.

Complexes

Fig. : Structure of solid basic zinc acetate, $[Zn4(\mu^4\text{-}O)(\eta^2\text{-}O_2CCH_3)_6]$.

The most common structure of zinc complexes is tetrahedral which is clearly connected with the fact that the octet rule is obeyed in these cases. Nevertheless, octahedral complexes comparable to those of the transition elements are not rare. Zn^{2+} is a class A acceptor in the classification of Ahrland, Chatt and Davies, and so forms stronger complexes with the first-row donor atoms oxygen or nitrogen than with second-row sulfur or phosphorus. In terms of HSAB theory Zn^{2+} is a hard acid.

In aqueous solution an octahedral complex, $[Zn(H_2O)_6]^{2+}$ is the predominant species. Aqueous solutions of zinc salts are mildly acidic because the aqua-ion is subject to hydrolysis with a pKaof around 5, depending on conditions.

$$[Zn(H_2O)_6]^{2+} \rightleftharpoons [Zn(H_2O)_5(OH)]^+ + H^+$$

Hydrolysis explains why basic salts such as basic zinc acetate and basic zinc carbonate, $Zn_3(OH)_4(CO_3).H_2O$ are easy to obtain. The reason for the hydrolysis is the high electrical charge density on the zinc ion, which pulls electrons away from an OH bond of a coordinated water molecule and releases a hydrogen ion. The polarizing effect of Zn^{2+} is part of the reason why zinc is found in enzymes such as carbonic anhydrase.

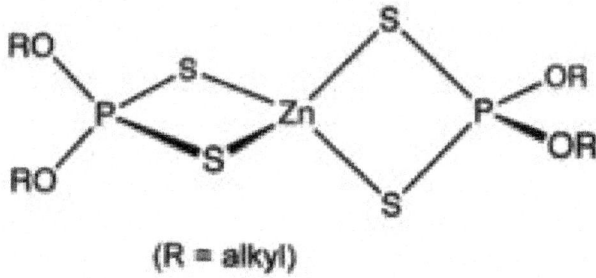

(R = alkyl)

Fig. : Structure of a monomeric zinc dialkyldithiophosphate.

No fluoro complexes are known, but complexes with the other halides and with pseodohalides, $[ZnX_3]$-and $[ZnX_4]^{2-}$ can be prepared. The case of the thiocyanate complex illustrates the class A character of the zinc ion as it is the N-bonded isomer, $[Zn(NCS)_4]^{2-}$ in contrast to $[Cd(SCN)_4]^{2-}$ which is S-bonded. Being a class-A acceptor does not preclude the formation of complexes with sulfur donors, as is shown by zinc dithiophosphate and the zinc finger complex.

The acetylacetonate complex, $Zn(acac)_2$ is interesting. As the ligand is bidentate a tetrahedral structure might be expected. However, the compound is in fact a trimer, $Zn3(acac)_6$ in which each Zn ion is co-ordinated by five oxygen atoms in a distorted trigonal bipyramidal structure. Other 5-co-ordinate structures can be engineered by choosing ligands which have specific stereo-chemical requirements. For example, terpyridine, which is a tridentate ligand forms the complex $[ZnCl_2]$. Another example would involve a tripodal ligand such as Tris(2-aminoethyl) amine. The compound zinc cyanide, $Zn(CN)_2$, is not 2-co-ordinate. It adopts a polymeric structure consisting of tetrahedral zinc centres linked by bridging cyanide ligands. The cyanide group shows head to tail disorder with any zinc atom having between 1 and 4 carbon atom neighbours and the remaining being nitrogen atoms. These two examples illustrate the difficulty of sometimes relating structure to stoichiometry.

A co-ordination number of 2 occurs in the amide Zn_2 (R^1=CMe$_3$, R^2=SiMe$_3$); the ligand is so bulky that there is not enough space for more than two of them.

Bio-complexes

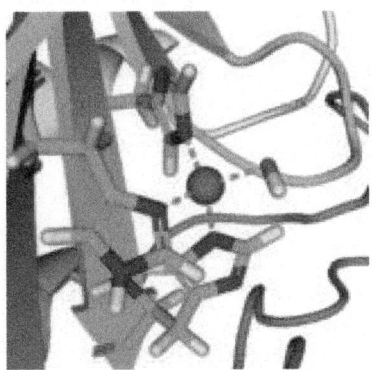

Fig. : Carbonic anhydrase : an hydroxide group (red) is shown attached to zinc (gray).

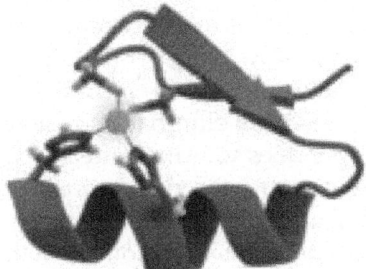

Fig. : Zinc finger motif. The zinc ion is co-ordinated by two histidine residues and two cysteine residues.

A very large number of metallo-enzymes contain zinc (II). Also many proteins contain zinc for structural reasons. The zinc ion is invariably 4-co-ordinate with at least three ligands that are amino-acid side-chains. The imidazole nitrogen of a histidine side-chain is a common ligand. The following are typical examples of the two kinds of zinc-protein complexes.

In the active site of resting Carbonic anhydrase a zinc ion is co-ordinated by three histidine residues. The fourth position is occupied by a water molecule, which is strongly polarized as in hydrolysis. When carbon dioxide enters the active site, it subject to nucleophilic attack by the oxygen atom which carries a partial negative charge, or indeed a full negative charge if the water molecule is dissociated. The CO_2 is rapidly converted into a bicarbonate ion.

$$[(\text{-hys})_3 Zn(H_2O)]^{2+} + CO_2 \rightarrow [(\text{-hys})_3 Zn]^{2+} + HCO_3^- + H+$$

Some peptidases, such as Glutamate carboxypeptidase II are thought to act in a similar way, with the zinc ion promoting the formation of a nucleophilic reagent.

The zinc finger motif is a rigid sub-structure in a protein which facilitates the binding of the protein to another molecule such as DNA. In this case all four co-ordination positions are occupied by the histidine and cysteine residues.

The tetrahedral geometry around the zinc ion constrains an α helix fragment and an antiparallel β sheet fragment to a particular orientation with respect to each other.

The magnesium ion, which has a higher concentration in biological fluids, cannot perform these functions as its complexes are much weaker than those of zinc.

Organometallic Compounds

Figure : Diethylzinc

Organozinc compounds contain zinc−carbon covalent bonds. Diethylzinc $((C_2H_5)_2Zn)$ was first reported in 1848. It was made by reaction of zinc and ethyl iodide and is the first compound known to contain a metal−carbon sigma bond. For a long time it was a mystery why copper (II) did not form an analogous compound. It was not until the 1980s that the reason was found : the zinc compound does not undergo the Beta-hydride elimination reaction whereas the compound of the transition metal copper does so. Alkyl and aryl zinc compounds are contain the linear C−Zn−C motif. Because the zinc centre is co-ordinatively unsaturated the compound are powerful electrophiles. In fact the low-molecular weight compounds will ignite spontaneously on contact with air and are immediately destroyed by reaction with water molecules. The use of zinc alkyls has been largely superseded by the use of the more easily handled Grignard reagents. This demonstrates yet another connection between the chemistries of zinc and magnesium.

Zinc cyanide, $Zn(CN)_2$, is used as a catalyst in some organic reactions.

Organometallic compounds of zinc(I) contain M−M bonds. decamethyldizincocene is now known.

ISOTOPES OF ZINC

Naturally occurring zinc (Zn) is composed of the 5 stable isotopes ^{64}Zn, ^{66}Zn, ^{67}Zn, ^{68}Zn, and ^{70}Zn with ^{64}Zn being the most abundant (48.6% natural abundance). Twenty-five radioisotopes have been characterised with the most abundant and stable being ^{65}Zn with a half-life of 244.26 days, and ^{72}Zn with a half-life of 46.5 hours. All of the remaining radioactive isotopes have half-lives that are less than 14 hours and the majority of these have half-lives that are less than 1 second. This element also has 10 meta states.

Zinc has been proposed as a "salting" material for nuclear weapons (cobalt is another, better-known salting material). A jacket of isotopically enriched ^{64}Zn, irradiated by the intense high-energy neutron flux from an exploding thermonuclear weapon, would transmute into the radioactive isotope ^{65}Zn with a half-life of 244 days and produce approximately 1.115 MeV of gamma radiation, significantly increasing the radioactivity of the weapon's fallout for several days. Such a weapon is not known to have ever been built, tested, or used.

Standard atomic mass : 65.409(4) u

Table

nuclide symbol	Z(p)	N(n)	isotopic mass (u)	half-life	decay mode(s)	daughter isotope(s)	nuclear spin	representative isotopic composition (mole fraction)	range of natural variation (mole fraction)
			excitation energy						
^{54}Zn	30	24	53.99295(43)#		2p	52Ni	0+		
^{55}Zn	30	25	54.98398(27)#	20# ms [>1.6 μs]	2p	^{53}Ni	5/2-#		
					β+	^{55}Cu			
^{56}Zn	30	26	55.97238(28)#	36(10) ms	β+	^{56}Cu	0+		
^{57}Zn	30	27	56.96479(11)#	38(4) ms	β+, p (65%)	^{56}Ni	7/2-#		
					β+ (35%)	^{57}Cu			
^{58}Zn	30	28	57.95459(5)	84(9) ms	β+, p (60%)	^{57}Ni	0+		
					β+ (40%)	^{58}Cu			
^{59}Zn	30	29	58.94926(4)	182.0(18) ms	β+ (99%)	^{59}Cu	3/2-		
					β+, p (1%)	^{58}Ni			
^{60}Zn	30	30	59.941827(11)	2.38(5) min	β+	^{60}Cu	0+		
^{61}Zn	30	31	60.939511(17)	89.1(2) s	β+	^{61}Cu	3/2-		
^{61}m1Zn			88.4(1) keV	<430 ms			1/2-		
^{61}m2Zn			418.10(15) keV	140(70) ms			3/2-		
^{61}m3Zn			756.02(18) keV	<130 ms			5/2-		
^{62}Zn	30	32	61.934330(11)	9.186(13) h	β+	^{62}Cu	0+		
^{63}Zn	30	33	62.9332116(17)	38.47(5) min	β+	^{63}Cu	3/2-		
^{64}Zn	30	34	63.9291422(7)	Observationally Stable			0+	0.48268(321)	
^{65}Zn	30	35	64.9292410(7)	243.66(9) d	β+	^{65}Cu	5/2-		
^{65}mZn			53.928(10) keV	1.6(6) μs			(1/2)-		
^{66}Zn	30	36	65.9260334(10)	Stable			0+	0.27975(77)	
^{67}Zn	30	37	66.9271273(10)	Stable			5/2-	0.04102(21)	
^{68}Zn	30	38	67.9248442(10)	Stable			0+	0.19024(123)	
^{69}Zn	30	39	68.9265503(10)	56.4(9) min	β-	^{69}Ga	1/2-		
^{69}mZn			438.636(18) keV β-(3.3%)	13.76(2) h 69Ga	IT	^{69}Zn	9/2+		
^{70}Zn	30	40	69.9253193(21)	Observationally Stable			0+	0.00631(9)	
^{71}Zn	30	41	70.927722(11)	2.45(10) min	β-	^{71}Ga	1/2-		
^{71}mZn			157.7(13) keV IT (.05%)	3.96(5) h 71Zn	β-(99.95%)	^{71}Ga	9/2+		

(Contd...)

(Contd...)

nuclide symbol	Z(p)	N(n)	isotopic mass (u)	half-life	decay mode(s)	daughter isotope(s)	nuclear spin	representative isotopic composition (mole fraction)	range of natural variation (mole fraction)
			excitation energy						
^{72}Zn	30	42	71.926858(7)	46.5(1) h	β-	^{72}Ga	0+		
^{73}Zn	30	43	72.92978(4)	23.5(10) s	β-	^{73}Ga	(1/2)-		
^{73}m1Zn			195.5(2) keV	13.0(2) ms			(5/2+)		
^{73}m2Zn			237.6(20) keV IT	5.8(8) s 73Zn	β-	^{73}Ga	(7/2+)		
^{74}Zn	30	44	73.92946(5)	95.6(12) s	β-	^{74}Ga	0+		
^{75}Zn	30	45	74.93294(8)	10.2(2) s	β-	^{75}Ga	(7/2+)#		
^{76}Zn	30	46	75.93329(9)	5.7(3) s	β-	^{76}Ga	0+		
^{77}Zn	30	47	76.93696(13)	2.08(5) s	β-	^{77}Ga	(7/2+)#		
^{77}mZn			772.39(12) keV β-(50%)	1.05(10) s 77Ga	IT (50%)	^{77}Zn	1/2-#		
^{78}Zn	30	48	77.93844(10)	1.47(15) s	β-	^{78}Ga	0+		
^{78}mZn			2673(1) keV	319(9) ns			(8+)		
^{79}Zn	30	49	78.94265(28)#	0.995(19) s	β-	^{79}Ga	(9/2+)		
					β-, n (1.3%)	^{78}Ga			
^{80}Zn	30	50	79.94434(18)	545(16) ms	β-(99%)	^{80}Ga	0+		
					β-, n (1%)	^{79}Ga			
^{81}Zn	30	51	80.95048(32)#	290(50) ms	β-	^{81}Ga	5/2+#		
					β-, n (7.5%)	^{80}Ga			
^{82}Zn	30	52	81.95442(54)#	100# ms [>300 ns]	β-	^{82}Ga	0+		
^{83}Zn	30	53	82.96103(54)#	80# ms [>300 ns]			5/2+#		

Abbreviations :

IT : Isomeric transition

1. Bold for stable isotopes
2. Final product of the silicon-burning process; its production is endothermic and accelerates the star's collapse
3. Believed to undergo β+β+ decay to ^{64}Ni with a half-life over 2.3×10^{18} a
4. Believed to undergo β-β- decay to ^{70}Ge with a half-life over 1.3×10^{16} a

Notes

- Values marked # are not purely derived from experimental data, but at least partly from systematic trends. Spins with weak assignment arguments are enclosed in parentheses.
- Uncertainties are given in concise form in parentheses after the corresponding last digits. Uncertainty values denote one standard deviation, except isotopic

composition and standard atomic mass from IUPAC which use expanded uncertainties.

- Nuclide masses are given by IUPAP Commission on Symbols, Units, Nomenclature, Atomic Masses and Fundamental Constants (SUNAMCO).
- Isotope abundances are given by IUPAC Commission on Isotopic Abundances and Atomic Weights.

Chapter 15

METAL COMPLEXES CONTAININGREDOX-ACTIVE LIGANDS

LIGAND

In co-ordination chemistry, a ligand is an ion or molecule (functional group) that binds to a central metal atom to form a co-ordination complex. The bonding between metal and ligand generally involves formal donation of one or more of the ligand's electron pairs. The nature of metal-ligand bonding can range from covalent to ionic. Furthermore, the metal-ligand bond order can range from one to three. Ligands are viewed as Lewis bases, although rare cases are known to involve Lewis acidic "ligand."

Metals and metalloids are bound to ligands in virtually all circumstances, although gaseous "naked" metal ions can be generated in high vacuum. Ligands in a complex dictate the reactivity of the central atom, including ligand substitution rates, the reactivity of the ligands themselves, and redox. Ligand selection is a critical consideration in many practical areas, including bio-inorganic and medicinal chemistry, homogeneous catalysis, and environmental chemistry.

Ligands are classified in many ways : their charge, their size (bulk), the identity of the co-ordinating atom(s), and the number of electrons donated to the metal (denticity or hapticity). The size of a ligand is indicated by its cone angle.

History

The composition of co-ordination complexes have been known since the early 1800s, *e.g.*, Prussian blue and copper vitriol. The key breakthrough occurred when Alfred Werner reconciled formulas and isomers. He showed, among other things, that the formulas of many cobalt (III) and chromium (III) compounds can be understood if the metal has six ligands in an octahedral geometry. The first to use the term "ligand" were Alfred Stock and Carl Somiesky, in relation to silicon chemistry. The theory allows one to understand the difference between co-ordinated and ionic chloride in the cobalt ammine chlorides and to explain many of the previously inexplicable isomers. He resolved the first co-ordination complex called hexol into optical isomers, overthrowing the theory that chirality was necessarily associated with carbon compounds.

Strong Field and Weak Field Ligands

Crystal Field Theory

Crystal field theory (CFT) is a model that describes the breaking of degeneracies of electron orbital states, usually d or f orbitals, due to a static electric field produced by a surrounding charge distribution (anion neighbors). This theory has been used to describe various spectroscopies of transition metal co-ordination complexes, in particular optical spectra (colours). CFT successfully accounts for some magnetic properties, colours, hydration enthalpies, and spinel structures of transition metal complexes, but it does not attempt to describe bonding. CFT was developed by physicists Hans Bethe and John Hasbrouck van Vleck in the 1930s. CFT was subsequently combined with molecular orbital theory to form the more realistic and complex ligand field theory (LFT), which delivers insight into the process of chemical bonding in transition metal complexes.

Overview of Crystal Field Theory Analysis

According to CFT, the interaction between a transition metal and ligands arises from the attraction between the positively charged metal cation and negative charge on the non-bonding electrons of the ligand. The theory is developed by considering energy changes of the five degenerate d-orbitals upon being surrounded by an array of point charges consisting of the ligands. As a ligand approaches the metal ion, the electrons from the ligand will be closer to some of the d-orbitals and farther away from others causing a loss of degeneracy. The electrons in the d-orbitals and those in the ligand repel each other due to repulsion between like charges. Thus the d-electrons closer to the ligands will have a higher energy than those further away which results in the d-orbitals splitting in energy. This splitting is affected by the following factors :

- The nature of the metal ion.
- The metal's oxidation state. A higher oxidation state leads to a larger splitting.
- The arrangement of the ligands around the metal ion.
- The nature of the ligands surrounding the metal ion. The stronger the effect of the ligands then the greater the difference between the high and low energy *d* groups.

The most common type of complex is octahedral; here six ligands form an octahedron around the metal ion. In octahedral symmetry the d-orbitals split into two sets with an energy difference, Δ_{oct} (the crystal-field splitting parameter) where the d_{xy}, d_{xz} and d_{yz} orbitals will be lower in energy than the d_{z^2} and $d_{x^2-y^2}$, which will have higher energy, because the former group is farther from the ligands than the latter and therefore experience less repulsion. The three lower-energy orbitals are collectively referred to as t_{2g}, and the two higher-energy orbitals as *e.g.* (These labels are based on the theory of molecular symmetry). Typical orbital energy diagrams are given below in the section High-spin and low-spin.

Tetrahedral complexes are the second most common type; here four ligands form a tetrahedon around the metal ion. In a tetrahedral crystal field splitting the *d*-orbitals again split into two groups, with an energy difference of Δ_{tet} where the lower energy orbitals will be d_z^2 and $d_{x^2-y^2}$, and the higher energy orbitals will be d_{xy}, d_{xz} and d_{yz} - opposite to the octahedral case. Furthermore, since the ligand electrons in tetrahedral symmetry are not oriented directly towards the *d*-orbitals, the energy splitting will be lower than in the octahedral case. Square planar and other complex geometries can also be described by CFT.

The size of the gap Δ between the two or more sets of orbitals depends on several factors, including the ligands and geometry of the complex. Some ligands always produce a small value of Δ, while others always give a large splitting. The reasons behind this can be explained by ligand field theory. The spectro-chemical series is an empirically-derived list of ligands ordered by the size of the splitting Δ that they produce:

$I^- < Br^- < S^{2-} < SCN^- < Cl^- < NO_3^- < N_3^- < F^- < OH^- < C_2O_4^{2-} < H_2O < NCS^- <$ $CH_3CN < py < NH_3 < en < 2,2'$-bipyridine $< phen < NO_2^- < PPh_3 < CN^- < CO$

It is useful to note that the ligands producing the most splitting are those that can engage in metal to ligand back-bonding.

The oxidation state of the metal also contributes to the size of Δ between the high and low energy levels. As the oxidation state increases for a given metal, the magnitude of Δ increases. A V^{3+} complex will have a larger Δ than a V^{2+} complex for a given set of ligands, as the difference in charge density allows the ligands to be closer to a V^{3+} ion than to a V^{2+} ion. The smaller distance between the ligand and the metal ion results in a larger Δ, because the ligand and metal electrons are closer together and therefore repel more.

High-spin and Low-spin

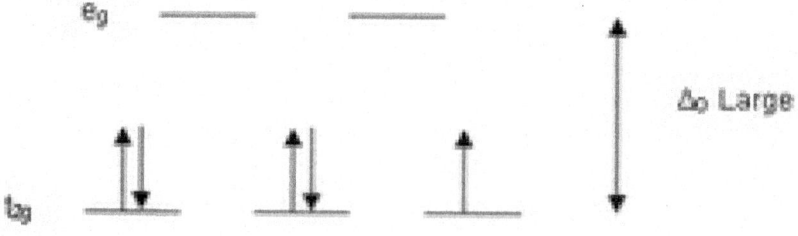

Fig. : Low Spin $[Fe(NO_2)_6]^{3-}$ crystal field diagram.

Ligands which cause a large splitting Δ of the d-orbitals are referred to as strong-field ligands, such as CN^- and CO from the spectrochemical series. In complexes with these ligands, it is unfavourable to put electrons into the high energy orbitals. Therefore, the lower energy orbitals are completely filled before population of the upper sets starts according to the *Aufbau principle*. Complexes such as this are called "low spin". For example, NO_2^- is a strong-field ligand and produces a large Δ. The octahedral ion $[Fe(NO_2)_6]^{3-}$, which has 5 d-electrons, would have the octahedral splitting diagram shown at right with all five electrons in the t_{2g} level.

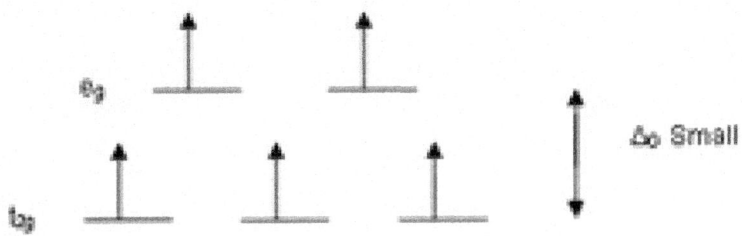

Fig. : High Spin $[FeBr_6]^{3-}$ crystal field diagram.

Conversely, ligands (like I^- and Br^-) which cause a small splitting Δ of the d-orbitals are referred to as weak-field ligands. In this case, it is easier to put electrons into the higher energy set of orbitals than it is to put two into the same low-energy orbital, because two electrons in the same orbital repel each other. So, one electron is put into each of the five d-orbitals before any pairing occurs in accord with Hund's rule and "high spin" complexes are formed. For example, Br^- is a weak-field ligand and produces a small Δ_{oct}. So, the ion $[FeBr_6]^{3-}$, again with five d-electrons, would have an octahedral splitting diagram where all five orbitals are singly occupied.

In order for low spin splitting to occur, the energy cost of placing an electron into an already singly occupied orbital must be less than the cost of placing the additional electron into an e_g orbital at an energy cost of Δ. As noted above, e_g refers to the d_z^2 and $d_{x^2-y^2}$ which are higher in energy than the t_{2g} in octahedral complexes. If the energy required to pair two electrons is greater than the energy cost of placing an electron in an e_g, Δ, high spin splitting occurs.

The crystal field splitting energy for tetrahedral metal complexes (four ligands) is referred to as Δ_{tet} and is roughly equal to $4/9\Delta_{oct}$ (for the same metal and same ligands). Therefore, the energy required to pair two electrons is typically higher than the energy required for placing electrons in the higher energy orbitals. Thus, tetrahedral complexes are usually high-spin.

The use of these splitting diagrams can aid in the prediction of the magnetic properties of co-ordination compounds. A compound that has unpaired electrons in its splitting diagram will be paramagnetic and will be attracted by magnetic fields, while a compound that lacks unpaired electrons in its splitting diagram will be diamagnetic and will be weakly repelled by a magnetic field.

Crystal Field Stabilization Energy

The crystal field stabilization energy (CFSE) is the stability that results from placing a transition metal ion in the crystal field generated by a set of ligands. It arises due to the fact that when the d-orbitals are split in a ligand field (as described above), some of them become lower in energy than before with respect to a spherical field known as the barycenter in which all five d-orbitals are degenerate. For example, in an octahedral case, the t_{2g} set becomes lower in energy than the orbitals in the barycenter. As a result of this, if there are any electrons occupying these orbitals, the metal ion is more stable in the ligand field relative to the barycenter by an amount known as the CFSE. Conversely, the e_g orbitals (in the octahedral case) are higher in energy than in the barycenter, so putting electrons in these reduces the amount of CFSE.

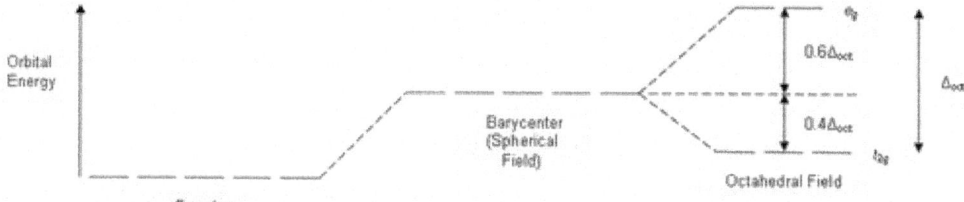

Fig. : Octahedral crystal field stabilization energy.

If the splitting of the d-orbitals in an octahedral field is Δ_{oct}, the three t_{2g} orbitals are stabilized relative to the barycenter by $^2/_5 \Delta_{oct}$, and the e_g orbitals are destabilized by $^3/_5 \Delta_{oct}$. As examples, consider the two d^5 configurations shown further up the page. The low-spin (top) example has five electrons in the t_{2g} orbitals, so the total CFSE is $5 \times ^2/_5 \Delta_{oct} = 2\Delta_{oct}$. In the high-spin example, the CFSE is $(3 \times ^2/_5 \Delta_{oct}) - (2 \times ^3/_5 \Delta_{oct}) = 0$ - in this case, the stabilization generated by the electrons in the lower orbitals is canceled out by the destabilizing effect of the electrons in the upper orbitals.

Crystal Field stabilization is applicable to transition-metal complexes of all geometries. Indeed, the reason that many d^8 complexes are square-planar is the very large amount of crystal field stabilization that this geometry produces with this number of electrons.

Explaining the colours of Transition Metal Complexes

The bright colors exhibited by many coordination compounds can be explained by Crystal Field Theory. If the d-orbitals of such a complex have been split into two sets as described above, when the molecule absorbs a photon of visible light one or more electrons may momentarily jump from the lower energy d-orbitals to the higher energy ones to transiently create an excited state atom. The difference in energy between the atom in the ground state and in the excited state is equal to the energy of the absorbed photon, and related inversely to the wavelength of the light. Because only certain wavelengths (λ) of light are absorbed–those matching exactly the energy difference - the compounds appears the appropriate complementary color.

As explained above, because different ligands generate crystal fields of different strengths, different colours can be seen. For a given metal ion, weaker field ligands create a complex with a smaller Δ, which will absorb light of longer λ and thus lower frequency v. Conversely, stronger field ligands create a larger Δ, absorb light of shorter λ, and thus higher v. It is, though, rarely the case that the energy of the photon absorbed corresponds exactly to the size of the gap Δ; there are other things (such as electron-electron repulsion and Jahn-Teller effects) that also affect the energy difference between the ground and excited states.

Which Colours are Exhibited?

Fig. : Colour wheel.

This colour wheel demonstrates which colour a compound will appear if it only has one absorption in the visible spectrum. For example, if the compound absorbs red light, it will appear green.

λ absorbed versus colour observed

400 nm Violet absorbed, Green-yellow observed (λ 560 nm)

450 nm Blue absorbed, Yellow observed (λ 600 nm)

490 nm Blue-green absorbed, Red observed (λ 620 nm)

570 nm Yellow-green absorbed, Violet observed (λ 410 nm)

580 nm Yellow absorbed, Dark blue observed (λ 430 nm)

600 nm Orange absorbed, Blue observed (λ 450 nm)

650 nm Red absorbed, Green observed (λ 520 nm).

Covalent Bond Classification

The Covalent Bond Classification is also referred to as the LXZ notation. It was published by *M. L. H. Green* in the mid-1990s as a solution for the need to describe covalent compounds such as *organometallic* complexes in a way that is not prone to limitations resulting from the definition of *oxidation state*. Instead of simply assigning a charge to an atom in the molecule (*i.e.* the oxidation state), the covalent bond classification method analyzes the nature of the *ligands* surrounding the atom of interest, which is often a *transition metal*. According to this method, there are three basic types of interactions that allow for co-ordination of the ligand. The three types of interaction are classified according to whether the ligating group donates two, one, or zero electrons. These three classes of ligands are respectively given the symbols L, X and Z.

Types of Ligands

X-type ligands are those that donate one electron to the metal and accept one electron from the metal when using the neutral ligand method of *electron counting*, or donate two electrons to the metal when using the donor pair method of electron counting. Regardless of whether it is considered neutral or anionic, these ligands yield normal *covalent bonds*. A few examples of this type of ligand are H, halogens (Cl, Br, F, etc.), OH⁻, CN, CH_3, and NO (bent).

L-type ligands are neutral ligands that donate two electrons to the metal center regardless of the electron counting method being used. These electrons can come from *lone pairs*, pi or sigma donors.[4] The bonds formed between these ligands and the metal are *dative covalent bonds*, which are also known as co-ordinate bonds. Examples of this type of ligand include CO, PR_3, NH_3, H_2O, carbenes, and alkenes.

Z-type ligands are those that accept two electrons from the metal center as opposed to the donation occurring with the other two types of ligands. However, these ligands also form dative covalent bonds like the L-type. This type of ligand is not usually used, because in certain situations it can be written in terms of L and X. For example if a Z ligand is accompanied by an L type, it can be written as X_2. Examples of these ligands are *Lewis acids*, such as BR_3.

Uses of the Notation

When given a metal complex and the trends for the ligand types, the complex can be written in a more simplified manner with the form $[ML_lX_xZ_z]^{Q\pm}$. The subscripts represent the numbers of each ligand type present in that complex, M is the metal center and Q is the overall charge on the complex. Some examples of this overall notation are as follows :

$[Mn(CO)_6]+ -> [ML_6]^+$

$[Ir(CO)(PPh_3)_2(Cl)(NO)]^{2+} -> [ML_3X_2]^{2+}$

$[Fe(CO)_2(CN)_4]^{2-} -> [ML_2X_4]^{2-}$

Also from this general form, the values for electron count, oxidation state, *co-ordination number*, number of d-electrons,[5] *valence number* and the *ligand bond number* can be calculated.

Electron Count = N + x + 2l–Q

Where N is the group number of the metal.

Oxidation State (OS) = x + Q

Co-ordination Number (CN) = x + 1

Number of d-electrons (dn) = N-OS = N-

Valence Number (VN) = x + 2z

Ligand Bond Number (LBN) = 1 + x + z

Other Uses

This template of writing a metal complex also allows for better comparison of molecules with different charges. This can happen when the assignment is reduced to its "equivalent neutral class". The equivalent neutral class is the classification of the complex if the charge was localized on the ligand as opposed to the metal center.[2] In other words, the equivalent neutral class is the representation of the complex as though there was no charge.

Polydentate and Polyhapto Ligand Motifs and Nomenclature

Denticity

Denticity refers to the number of *atoms* in a single *ligand* that bind to a central atom in a *co-ordination complex*. In many cases, only one atom in the ligand binds to the metal, so the denticity equals one, and the ligand is said to be **monodentate** (sometimes called **unidentate**). Ligands with more than one bonded atom are called **polydentate** or **multi-dentate**. The word *denticity* is derived from *dentis*, the Latin word for tooth. The ligand is thought of as biting the metal at one or more linkage points. The denticity of a ligand is described with the Greek letter κ ('kappa'). For example, $κ^6$-*EDTA* describes an EDTA ligand that co-ordinates through 6 **non-contiguous** atoms.

Denticity is different from *hapticity* because hapticity refers exclusively to ligands where the co-ordinating atoms are **contiguous.** In these cases the η notation is used. *Bridging ligands* use the μ ('mu') notation.

Classes of Denticity

Polydentate ligands are *chelating agents* and classified by their denticity. Some atoms cannot form the maximum possible number of bonds a ligand could make. In that case one or more *binding sites* of the ligand are unused. Such sites can be used to form a bond with another *chemical species*.

- Bidentate (also called didentate) ligands bind with two atoms, an example being *ethylenediamine.*
- Tridentate ligands bind with three atoms, an example being *terpyridine.* Tridentate ligands usually bind *via* two kinds of connectivity, called "mer" and "fac." Cyclic tridentate ligands such as *TACN* and *9-ane-S3* bind in a facial manner.

Fig. : Structure of the pharmaceutical *Oxaliplatin*, which features
two different bidentate ligands.

- Tetradentate ligands bind with four atoms, an example being *triethylenetet-ramine* (abbreviated trien). Tetradentate ligands bind *via* three connectivities depending on their topology and the geometry of the metal center. For octahedral metals, the linear tetradentate trien can bind *via* three geometries. Tripodal tetradentate ligands, *e.g. tris(2-aminoethyl) amine*, are more constrained, and on octahedra leave two cis sites. Many naturally occurring *macrocyclic* ligands are tetradentative, an example being the *porphyrin* in *heme*.

- Pentadentate ligands bind with five atoms, an example being ethylenediaminetriacetic acid.

- Hexadentate ligands bind with six atoms, an example being *EDTA* (although it can bind in a tetradentate manner).

- Ligands of denticity greater than 6 are well known. The ligands *1,4,7,10-tetraazacyclododecane-1,4,7,10-tetraacetate* (DOTA) and *diethylene triamine pentaacetate* (DTPA) are octadentate. They are particularly useful for binding lanthanide ions, which typically have co-ordination numbers greater than 6.

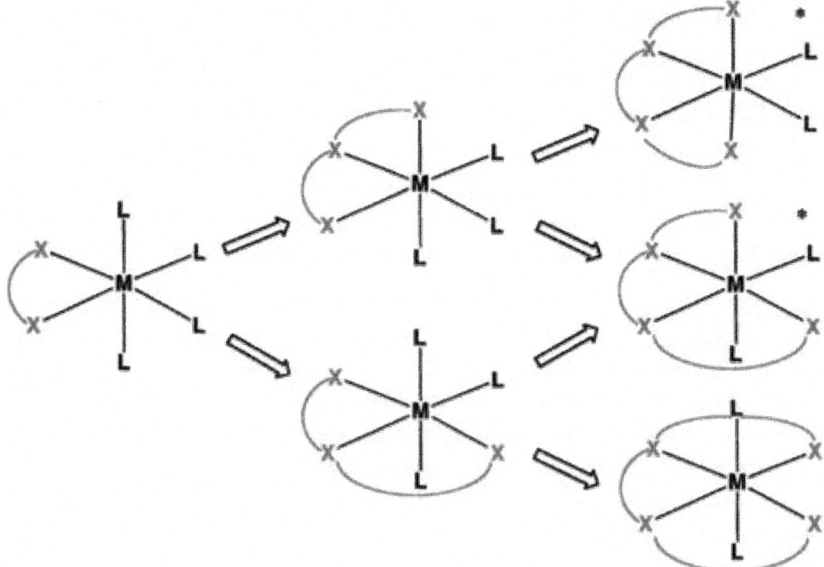

Fig. : Relationship between "linear" bi-, tri-and tetradentate ligands (red) bound to an octahedral metal center. The structures marked with * are chiral owing to the backbone of the tetradentate ligand.

Stability Constants

In general, the stability of a metal complex correlates with the denticity of the ligands. The stability of a complex is represented quantitatively in the form of *Stability constants*. Hexadentate ligands tend to bind metal ions more strongly than ligands of lower denticity.

Chelation

Chelation describes a particular way that *ions* and molecules bind metal ions. According to the *International Union of Pure and Applied Chemistry*, chelation involves the formation or presence of two or more separate *co-ordinate bonds* between a *polydentate* (multiple bonded) *ligand* and a single central atom. Usually these *ligands* are *organic compounds*, and are called chelants, chelators, chelating agents, or sequestering agents.

Chelate Effect

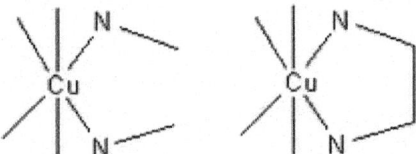

Fig. : *Ethylenediamine ligand,* binding to a central atom with two bonds.

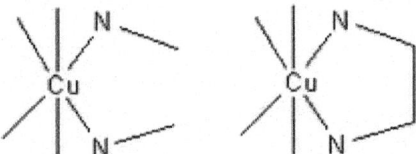

Fig. : Cu^{2+} *complexes* with *methylamine* (left) *and ethylenediamine.*

The chelate effect describes the enhanced affinity of chelating ligands for a metal ion compared to the affinity of a collection of similar nonchelating (monodentate) ligands for the same metal.

Consider the two equilibria, in aqueous solution, between the *copper (II) ion,* Cu^{2+} and *ethylenediamine* (en) on the one hand and *methylamine,* $MeNH_2$ on the other.

$$Cu^{2+} + en \rightleftharpoons [Cu(en)]^{2+} \quad (1)$$

$$Cu^{2+} + 2\ MeNH_2 \rightleftharpoons [Cu_2]^{2+} \quad (2)$$

In (1) the *bidentate* ligand ethylene diamine forms a chelate complex with the copper ion. Chelation results in the formation of a five–membered ring. In (2) the bidentate ligand is replaced by two *monodentate* methylamine ligands of approximately the same donor power, meaning that the *enthalpy* of formation of $Cu-N$ bonds is approximately the same in the two reactions. Under conditions of equal copper concentrations and when the concentration of methylamine is twice the concentration of ethylenediamine, the concentration of the complex (1) will be greater than the concentration of the complex (2). The effect increases with the number of chelate rings so the concentration of the *EDTA* complex, which has six chelate rings, is much much higher than a corresponding complex with two monodentate nitrogen donor ligands and four monodentate carboxylate ligands. Thus, the *phenomenon* of the chelate effect is a firmly established *empirical* fact.

The *thermodynamic* approach to explaining the chelate effect considers the *equilibrium constant* for the reaction : the larger the equilibrium constant, the higher the concentration of the complex.

$$[Cu(en)] = \beta_{11}[Cu][en]$$

$$[Cu_2] = \beta_{12}[Cu][MeNH_2]^2$$

Electrical charges have been omitted for simplicity of notation. The square brackets indicate concentration, and the subscripts to the *stability constants*, β, indicate the *stoichiometry* of the complex. When the *analytical concentration* of methylamine is twice that of ethylenediamine and the concentration of copper is the same in both reactions, the concentration $[Cu(en)]$ is much higher than the concentration $[Cu_2]$ because $\beta_{11} \gg \beta_{12}$.

An equilibrium constant, K, is related to the standard *Gibbs free energy*, $\Delta G^{\oplus}$ by

$$\Delta G^{\oplus} = -RT \ln K = \Delta H^{\oplus} - T\Delta S^{\oplus}$$

where R is the *gas constant* and T is the temperature in *Kelvin*. $\Delta H^{\oplus}$ is the standard *enthalpy* change of the reaction and $\Delta S^{\oplus}$ is the standard *entropy* change.

It has already been posited that the enthalpy term should be approximately the same for the two reactions. Therefore, the difference between the two stability constants is due to the entropy term. In equation (1) there are two particles on the left and one on the right, whereas in equation (2) there are three particles on the left and one on the right. This means that less *entropy of disorder* is lost when the chelate complex is formed than when the complex with monodentate ligands is formed. This is one of the factors contributing to the entropy difference. Other factors include solvation changes and ring formation. Some experimental data to illustrate the effect are shown in the following table.

Equilibrium	$\log \beta$	$\Delta G^{\oplus}$	$\Delta H^{\oplus}/kJ\ mol^{-1}$	$-T\Delta S^{\oplus}/kJ\ mol^{-1}$
$Cd^{2+} + 4\ MeNH_2 \rightleftharpoons Cd_4^{2+}$	6.55	−37.4	−57.3	19.9
$Cd^{2+} + 2\ en \rightleftharpoons Cd(en)_2^{2+}$	10.62	−60.67	−56.48	−4.19

These data show that the standard enthalpy changes are indeed approximately equal for the two reactions and that the main reason for the greater stability of the chelate complex is the entropy term, which is much less unfavourable. In general it is difficult to account precisely for thermodynamic values in terms of changes in solution at the molecular level, but it is clear that the chelate effect is predominantly an effect of entropy.

Other explanations, including that of Schwarzenbach, are discussed in Greenwood and Earnshaw (*loc.cit*).

In Nature

Virtually all biochemicals exhibit the ability to dissolve certain metal *cations*. Thus, *proteins, polysaccharides,* and polynucleic acids are excellent polydentate ligands for many metal ions. Organic compounds such as the amino acids *glutamic acid* and *histidine,* organic diacids such as *malate,* and polypeptides such as *phytochelatin* are also typical chelators. In addition to these adventitious chelators, several biomolecules are specifically produced to bind certain metals.

In Biochemistry and Microbiology

Virtually all metalloenzymes feature metals that are chelated, usually to peptides or co-factors and prosthetic groups. Such chelating agents include the *porphyrin* rings in *hemoglobin* and *chlorophyll.* Many microbial species produce water-soluble pigments that serve as chelating agents, termed *siderophores.* For example, species of *Pseudomonas* are known to secrete *pyocyanin* and *pyoverdin* that bind iron. *Enterobactin,* produced by *E. coli,* is the strongest chelating agent known.

In Geology

In earth science, chemical *weathering* is attributed to organic chelating agents, *e.g.* *peptides* and *sugars,* that extract metal ions from minerals and rocks. Most metal complexes in the environment and in nature are bound in some form of chelate ring, *e.g.* with a *humic acid* or a protein. Thus, metal chelates are relevant to the mobilization of *metals* in the *soil,* the uptake and the accumulation of *metals* into *plants* and *microorganisms.* Selective chelation of *heavy metals* is relevant to bioremediation, *e.g.* removal of *Cs* from radioactive waste.

Applications

Chelators are used in producing nutritional supplements, *fertilizers, chemical analysis,* as *water softeners,* commercial products such as *shampoos* and food *preservatives,* medicine, heavy metal detox, and industrial applications.

In 2010, the *Asia-Pacific* region was the largest outlet, generating about 45% of worldwide demand for chelating agents. The region was followed by Western Europe and North America. The global chelating agents market is expected to reach more than 5 million *tonnes* in 2018.

Nutritional Supplements

In the 1960s, scientists developed the concept of chelating a metal ion prior to feeding the element to the animal. They believed that this would create a neutral compound, protecting the mineral from being complexed with insoluble salts within the stomach, which would render the metal unavailable for absorption. Amino acids, being effective metal binders, were chosen as the prospective ligands, and research was conducted on the metal-amino acid combinations. The research supported that the metal-amino acid chelates were able to enhance mineral absorption.

During this period, synthetic chelates were also being developed. An example of such synthetics is ethylenediaminetetraacetic acid (EDTA). These synthetics applied the same concept of chelation and did create chelated compounds; however, these synthetics were too stable and not nutritionally viable. If the mineral was taken from the EDTA ligand, the ligand could not be used by the body and would be expelled. During the expulsion process the EDTA ligand will randomly chelate and strip another mineral from the body.

According to the Association of American Feed Control Officials, a metal amino acid chelate is defined as the product resulting from the reaction of a metal ion from a soluble metal salt with a mole ratio of one to three (preferably two) moles of amino acids. The average weight of the hydrolyzed amino acids must be approximately 150 and the resulting molecular weight of the chelate must not exceed 800 Da.

Since the early development of these compounds, much more research has been conducted, and has been applied to human nutrition products in a similar manner to the animal nutrition experiments that pioneered the technology. Ferrous bis-glycinate is an example of one of these compounds that has been developed for human nutrition.

Fertilizers

Metal chelate compounds are common components of fertilizers to provide micro-nutrients. These micro-nutrients (manganese, iron, zinc, copper) are required for the overall health of the plants. Most fertilizers contain phosphate salts that, in the absence of chelating agents, typically convert these metal ions into insoluble solids that are of no nutritional value to the plants. *EDTA* is the typical chelating agent for this purpose.

Heavy Metal Detoxification

Chelation therapy is the use of chelating agents to detoxify *poisonous* metal agents such as *mercury, arsenic,* and *lead* by converting them to a chemically inert form that can be excreted without further interaction with the body, and was approved by the *U.S. Food and Drug Administration* in 1991. In *alternative medicine,* chelation is used as a *treatment* for *autism,* although this practice is controversial due to its potentially deadly side-effects, the absence of scientific plausibility, and the lack of FDA approval.

Although they can be beneficial in cases of heavy metal poisoning, chelating agents can also be dangerous. Use of disodium EDTA instead of calcium EDTA has resulted in fatalities due to *hypocalcemia.*

Other Medical Applications

Chelation in the intestinal tract is a cause of numerous interactions between drugs and metal ions (also known as *"minerals"* in nutrition). As examples, *antibiotic drugs* of the *tetracycline* and *quinolone* families are chelators of Fe^{2+}, Ca^{2+} and Mg^{2+} ions.

EDTA is also used in *root canal treatment* as an intracanal irrigant. EDTA softens the dentin which may improve access to the entire canal length and is utilized as an irrigant to assist in the removal of the smear layer.

Chelate complexes of *gadolinium* are often used as *contrast agents* in *MRI scans*.

Chemical Applications

Homogeneous catalysts are often chelated complexes. A typical example is the *ruthenium (II) chloride* chelated with *BINAP* (a bidentate *phosphine*) used in *e.g.* *Noyori asymmetric hydrogenation* and asymmetric isomerization. The latter has the practical use of manufacture of synthetic *(–)-menthol*.

Citric acid is used to *soften water* in *soaps* and laundry *detergents*. A common synthetic chelator is *EDTA*. *Phosphonates* are also well-known chelating agents. Chelators are used in water treatment programs and specifically in *steam engineering, e.g., boiler water treatment system : Chelant Water Treatment system*.

Products such as Bio-Rust and Evapo-Rust are chelating agents sold for the removal of rust from iron and steel.

Hapticity

Hapticity is the *co-ordination* of a *ligand* to a metal center *via* an uninterrupted and contiguous series of *atoms*. The hapticity of a ligand is described with the Greek letter η. For example, η^2 describes a ligand that co-ordinates through 2 contiguous atoms. In general the η-notation only applies when multiple atoms are co-ordinated (otherwise the κ-*notation* is used). In addition, if the ligand co-ordinates through multiple atoms that are not contiguous then this is considered *denticity* (not hapticity), and the κ-notation is used once again. Lastly, *bridging ligands* are described with the μ ('mu') notation.

History

The need for additional nomenclature for organometallic compounds became apparent in the mid-1950s when Dunitz, *Orgel,* and Rich described the structure of the *"sandwich complex" ferrocene* by *X-ray crystallography* where an *iron* atom is *"sandwiched"* between two parallel *cyclopentadienyl* rings. *Cotton* later proposed the term *hapticity* derived from the adjectival prefix hapto (from the Greek *haptein,* to fasten, denoting contact or combination) placed before the name of the olefin, where the Greek letter η (eta) is used to denote the number of contiguous atoms of a ligand that bind to a metal center. The term is usually employed to refer to ligands containing extended π-systems or where *agostic bonding* is not obvious from the formula.

Historically Important Compounds where the Ligands are Described with Hapticity

- *Ferrocene*-bis(η^5-*cyclopentadienyl*)*iron*
- *Uranocene*-bis(η^8-1,3,5,7-*cyclooctatetraene*)*uranium*
- $W(CO)_{32}(\eta^2$-$H_2)$-the first compound to be synthesized with a *dihydrogen* ligand.
- $IrCl(CO)[P(C_6H_5)_3]_2$-the *dioxygen* derivative which forms reversibly upon oxygenation of *Vaska's complex*.

Examples

The η-notation is encountered in many co-ordination compounds :

- Side-on bonding of molecules containing σ-bonds like H_2 :
  - $W(CO)_{32}$
- Side-on bonded ligands containing multiple bonded atoms, *e.g. ethylene* in *Zeise's salt* or with *fullerene,* which is bonded through donation of the π-bonding electrons :
  - $K[PtCl_3(\eta^2$-$C_2H_4)]\cdot H_2O$
- Related complexes containing bridging π-ligands :
  - $(\mu$-η^2:η^2-$C_2H_2)Co_2(CO)_6$ and $(Cp^*_{2sm})_2(\mu$-η^2:η^2-$N_2)$
  - *Dioxygen* in bis{(trispyrazolylborato)copper (II)}$(\mu$-η^2:η^2-$O_2)$,

   Note that with some *bridging ligands,* an alternative bridging mode is observed, *e.g.* κ^1,κ^1, like in $(Me_3SiCH_2)_3V(\mu$-N_2-$\kappa^1(N),\kappa^1(N'))V(CH_2SiMe_3)_3$ contains a bridging dinitrogen molecule, where the molecule is end-on co-ordinated to the two metal centers.

- The bonding of π-bonded species can be extended over several atoms, *e.g.* in *allyl, butadiene* ligands, but also in *cyclopentadienyl* or *benzene* rings can share their electrons.
- Apparent violations of the *18-electron rule* sometimes are explicable in compounds with unusual hapticities :
  - The 18-VE complex $(\eta^5$-$C_5H_5)Fe(\eta^1$-$C_5H_5)(CO)_2$ contains one η^5 bonded cyclopentadienyl, and one η^1 bonded cyclopentadienyl.
  - *Reduction* of the 18-VE compound $[Ru(\eta^6$-$C_6Me_6)_2]^{2+}$ (where both aromatic rings are bonded in an η^6-co-ordination), results in another 18VE compound : $[Ru(\eta^6$-$C_6Me_6)(\eta^4$-$C_6Me_6)]$.
- Examples of polyhapto co-ordinated heterocyclic and inorganic rings : $Cr(\eta^5$-$C_4H_4S)(CO)_3$ contains the *sulfur* heterocycle *thiophene* and $Cr(\eta^6$-$B_3N_3Me_6)(CO)_3$ contains a co-ordinated inorganic ring (B_3N_3 ring).

Electrons Donated by "π-ligands" vs. Hapticity

Ligand	Electrons contributed (neutral counting)	Electrons contributed (ionic counting)
η^1 Allyl	1	2
η^3-Allyl cyclopropenyl	3	4
η^3-Allenyl	3	4
η^2-Butadiene	2	2
η^4-Butadiene	4	4
η^1-cyclopentadienyl	1	2
η^3-cyclopentadienyl	3	4
η^5-cyclopentadienyl pentadienyl cyclohexadienyl	5	6
η^2-Benzene	2	2
η^4-Benzene	4	4
η^6-Benzene	6	6
η^7-Cycloheptatrienyl	7	6
η^8-Cyclooctatetraenyl	8	10

Changes in Hapticity

The hapticity of a ligand can change in the course of a reaction. *E.g.* in a redox reaction :

$$[Ru(\eta^6\text{-}C_6H_6)_2]^{2+} \qquad Ru(\eta^6\text{-}C_6H_6)(\eta^4\text{-}C_6H_6)$$

Here one of the η^6-benzene rings changes to a η^4-benzene.

Similarly hapticity can change during a substitution reaction :

Here the η^5-cyclopentadienyl changes to an η^3-cyclopentadienyl, giving room on the metal for an extra 2-electron donating ligand 'L'. Removal of one molecule of CO and again donation of two more electrons by the cyclopentadienyl ligand restores the η^5-cyclopentadienyl. The so-called *indenyl effect* also describes changes in hapticity in a substitution reaction.

Hapticity vs. Denticity

Hapticity must be distinguished from *denticity*. Polydentate ligands co-ordinate *via* multiple co-ordination sites within the ligand. In this case the co-ordinating atoms are identified using the κ-notation, as for example seen in co-ordination of *1,2-bis(diphenylphosphino)ethane* ($Ph_2PCH_2CH_2PPh_2$), to *NiCl$_2$* as dichloro [ethane-1,2-diylbis(diphenylphosphane)-κ^{2}P] nickel (II). If the co-ordinating atoms are contiguous (connected to each other), the η-notation is used, as *e.g.* in *titanocene dichloride* : dichlorobis(η^5-2,4-cyclopentadien-1-yl)titanium.

Hapticity and Fluxionality

Molecules with polyhapto ligands are often "fluxional", also known as stereochemically non-rigid. Two classes of fluxionality are prevalent for organometallic complexes of polyhapto ligands :

- *Case 1, typically* : When the hapticity value is less than the number of sp^2 carbon atoms. In such situations, the metal will often migrate from carbon to carbon, maintaining the same net hapticity. The η^1-C_5H_5 ligand in (η^5-C_5H_5)Fe(η^1-C_5H_5)(CO)$_2$ re-arranges rapidly in solution such that Fe binds alternatingly to each carbon atom in the η^1-C_5H_5 ligand. This reaction is *degenerate* and, in the jargon of *organic chemistry*, it is an example of a *sigmatropic re-arrangement*.

- *Case 2, typically* : Complexes containing cyclic polyhapto ligands with maximized hapticity. Such ligands tend to rotate. A famous example is *ferrocene*, Fe(η^5-C_5H_5)$_2$, wherein the Cp rings rotate with a low energy barrier about the *principal axis* of the molecule that "skewers" each ring. This "ring whizzing" explains, inter alia, why only one isomer can be isolated for Fe(η^5-C_5H_4Br)$_2$. In this case, the rotamers are not necessarily degenerate, but the rotational barriers have low energies of activation.

Ligand Motifs

Outer-sphere Ligands

In co-ordination chemistry, the ligands that are directly bonded to the metal (that is, share electrons), form part of the *first co-ordination sphere* and are sometimes called "inner sphere" ligands. "*Outer-sphere*" ligands are not directly attached to the metal, but are bonded, generally weakly, to the first co-ordination shell, affecting the inner sphere in subtle ways. The complex of the metal with the inner sphere ligands is then called a co-ordination complex, which can be neutral, cationic, or *anionic*. The complex, along with its *counterions*, is called a *co-ordination compound*.

Trans-spanning Ligand

Trans-spanning ligands are bidentate *ligands* that can span opposite sites of a complex with square-planar geometry. A wide variety of ligands that chelate in the cis fashion already exist, but very few can link opposite vertices on a co-ordination polyhedron. Early attempts to generate trans-spanning bidentate ligands relied on *polymethylene* chains to link the donor functionalities, but such ligands often lead to *co-ordination polymers*.

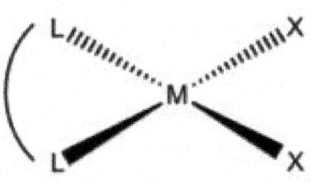

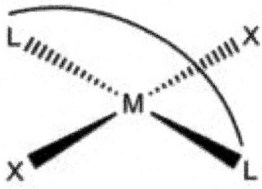

cis-chelating ligand   trans-chelating ligand

History

A diphosphane linked with *pentamethylene* was claimed to span across a square planar complex. This early attempt was followed by ligands with more rigid backbones. "TRANSPHOS" was the first trans-spanning diphosphane ligand that usually co-ordinates to palladium (II) and platinum (II) in a trans manner. TRANSPHOS features benzo [c] phenanthrene substituted by diphenylphosphinomethyl (Ph_2PCH_2) groups at the 1 and 11 positions. The polycyclic framework suffers sterically clashing hydrogen centers.

Xantphos, SPANphos, TRANSDIP and Related Ligands

Xantphos is a trans-spanning ligand, without the steric problems associated with TRANSPHOS. *SPANphos* is comparable to XANTPHOS but more reliably transspanning. TRANSDIP, based on a α-*cyclodextrin*, is the first ligand to give exclusively trans-spanned complexes, even with d^8 metal ion halides.

Linkage Isomerism

Linkage isomerism is the existence of *co-ordination compounds* that have the same composition differing with the connectivity of the metal to a *ligand*.

       Typical ligands that give rise to linkage *isomers* are :

- *thiocyanate*, SCN^-–*isothiocyanate*, NCS^-
- *selenocyanate*, $SeCN^-$–*isoselenocyanate*, $NCSe^-$
- *nitrite*, NO_2^-
- *sulfite*, SO_3^{2-}

       Examples of linkage isomers are violet-coloured $[(NH_3)_5Co\text{-}SCN]^{2+}$ and orange-coloured $[(NH_3)_5Co\text{-}NCS]^{2+}$. The isomerization of the S-bonded isomer to the N-bonded isomer occurs intra-molecularly. In the complex, *dichlorotetrakis (dimethyl sulfoxide) ruthenium (II)*, linkage isomerism of dimethyl sulfoxide ligands can be observed in the NMR spectrum due to the effect of S *vs.* O bonding on the methyl groups of DMSO. The proper notation for linkage isomerism is the kappa notation where the atom directly bonding to the metal is proceeded by the lowercase Greek letter kappa; κ. For example, NO_2^- is represented as nitrito-κ-N and nitrito-κ-O, replacing the old system of trivial names such as nitro and nitroso.

History

The first reported example of linkage isomerism had the formula $[Co(NH_3)_5(NO_2)]$ Cl_2. The *cationic* cobalt *complex* exists in two separable linkage isomers. In the yellow-coloured isomer, the nitro ligand is bound through nitrogen. In the red linkage isomer, the nitrito is bound through one oxygen atom. The O-bonded isomer is often written as $[Co(NH_3)_5(ONO)]^{2+}$. Although the existence of the isomers had been known since the late 1800s, only in 1907 was the structural difference explained. It was later shown that the red isomer converted to the yellow isomer upon UV-irradiation. In this particular example, the formation of the nitro isomer $(Co-NO_2)$ from the nitrito isomer $(Co-ONO)$ occurs through the re-arrangement of the molecular structure. Thus, no bonds are broken during isomerization.

Fig. : Structures of the two linkage isomers of $[Co(NH_3)_5(NO_2)]^{2+}$.

Bridging Ligand

Fig. : An example of a μ bridging ligand.

A **bridging ligand** is a *ligand* that connects two or more atoms, usually metal ions. The ligand may be atomic or polyatomic. Virtually all complex organic compounds can serve as bridging ligands, so the term is usually restricted to small ligands such as pseudohalides or to ligands that are specifically designed to link two metals.

In naming a complex wherein a single atom bridges two metals, the bridging ligand is preceded by the *Greek* character 'mu', μ, with a superscript number denoting the number of metals bound to the bridging ligand. μ is often denoted simply as μ. When describing co-ordination complexes care should be taken not to confuse μ with η, which relates to *hapticity*. Ligands that are not bridging, are called **terminal ligands**.

List of Bridging Inorganic Ligands

Virtually all ligands are known to bridge, with the exception of amines and ammonia. Common inorganic bridging ligands include most of the common anions.

Bridging Ligand	Name	Example
OH	hydroxide	$[Fe_2(OH)_2(H_2O)_8]$
O	oxide	$[Cr_2O_7]$
SH	hydrosulfido	$Cp_2Mo_2(SH)_2S_2$
NH_2	amido	$HgNH_2Cl$
N	nitride	$[Ir_3N(SO_4)_6(H_2O)_3]$
CO	carbonyl	$Fe_2(CO)_9$
Cl	Chloride	Nb_2Cl_{10}
H	Hydride	B_2H_6
CN	Cyanide	approx. $Fe_7(CN)_{18}$

Many simple organic ligands form strong bridges between metal centers. Many common examples include organic derivatives of the above inorganic ligands (R = alkyl, aryl) : OR, SR, NR_2, NR, PR_2 (phosphido, note the ambiguity with the preceding entry), PR (phosphinidino), and many more.

Bonding

For doubly bridging (μ-) ligands, two limiting representation are 4e and 2e bonding interactions. These cases are illustrated in main group chemistry by $[Me_2Al\mu\text{-}Cl]2$ and $[Me_2Al(\mu\text{-}Me]_2$. Complicating this analysis is the possibility of metal-metal bonding. Computational studies suggest that metal-metal bonding is absent in many compounds where the metals are separated by bridging ligands. For example, calculations suggest that $Fe_2(CO)_9$ lacks a Fe-Fe bond by virtue of a 3-center, 2-electron bond involving one of three bridging CO ligands.

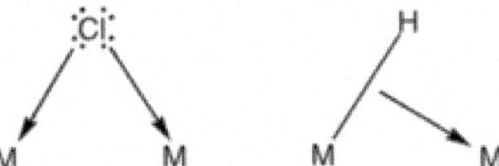

Fig. : Representations of two kinds of M-bridging ligand interactions, 3-center, 4 electron bond (left) and 3-center, 2 electron bonding.

Polyfunctional Ligands

Polyfunctional ligands can attach to metals in many ways and thus can bridge metals in diverse ways, including sharing of one atom or using several atoms.

Examples of such polyatomic ligands are the oxoanions CO_3 and the related *Carboxylate*, PO_4, and the *polyoxometallates*. Several organophosphorus ligands have been developed that bridge pairs of metals, a well-known example being $Ph_2PCH_2PPh_2$.

Metal–ligand Multiple Bond

In Chemistry, a **metal–ligand multiple bond** describes the interaction of certain *ligands* with a metal with a *bond order* greater than one. *Co-ordination complexes* featuring multiply bonded ligands are of both scholarly and practical interest. *Transition metal carbene complexes* catalyze the *olefin metathesis* reaction. Metal oxo intermediates are pervasive in oxidation catalysis. *oxygen evolving complex.*

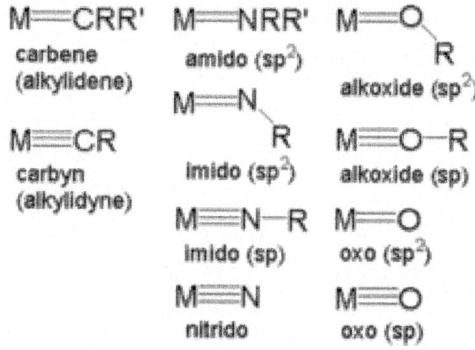

Fig. : Most common classes of complexes showing metal ligand multiple bonds.

As a cautionary note, the classification of a metal ligand bond as being "multiple" bond order is ambiguous and even arbitrary because *bond order* is a formalism. Furthermore, the usage of multiple bonding is not uniform. Symmetry arguments suggest that most ligands engage metals *via* multiple bonds. The term 'metal ligand multiple bond" is often reserved for ligands of the type CR_n and NR_n (n = 0, 1, 2) and OR_n (n = 0, 1) where R is H or an organic substituent, or pseudohalide. Historically, CO and NO are not included in this classification, nor are halides.

Pi-donor Ligands

In *co-ordination chemistry*, a pi-donor ligand is a kind of ligand endowed with filled non-bonding orbitals that overlap with metal-based orbitals. Their interaction is complementary to the behaviour of *pi-acceptor ligands*. The existence of terminal *oxo ligands* for the early transition metals is one consequence of this kind of bonding. Classic pi-donor ligands are oxide (O), nitride (N), imide (RN), alkoxide (RO), amide (R_2N), and fluoride. For late transition metals, strong pi-donors form anti-bonding interactions with the filled d-levels, with consequences for spin state, redox potentials, and ligand exchange rates. Pi-donor ligands are low in the *spectro-chemical series*.

Multiple Bond Stabilization

Metals bound to so-called *triply bonded carbyne, imide, nitride (nitrido)*, and *oxide (oxo)* ligands are generally assigned to high oxidation states with low d electron counts. The high *oxidation state* stabilizes the highly reduced ligands. The low *d electron count* allow for many bonds between ligands and the metal center. A d metal center can accommodate up to 9 bonds without violating the *18 electron rule*, whereas a d species can only accommodate 6 bonds.

Reactivity Explained Through Ligand Hybridization

A ligand described in ionic terms can bond to a metal through however many lone pairs it has available. For example many alkoxides use one of their three lone pairs to make a single bond to a metal center. In this situation the oxygen is sp *hybridized* according to *valence bond theory*. Increasing the bond order to two by involving another lone pair changes the hybridization at the oxygen to an sp center with an expected expansion in the M-O-R bond angle and contraction in the M-O bond length. If all three lone pairs are included for a bond order of three than the M-O bond distance contracts further and since the oxygen is a sp center the M-O-R bond angle is 180° or linear. Similarly with the imidos are commonly referred to as either bent (sp) or linear (sp). Even the oxo can be sp or sp hybridized. The triply bonded oxo, similar to *carbon monoxide*, is partially positive at the oxygen atom and unreactive towards *bronsted acids* at the oxygen atom. When such a complex is reduced, the triple bond can be converted to a *double bond* at which point the oxygen no longer bears a partial positive charge and is reactive towards acid.

Conventions

Bonding Representations

Imido ligands, also known as imides or nitrenes, most commonly form "linear six electron bonds" with metal centers. Bent imidos are a rarity limited by complexes electron count, orbital bonding availability, or some similar phenomenon. It is common to draw only two lines of bonding for all imidos, including the most common linear imidos with a six electron bonding interaction to the metal center. Similarly amido complexes are usually drawn with a single line even though most amido bonds involve four electrons. Alkoxides are generally drawn with a single bond although both two and four electron bonds are common. Oxo can be drawn with two lines regardless of whether four electrons or six are involved in the bond, although it is not uncommon to see six electron oxo bonds represented with three lines.

Representing Oxidation States

There are two motifs to indicate a metal oxidation state based around the actual charge separation of the metal center. Oxidation states up to +3 are believed to be an accurate representation of the charge separation experienced by the metal center. For oxidation states of +4 and larger, the oxidation state becomes more of a formalism with much of the positive charge distributed between the ligands. This distinction can be expressed by using a Roman numeral for the lower oxidation states in the upper right of the metal atomic symbol and an Arabic number with a plus sign for the higher oxidation states. This formalism is not rigorously followed and the use of Roman numerals to represent higher oxidation states is common.

$$[ML_n] \; vs. \; [O{=}ML_n]$$

Spectator Ligand

In *co-ordination chemistry*, a **spectator ligand** is a *ligand* that does not participate in chemical reactions of the complex. Instead, spectator ligands (vs "actor ligands") occupy co-ordination sites. Spectator ligands tend to be of *polydentate*, such that the M-spectator ensemble is inert kinetically. Although they do not participate in reactions of the metal, spectator ligands influence the reactivity of the metal center to which they are bound.

There are several different classes of ligand that can be considered spectator ligands. A few examples of common spectator ligands include *trispyrazolylborates* (Tp), *cyclopentadienyl* ligands (Cp), and many chelating diphosphines such as *1,2-bis(diphenylphosphino)ethane* ligands (dppe). Varying the substituents on the spectator ligands greatly influences the solubility, stability, electronic and steric properties of the metal complex.

Specialized Ligand Types

Bulky Ligands

Bulky ligands are used to control the steric properties of a metal center. They are used for many reasons, both practical and academic. On the practical side, they influence the selectivity of metal catalysts, *e.g.*, in *hydroformylation*. Of academic interest, bulky ligands stabilize unusual co-ordination sites, *e.g.*, reactive coligands or low co-ordination numbers. Often bulky ligands are employed to simulate the steric protection afforded by proteins to metal-containing active sites. Of course excessive steric bulk can prevent the co-ordination of certain ligands.

Chiral Ligand

In *chemistry* a **chiral ligand** is a specially adapted *ligand* used for *asymmetric synthesis*. This ligand is an *enantiopure organic compound* that combines with a *metal center* by *chelation* to form an *asymmetric catalyst*. This *catalyst* engages in a *chemical reaction* and transfers its chirality to the reaction product which as a result also becomes chiral. In an ideal reaction one equivalent of catalyst can *turn over* many more equivalents of reactant which enables the synthesis of a large amount of a chiral compound from achiral precursors with the aid of a very small (often expensive) chiral ligand.

First Discovery

The first such ligand, the *diphosphine DiPAMP* was developed in 1968 by *William S. Knowles* of *Monsanto Company*, who won the 2001 *Nobel Prize in Chemistry*, and ultimately used in the industrial production of *L-DOPA*.

Privileged Ligands

Many thousands of chiral ligands have been prepared and tested since then but only several compound classes have been found to have a general scope. These ligands are therefore called **privileged ligands**. Important members depicted below are *BINOL, BINAP, TADDOL, DIOP, BOX* and *DuPhos* (a *phosphine ligand*), all available as enantiomeric pairs.

Other members are *Salen, cinchona alkaloids* and *phosphoramidites*. Many of these ligands possess *C2 symmetry* which limits the number of possible reaction pathways and thereby increasing *enantioselectivity*.

Chiral Fence

Chiral ligands work *asymmetric induction* somewhere along the *reaction co-ordinate*. The image depicted on the right gives a general idea how a chiral ligand may induce an enantioselective reaction. The ligand (in green) has C2 symmetry with its nitrogen, oxygen or phosphorus atoms hugging a central metal atom (in red). In this particular ligand the right side is sticking out and its left side points away. The substrate in this reduction is *acetophenone* and the reagent (in blue) a *hydride* ion. In absence of the metal and the ligand the *re face* approach of the hydride ion gives the (S)-enantiomer and the *si face* approach the (R)-enantiomer in equal amounts (a racemic mixture like expected). The ligand/metal presence changes all that. The *carbonyl* group will co-ordinate with the metal and due to the *steric bulk* of the *phenyl* group it will only be able to do so with its si face exposed to the hydride ion with in the ideal situation exclusive formation of the (R) enantiomer. The re face will simply hit the **chiral fence**. Note that when the ligand is replaced by its mirror image the other enantiomer will form and that a racemic mixture of ligand will once again yield a racemic product. Also note that if the steric bulk of both carbonyl substituents is very similar the strategy will fail.

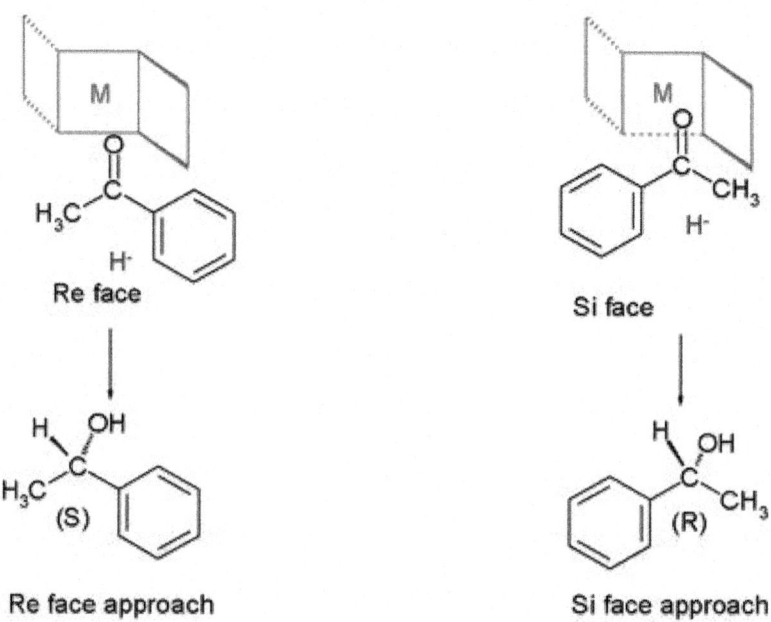

Re face   Si face

Re face approach   Si face approach

Chiral Counterions

So-called **chiral ions** team up with traditional cationic catalysts in asymmetric synthesis as demonstrated in this *allene* hydroxyalkoxylation in which the active catalyst is a salt of gold (I) and a *phosphate* of a chiral *binaphthol* :

Hemilability

In *co-ordination chemistry* and *catalysis* **hemilability** (*hemi*-half, *lability*-a suscepti-bility to change) refers to a property of many *polydentate ligands* which contain at least two electronically different co-ordinating groups, such as *hard and soft donors*. These **hybrid** or **heteroditopic** ligands form *complexes* where one co-ordinating group is *easily displaced* from the metal centre while the other group remains firmly bound; a behaviour which has been found to increase the reactivity of *catalysts* when compared to the use of more traditional ligands.

Overview

In general, catalytic cycles can be divided into 3 stages :

1. Co-ordination of the starting material(s)
2. Catalytic transformation of the starting material(s) to the product(s)
3. Displacement of the product(s) to regain the catalyst (or *pre-catalyst*).

Traditionally the focus of catalytic research has been on the reaction taking place in the second stage, however there will be energy changes associated with the beginning and end steps due to their effect on the *co-ordination sphere* and *geometry* of the complex, as well as its *oxidation number* in cases of *oxidative addition and reductive elimination*. When these energy changes are large they can dictate the *turn-over rate* of the catalyst and hence its effectiveness.

Hemilabile ligands reduce the *activation energy* of these changes by readily undergoing partial and reversible displacement from the metal centre. Hence a co-ordinately saturated hemilabile complex will readily re-organise to allow the co-ordination of reagents but will also promote the ejection of products due to re-co-ordination of the labile section of the ligand. The low energy barrier between the fully and hemi co-ordinated states results in frequent inverconvertion between the two, which promotes a fast catalytic turn-over rate.

Examples

- The oxidative addition of MeI to Ir (I) complexes was shown to proceed about 100 times faster with a hemilabile *phosphane* ligand compared to a very similar non-labile ligand.

- *Hydrovinylation (olefin dimerisation)*, which is typically difficult to carry out *enantioselectively*, has been shown to proceed with high *enantiomeric excess* when using a *chiral phosphine* ligand with an appropriately placed hemilabile co-ordinating group. The *Pauson-Khand reaction*, which is conceptually similar, has also been shown to give improved results when hemilabile P,S type hybrid ligands were used.

- Iridium (I) complexes incorporating hemilabile ligands which contain *methoxy*, *dimethylamino*, and *pyridine* as donor functions have been shown to be effective catalysts for *transfer hydrogenation*.

Non-innocent Ligand

In *chemistry*, a **non-innocent ligand** is a *ligand* in a *metal complex* where the oxidation state is unclear. Typically, complexes containing non-innocent ligands are *redox* active at mild *potentials*. The concept assumes that redox reactions in *metal complexes* are either metal or ligand localized, which is a simplification, albeit a useful one.

Redox Reactions of Complexes Containing Innocent vs. Non-innocent Ligands

Conventionally, redox reactions of co-ordination complexes are assumed to be metal-centered. The reduction of MnO_4 to MnO_4 is described by the change in oxidation state of *manganese* from 7+ to 6+. The *oxide* ligands do not change in oxidation state, remaining 2-(a more careful examination of the electronic structure of the redox partners reveals however that the oxide ligands are affected by the redox change). Oxide is an innocent ligand. Another example of conventional metal-centered *redox couple* is $[Co(NH_3)_6]/[Co(NH_3)_6]$. Ammonia is innocent in this transformation.

A clear example of redox non-innocent behaviour of ligands is observed for $[Ni(S_2C_2Ph_2)_2]$, which exists in three *oxidation states* : z = 2-, 1-, and 0. If the ligands are always considered to be dianionic (as is done in formal oxidation state counting), then z = 0 requires that that nickel has a formal oxidation state of +IV. The formal oxidation state of the central nickel atom therefore, ranges from +II to +IV in the above transformations. However, the formal oxidation state is different from the real (spectroscopic) oxidation state based on the (spectroscopic) metal d-electron configuration. The stilbene-1,2-dithiolate behaves as a redox non-innocent ligand, and the oxidation processes actually take place at the ligands rather than the metal. This leads to the formation of ligand radical complexes. The charge-neutral complex (z =0) is therefore best described as a Ni derivative of $S_2C_2Ph_2$. The *diamagnetism* of this complex arises from anti-ferromagnetic coupling between the unpaired electrons of the two ligand radicals.

The complex Cr(2,2'-bipyridine)$_3$ is a derivative of Cr (III) bound to three radical anions of 2,2'bipyridine, which is in this case also behaving as a redox non-innocent ligand. On the other hand, one-electron oxidation of [Ru(2,2'-bipyridine)$_3$] is localized on Ru and the bipyridine is behaving as a normal, innocent ligand in this case.

History

C.K. Jørgenson (Cologny-Geneva) described ligands as "innocent" and "suspect" : "Ligands are innocent when they allow oxidation states of the central atoms to be defined. The simplest case of a suspect ligand is *NO*..."

Redox non-innocent ligands have been intensively investigated spectro-scopically by the groups of K. Wieghardt (MPI Mülheim a/d Ruhr) and W. Kaim (Stuttgart) over the past years. Quite recently it became obvious that redox non-innocent ligands are not just a spectroscopic curiosity, as the radical reactivity of redox non-innocent ligands was demonstrated to play a crucial role in the mechanism of bio-catalytic processes mediated by several metallo-enzymes (*e.g.* Gallactose Oxidase, Cytochrome P450, methane mono-oxygenase). More recently, some synthetic research groups have started to systematically investigate the (catalytic) reactivity of transition metal complexes with redox non-innocent ligands in organometallic chemistry.

Typical Ligands that often Behave as Redox Non-innocent Ligands

Fig. : The pyridine-2,6-diimine ligand can be easily reduced by one or two electrons.

- O_2 *and NO* : Ligands with extended pi-delocalization such as *porphyrins* and *phthalocyanines*, ligands with the generalised formulas [D-CR=CR-D] or D=CR-CR=D (D = O, S, NR' and R, R' = *alkyl* or *aryl*), and similar related systems are often non-innocent. For example :
- Dioxalenes, such as *catecholates*.

- *Dithiolenes,* such as 1,2-maleonitriledithiolate
- Diimines such as derivatives of 1,2-diaminobenzene, *α-diimines,* and *dimethylglyoxime.*
- Pyridine-2,6-diimine ligands (relevant in polymerisation and hydrogenation catalysis).

Redox Non-innocent Ligands in Organometallic Chemistry and Catalysis

In paramagnetic organometallic complexes of Rh and Ir (metallo-radicals), ethene ligands, amido ligands, and (reactive) carbene ligands are sometimes also behaving as 'redox non-innocent' ligands :

- Solvent co-ordination to some metallo-radical Ir(ethene) species transfers the spin-density from the metal to the **redox non-innocent ethene** ligand, after which direct radical coupling reactions with the olefinic ligand radical become possible.
- Oxidation of certain Rh-amido and Ir-amido complexes does not lead to the expected M-amido species. Instead the unexpected **M-aminyl radical complexes** are formed.
- Carbene formation from diazo compounds at metallo-radical Ir species unexpectedly leads to formation of '**carbene radicals**'. This is a result of the redox non-innocent character of Fischer-type carbenes, where one-electron reduction of the carbene ligand by Ir leads to formation of carbon centered 'carbene radicals' co-ordinated to Ir. These 'carbene radicals' reveal interesting radical-type reactivities.
- The application of redox non-innocent ligands in homogeneous catalysis was recently reviewed.

Redox Non-innocent Ligands in Biology

Metalloenzymes often feature non-innocent ligands. A common non-innocent ligand is found in metalloporphyrins. In the enzyme *cytochrome P450,* the porphyrin ligand sustains oxidation during the catalytic cycle. In other *heme* proteins, such as *myoglobin,* ligand-centered redox does not occur and the porphyrin is innocent.

Galactose Oxidase provides a seminal example for the involvement of reactive non-innocent ligands in bio-catalytic turnover. GOase converts chemo-selectively primary alcohols with O_2 into aldehydes and H_2O_2, with impressive turnover frequencies. The active site of the enzyme GOase contains a tyrosinyl radical which is co-ordinated to a Cu ion. In the key steps of the catalytic cycle, a co-operative Brønsted-basic ligand-site deprotonates the alcohol, and subsequently the oxygen atom of the tyrosinyl radical abstracts a hydrogen atom from the alpha-CH functionality of the co-ordinated alcoholate substrate. Thus, the tyrosinyl radical is a reactive fragment in the catalytic cycle which cooperates with the Cu site. This is essential for the function of the enzyme, because the Cu-ion is only capable of one-electron transformations. It is the interplay of the 1e reactivity of the ligand radical and the 1e reactivity of the metal which makes the overall process possible. The radical abstraction nature of the process makes the process extremely fast. Anti-ferromagnetic coupling between the unpaired spins of the tyrosine radical ligand and the d Cu ion (open-shell singlet ground state) explains the observed diamagnetic nature of the resting state of the enzyme, as was confirmed by synthetic model studies.

The oxygen molecule in *oxyhemoglobin* (or oxymyoglobin) would appear to satisfy the definition of a non-innocent ligand. Deoxyhemoglobin is ferrous and penta-co-ordinate, the (innocent) ligands being four N of the porphyrin and Nε of the proximal histidine. An O2 molecule binds the sixth co-ordination position.

There is evidence from a number of lines that partial electron transfer from the Fe to O_2 occurs, so that the complex is better described as superoxide anion bound to ferric heme, although spin coupling makes the complex diamagnetic. The change in oxidation and spin state of the Fe results in a change of bond length to the five innocent ligands which results in the heme switching from a "domed" to a planar conformation, which in turn drives conformational changes in the protein responsible for the cooperativity of O2 binding. This cooperativity is essential for efficient oxygen transport, so in a way we all owe our lives to the suspect nature of the O_2 ligand!

Common Ligands

Virtually every molecule and every ion can serve as a ligand for (or "co-ordinate to") metals. Monodentate ligands include virtually all anions and all simple Lewis bases. Thus, the *halides* and *pseudohalides* are important anionic ligands whereas *ammonia, carbon monoxide,* and *water* are particularly common charge-neutral ligands. Simple organic species are also very common, be they anionic (*RO* and *RCO$_2$*) or neutral (*R_2O, R_2S, $R_{3-x}NH_x$,* and *R_3P*). The steric properties of some ligands are evaluated in terms of their *cone angles*.

Beyond the classical Lewis bases and anions, all unsaturated molecules are also ligands, utilizing their π-electrons in forming the co-ordinate bond. Also, metals can bind to the σ bonds in for example *silanes, hydrocarbons,* and *dihydrogen.*

In complexes of *non-innocent ligands,* the ligand is bonded to metals *via* conventional bonds, but the ligand is also redox-active.

Examples of Common Ligands (by Field Strength)

In the following table the ligands are sorted by field strength (weak field ligands first) :

Ligand	Formula (bonding atom(s) in bold)	Charge	Most common denticity	Remark(s)
Iodide (iodo)	I	monoanionic	*monodentate*	
Bromide (bromido)	Br	monoanionic	monodentate	
Sulfide (thio or less commonly "bridging thiolate")	S	dianionic	monodentate (M=S), or bidentate bridging (M-S-M')	
Thiocyanate (S-thiocyanato)	S-CN	monoanionic	monodentate	ambidentate
Chloride (chlorido)	Cl	monoanionic	monodentate	also found bridging
Nitrate (nitrato)	O-NO$_2$	monoanionic	monodentate	
Azide	N-N$_2$	monoanionic	monodentate	
Fluoride (fluoro)	F	monoanionic	monodentate	

(Contd...)

(*Contd...*)

Ligand	Formula (bonding atom(s) in bold)	Charge	Most common denticity	Remark(s)
Hydroxide (hydroxo)	O-H	monoanionic	monodentate	often found as a bridging ligand
Oxalate (oxalato)	[O-C(=O)-C(=O)-O]	dianionic	bidentate	
Water (aqua)	H-O-H	neutral	monodentate	monodentate
Nitrite (nitrito)	O-N-O	monoanionic	monodentate	ambidentate
Isothiocyanate (isothiocy-anato)	N=C=S	monoanionic	monodentate	ambidentate
Acetonitrile (acetonitrilo)	CH$_3$CN	neutral	monodentate	
Pyridine	C$_5$H$_5$N	neutral	monodentate	
Ammonia (ammine or less commonly "ammino")	NH$_3$	neutral	monodentate	
Ethylenediamine	en	neutral	bidentate	
2,2'-Bipyridine	bipy	neutral	bidentate	easily reduced to its (radical) anion or even to its dianion
1,10-Phenanthroline	phen	neutral	bidentate	
Nitrite	N-O$_2$	monoanionic	monodentate	ambidentate
Triphenylphosphine	PPh$_3$	neutral	monodentate	
Cyanide	CN	monoanionic	monodentate	can bridge be-tween metals (both metals bound to C, or one to C and one to N)
Carbon monoxide (carbonyl)	CO	neutral	monodentate	can bridge be-tween metals (both metals bound to C)

Note : The entries in the table are sorted by field strength, binding through the stated atom (*i.e.* as a terminal ligand), the 'strength'' of the ligand changes when the ligand binds in an alternative binding mode (*e.g.*, when it bridges between metals) or when the conformation of the ligand gets distorted (*e.g.*, a linear ligand that is forced through steric interactions to bind in a non-linear fashion).

Other General Encountered Ligands (Alphabetical)

In this table other common ligands are listed in alphabetical order.

Ligand	formula (bonding atom(s) in bold)	Charge	Most common denticity	Remark(s)
Acetylacetonate (Acac)	CH$_3$-C(O)-CH$_2$-C(O)-CH$_3$	monoanionic	bidentate	In general bidentate, bound through both oxygens, but sometimes bound through the central carbon only
Alkenes	R$_2$C=CR$_2$	neutral		compounds with a C-C double bond
Benzene	C$_6$H$_6$	neutral		and other arenes
1,2-Bis (diphe-nylphosphino) ethane (dppe)	Ph$_2$PC$_2$H$_4$PPh$_2$	neutral	bidentate	
1,1-Bis (diphe-nylphosphino) meth-ane (dppm)	C$_{25}$H$_{22}$P$_2$	neutral		Can bond to 2 metal atoms at once, forming dimers
Corroles			tetradentate	
Crown ethers		neutral		primarily for alkali and alka-line earth metal cations
2,2,2-crypt			hexadentate	primarily for alkali and alka-line earth metal cations
Cryptates		neutral		
Cyclopentadienyl (Cp)	[C$_5$H$_5$]	monoanionic		Although monoanionic, by the nature of its occupied MO's, it is capable of acting as a tri-dentate ligand.
Diethylenetriamine (dien)	C$_4$H$_{13}$N$_3$	neutral	tridentate	related to TACN, but not con-strained to facial complexation
Dimethylglyoximate (dmgH)		monoanionic		
Ethylenediaminetet-raacetate (EDTA)	(HOOC-CH$_2$)$_2$N-(CH$_2$)$_2$-N(CH$_2$-COOH)$_2$	tetra-anionic	hexadentate	actual ligand is the tetra-anion
Ethylenediamin-etriacetate		trianionic	pentaden-tate	actual ligand is the trianion

(Contd...)

(*Contd...*)

Ligand	formula (bonding atom(s) in bold)	Charge	Most common denticity	Remark(s)
Ethyleneglycol-bis (oxyethylenenitrilo)-tetraacetate (EGTA)	(H**OO**C-CH$_2$)$_2$N-(CH$_2$)$_2$-**O**-(CH$_2$)$_2$-**O**-(CH$_2$)$_2$-N(CH$_2$-C**OO**H)$_2$	tetra-anionic	octodentate	
glycinate (Glycinato)		monoanionic	bidentate	other α-amino acid anions are comparable (but chiral)
Heme		dianionic	tetradentate	macrocyclic ligand
Nitrosyl	N**O**	cationic		bent (1e) and linear (3e) bonding mode
Oxo	**O**	dianion	monodentate	sometimes bridging
Pyrazine	**N**$_2$C$_4$H$_4$	neutral	ditopic	sometimes bridging
Scorpionate ligand			tridentate	
Sulfite		monoanionic	monodentate	ambidentate
2,2′;6′,2″-Terpyridine		neutral	tridentate	meridional bonding only
Triazacyclononane (tacn)	(C$_2$H$_4$)$_3$(**N**R)$_3$	neutral	tridentate	macrocyclic ligand
Tricyclohexylphosphine	$_3$**P** or (**P**Cy$_3$)	neutral	monodentate	
Triethylenetetramine		neutral	tetradentate	
Trimethylphosphine	**P**Me$_3$	neutral	monodentate	
Tri(*o*-tolyl)phosphine	**P**(*o*-tolyl)$_3$	neutral	monodentate	
Tris(2-aminoethyl) amine (tren)	(**N**H$_2$CH$_2$CH$_2$)$_3$**N**	neutral	tetradentate	
Tris(2-diphenylphosphineethyl) amine (np$_3$)		neutral	tetradentate	
Terpyridine	C$_{15}$H$_{11}$N$_3$	neutral	tridentate	
Tropylium	C$_7$H$_7$	cationic		
Carbon dioxide	CO$_2$			

LIGAND EXCHANGE

A **Ligand exchange** (also **ligand substitution**) is a type of *chemical reaction* in which a ligand in a compound is replaced by another. One type of pathway for substitution is the *ligand dependent pathway*. In organometallic chemistry this can take place *via associative substitution* or by *dissociative substitution*. Another form of ligand exchange is seen in the *nucleophilic abstraction* reaction.

LIGAND-PROTEIN BINDING DATABASE

BioLiP is a comprehensive ligand-protein interaction database, with the 3D structure of the ligand-protein interactions taken from the *Protein Data Bank*.

Chapter 16

FULLERENE CHEMISTRY

Fullerene chemistry is a field of *organic chemistry* devoted to the chemical properties of *fullerenes*. Research in this field is driven by the need to functionalize fullerenes and tune their properties. For example, fullerene is notoriously insoluble and adding a suitable group can enhance solubility. By adding a polymerizable group, a fullerene polymer can be obtained. Functionalized fullerenes are divided into two classes : *exohedral fullerenes* with substituents outside the cage and *endohedral fullerenes* with trapped molecules inside the cage.

CHEMICAL PROPERTIES OF FULLERENES

Fullerene or C_{60} is *soccer-ball-shaped* or I_h with 12 pentagons and 20 hexagons. According to *Euler's theorem* these 12 pentagons are required for closure of the carbon network consisting of n hexagons and C_{60} is the first stable fullerene because it is the smallest possible to obey this rule. In this structure none of the pentagons make contact with each other. Both C_{60} and its relative C_{70} obey this so-called **isolated pentagon rule (IPR)**. The next homologue C_{84} has 24 IPR isomers of which several are isolated and another 51,568 non-IPR isomers. Non-IPR fullerenes have thus far only been isolated as endohedral fullerenes such as $Tb_3N@C_{84}$ with two fused pentagons at the apex of an egg-shaped cage. or as fullerenes with exohedral stabilization such as $C_{50}Cl_{10}$ and reportedly $C_{60}H_8$.

Because of the molecule's spherical shape the carbon atoms are highly *pyramidalized*, which has far-reaching consequences for reactivity. It is estimated that *strain energy* constitutes 80% of the *heat of formation*. The conjugated carbon atoms respond to deviation from planarity by *orbital rehybridization* of the sp^2 orbitals and *pi orbitals* to a sp orbital with a gain in p-character. The p lobes extend further outside the surface than they do into the interior of the sphere and this is one of the reasons a fullerene is *electronegative*. The other reason is that the empty low-lying pi orbitals also have a high s character.

The double bonds in fullerene are not all the same. Two groups can be identified : 30 so-called [6,6] double bonds connect two hexagons and 60 [5,6] bonds connect a hexagon and a pentagon. Of the two the [6,6] bonds are shorter with more double-bond character and therefore, a hexagon is often represented as a *cyclohexatriene* and a pentagon as a pentalene or [5] *radialene*. In other words, although the carbon atoms in fullerene are all conjugated the superstructure is not a super *aromatic compound*. The *X-ray diffraction bond length* values are 135.5 *pm* for the [6,6] bond and 146.7 pm for the [5,6] bond.

C_{60} fullerene has 60 pi electrons but a *closed shell configuration* requires 72 electrons. The fullerene is able to acquire the missing electrons by reaction with *potassium* to form first the K_6C_6-60 salt and then the $K_{12}C_{12}-60$ In this compound the bond length alternation observed in the parent molecule has vanished.

FULLERENE REACTIONS

Fullerenes tend to react as electrophiles. An additional driving force is relief of *strain* when double bonds become saturated. Key in this type of reaction is the level of functionalization *i.e.* monoaddition or multiple additions and in case of multiple additions their topological relationships (new substituents huddled together or evenly spaced). In conformity with *IUPAC* rules, the terms *methanofullerene* are used to indicate the ring-closed (*cyclopropane*) *fullerene* derivatives, and fulleroid to ring-open (*methanoannulene*) structures.

Nucleophilic Addition

Fullerenes react as *electrophiles* with a host of nucleophiles in *nucleophilic additions*. The intermediary formed *carbanion* is captured by another electrophile. Examples of nucleophiles are *Grignard reagents* and *organolithium reagents*. For example the reaction of C_{60} with *methylmagnesium chloride* stops quantitatively at the penta-adduct with the methyl groups centered around a cyclopentadienyl anion which is subsequently protonated. Another nucleophilic reaction is the *Bingel reaction*. Fullerene reacts with *chlorobenzene* and *aluminium chloride* in a *Friedel-Crafts alkylation* type reaction. In this hydroarylation the reaction product is the 1,2-addition adduct (Ar-CC-H).

Pericyclic Reactions

The [6,6] bonds of fullerenes react as dienes or dienophiles in *cycloadditions* for instance *Diels-Alder reactions*. 4-membered rings can be obtained by [2+2]cycloadditions for instance with *benzyne*. An example of a *1,3-dipolar cycloaddition* to a 5-membered ring is the *Prato reaction*.

Hydrogenation

Fullerenes are easily hydrogenated by several methods. Examples of hydrofullerenes are $C_{60}H_{18}$ and $C_{60}H_{36}$. However, completely hydrogenated $C_{60}H_{60}$ is only hypothetical because of large strain. Highly hydrogenated fullerenes are not stable, as prolonged hydrogenation of fullerenes by direct reaction with hydrogen gas at high temperature conditions results in cage fragmentation. At the final reaction stage this causes collapse of cage structure with formation of polycyclic aromatic hydrocarbons.

Oxidation

Although more difficult than reduction, oxidation of fullerene is possible for instance with oxygen and *osmium tetraoxide*.

Hydroxylation

Fullerenes can be hydroxylated to *fullerenols* or *fullerols*. Water solubility depends on the total number of hydroxyl groups that can be attached. One method is fullerene reaction in diluted *sulfuric acid* and *potassium nitrate* to $C_{60}(OH)_{15}$. Another method is reaction in diluted *sodium hydroxide* catalysed by *TBAH* adding 24 to 26 hydroxyl groups. Hydroxylation has also been reported using solvent-free NaOH/*hydrogen peroxide*. $C_{60}(OH)_8$ was prepared using a multi-step procedure starting from a mixed peroxide fullerene. The maximum number of *hydroxyl* groups that can be attached (hydrogen peroxide method) stands at 36–40.

Electrophilic Addition

Fullerenes react in *electrophilic additions* as well. The reaction with *bromine* can add up to 24 bromine atoms to the sphere. The record holder for fluorine addition is $C_{60}F_{48}$. According to *in silico* predictions the as yet elusive $C_{60}F_{60}$ may have some of the fluorine atoms in endo positions (pointing inwards) and may resemble a tube more than it does a sphere.

Retro Additions

Protocols have been investigated for removing substituents *via* retroadditions after they have served their purpose. Examples are the *retro-Bingel reaction* and the *retro-Prato reaction*.

Carbene Additions

Fullerenes react with *carbenes* to methano-fullerenes. The reaction of fullerene with *dichlorocarbene* (created by *sodium trichloroacetate* pyrolysis) was first reported in 1993. A single addition takes place along a [6,6] bond.

Radical Additions

Fullerenes can be considered *radical scavengers*. With a simple hydrocarbon radical such as the *t-butyl* radical obtained by *thermolysis* or *photolysis* from a suitable precursor the tBuC60 radical is formed that can be studied. The unpaired electron does not delocalize over the entire sphere but takes up positions in the vicinity of the tBu substituent.

FULLERENES AS LIGANDS

Fullerene is a *ligand* in *organometallic chemistry*. The [6,6] double bond is electron-deficient and usually forms metallic bonds with η = 2 *hapticity*. Bonding modes such as η = 5 or η = 6 can be induced by modification of the *co-ordination sphere*.

- C_{60} fullerene reacts with *tungsten hexacarbonyl* $W(CO)_6$ to the $(\eta^2\text{-}C_{60})W(CO)_5$ complex in a *hexane* solution in direct sunlight.

MULTI-STEP FULLERENE SYNTHESIS

Although the procedure for the synthesis of the C_{60} fullerene is well established (generation of a large current between two nearby graphite electrodes in an inert atmosphere) a 2002 study described an *organic synthesis* of the compound starting from simple organic compounds.

FVP 1100°C 0.01 mm

0.1 - 1%

In the final step a large *polycyclic aromatic hydrocarbon* consisting of 13 hexagons and three pentagons was submitted to *flash vacuum pyrolysis* at 1100°C and 0.01 *Torr*. The three *carbon chlorine bonds* served as *free radical* incubators and the ball was stitched up in a no-doubt complex series of *radical reactions*. The *chemical yield* was low : 0.1 to 1%. A small percentage of fullerenes is formed in any process which involves burning of hydrocarbons, *e.g.* in candle burning. The yield through a combustion method is often above 1%. The method proposed above does not provide any advantage for synthesis of fullerenes compared to the usual combustion method, and therefore, the organic synthesis of fullerenes remains a challenge for chemistry.

A similar exercise aimed at construction of a C78 cage in 2008 (but leaving out the precursor's halogens) did not result in a sufficient yield but at least the introduction of *Stone Wales defects* could be ruled out. C60 synthesis through a fluorinated fullerene precursor was reported in 2013

MULTI-STEP NANORIBBON SYNTHESIS

In the field of *graphene nanoribbons* a bottom-up approach has also been investigated

OPEN-CAGE FULLERENES

A part of fullerene research is devoted to so-called **open-cage fullerenes** whereby one or more bonds are removed chemically exposing an orifice. In this way it is possible to insert into it small molecules such as hydrogen, helium or lithium. The first such open-cage fullerene was reported in 1995. In *endohedral hydrogen fullerenes* the opening, hydrogen insertion and closing back up has already been demonstrated.

HETERO-FULLERENES

In **hetero-fullerenes** at least one carbon atom is replaced by another element. Based on *spectroscopy*, substitutions have been reported with *boron* (**bora-fullerenes**), *nitrogen* (**aza-fullerenes**), *oxygen, arsenic, germanium, phosphorus, silicon, iron, copper, nickel, rhodium* and *iridium*. Reports on isolated hetero-fullerenes are limited to those based on nitrogen and oxygen.

FULLERENE DIMERS

The C_{60} fullerene dimerizes in a formal [2+2] *cycloaddition* to a C_{120} bucky dumbbell in the solid state by *mechano-chemistry* (high-speed vibration milling) with *potassium cyanide* as a catalyst. The trimer has also been reported using *4-aminopyridine* as catalyst (4% yield) and observed with *scanning tunneling microscopy* as a *monolayer*.

NANOTUBE CHEMISTRY

Carbon nanotubes, also part of the fullerene family, can be described as *graphene* sheets rolled into a cylindrical tube. Unlike the spherical fullerenes made up of hexagons and pentagons, nanotubes only have hexagons present but in terms of reactivity both systems have much in common. Due to electrostatic forces nanotubes have a nasty tendency to cluster together into bundles and many potential applications require an *exfoliation* process. One way to do this is by chemical surface modification.

A useful tool for the analysis of derivatised nanotubes is *Raman spectroscopy* which shows a **G-band** (G for *graphite*) for the native nanotubes at 1580 cm and a **D-band** (D for *defect*) at 1350 cm when the graphite lattice is disrupted with conversion of sp^2 to sp^3 *hybridized carbon*. The ratio of both peaks I_D/I_G is taken as a measure of functionalization. Other tools are *UV spectroscopy* where pristine nanotubes show distinct *Van Hove singularities* where functionalized tubes do not, and simple *TGA analysis*.

In one type of chemical modification, *aniline* is oxidized to a *diazonium* intermediate. After expulsion of nitrogen, it forms a covalent bond as an *aryl radical* :

H2O, 80°C, 12 hrs.

Also known are protocols for *cycloadditions* such as *Diels-Alder reactions,* 1,3-dipolar cycloadditions of azomethine ylides and azide–alkyne cycloaddition reactions. One example is a DA reaction assisted by *chromium hexacarbonyl* and high pressure. The I_D/I_G ratio for reaction with *Danishefsky's diene* is 2.6.

Nanotubes can also be alkylated with *alkyl halides* using lithium or sodium metal and liquid ammonia (*Birch reduction* conditions). The initial nanotube salt can function as an polymerization initiator and can react with peroxides to form alkoxy functionalized nanotubes

FULLERENE PURIFICATION

Fullerene purification is the process of obtaining a *fullerene* compound free of contamination. In fullerene production mixtures of C_{60}, C_{70} and higher *homologues* are always formed. Fullerene purification is key to *fullerene* science and determines fullerene prices and the success of practical applications of fullerenes. The first available purification method for C_{60} fullerene was by *HPLC* from which small amounts could be generated at large expense.

A practical laboratory-scale method for purification of soot enriched in C_{60} and C_{70} starts with *extraction* in *toluene* followed by *filtration* with a paper filter. The solvent is evapourated and the residue (the toluene-soluble soot fraction) redissolved in toluene and subjected to *column chromatography*. C_{60} elutes first with a purple colour and C_{70} is next displaying a reddish-brown colour.

In nanotube processing the established purification method for removing amorphous carbon and metals is by competitive oxidation (often a *sulfuric acid/ nitric acid* mixture). It is assumed that this oxidation creates oxygen containing groups (*hydroxyl, carbonyl, carboxyl*) on the nanotube surface which electrostatically stabilize them in water and which can later be utilized in chemical functionalization. One report reveals that the oxygen containing groups in actuality combine with carbon contaminations absorbed to the nanotube wall that can be removed by a simple base wash. Cleaned nanotubes are reported to have reduced D/G ratio indicative of less functionalization, and the absence of oxygen is also apparent from *IR spectroscopy* and *X-ray photoelectron spectroscopy.*

Experimental Purification Strategies

A recent kilogram-scale fullerene purification strategy was demonstrated by Nagata *et. al.* In this method C_{60} was separated from a mixture of C_{60}, C_{70} and higher fullerene compounds by first adding the *amidine* compound *DBU* to a solution of the mixture in *1,2,3-trimethylbenzene.* DBU as it turns out only reacts to C_{70} fullerenes and higher which reaction products separate out and can be removed by filtration. C_{60} fullerenes do not have any affinity for DBU and are subsequently isolated. Other diamine compounds like *DABCO* do not share this selectivity.

C_{60} but not C_{70} forms a 1 : 2 *inclusion compound* with *cyclodextrin* (CD). A separation method for both fullerenes based on this principle is made possible by anchoring cyclodextrin to *colloidal gold* particles through a *sulfur*-sulfur bridge. The Au/CD compound is very stable and soluble in water and selectively extracts C_{60} from the insoluble mixture after *refluxing* for several days. The C_{70} fullerene component is then removed by simple *filtration.* C_{60} is driven out from the Au/CD compound by adding *adamantol* which has a higher affinity for the cyclodextrin cavity. Au/CD is completely *recycled* when adamantol in turn is driven out by adding *ethanol* and ethanol removed by evapouration; 50 *mg* of Au/CD captures 5 mg of C_{60} fullerene.

CIS–TRANS ISOMERISM

Fig. : *cis*-but-2-ene.

Fig. : *trans*-but-2-ene.

In *organic chemistry*, **cis/trans** isomerism (also known as **geometric isomerism**) is a form of *stereo-isomerism* describing the *relative* orientation of *functional groups* within a molecule. It is not to be confused with **E/Z isomerism**, which is the related *absolute* stereo-chemical description, only to be used with alkenes. In general, such isomers contain *double bonds*, which cannot rotate, but they can also arise from ring structures, wherein the rotation of bonds is greatly restricted. *Cis* and *trans isomers* occur both in organic molecules and in inorganic co-ordination complexes. *Cis* and *trans* descriptors are not used for cases of *conformational isomerism* where the two geometric forms easily interconvert, such as most open-chain single-bonded structures; instead, the terms *syn* and *anti* would be used.

The terms *cis* and *trans* are from Latin, in which *cis* means "on the same side" and *trans* means "on the other side" or "across". The term "geometric isomerism" is considered an obsolete synonym of "*cis/trans* isomerism" by *IUPAC*.

Organic Chemistry

When the *substituent groups* are oriented in the same direction, the *diastereomer* is referred to as *cis*, whereas, when the substituents are oriented in opposing directions, the diastereomer is referred to as *trans*. An example of a small hydrocarbon displaying *cis/trans* isomerism is *2-butene*.

Alicyclic compounds can also display *cis/trans* isomerism. As an example of a geometric isomer due to a ring structure, consider 1,2-dichlorocyclohexane :

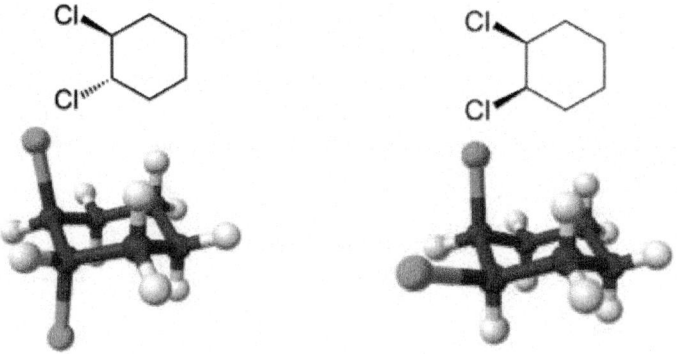

Fig. : *trans*-1,2-dichlorocyclohexane.   **Fig. :** *cis*-1,2-dichlorocyclohexane.

Comparison of Physical Properties

Cis and *trans* isomers often have different physical properties. Differences between isomers, in general, arise from the differences in the shape of the molecule or the overall *dipole moment*.

Fig. : *cis*-2-pentene.   **Fig. :** *trans*-2-pentene.

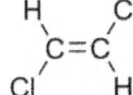

Fig. : cis-1,2-dichloroethene.

Fig. : trans-1,2-dichloroethene.

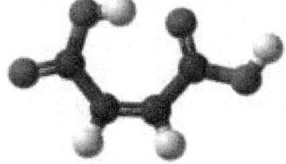

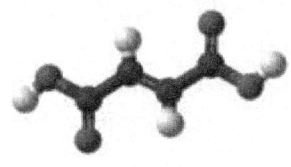

Fig. : cis-butenedioic acid (maleic acid). Fig. : trans-butenedioic acid (fumaric acid).

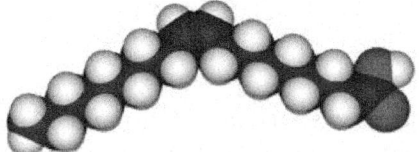

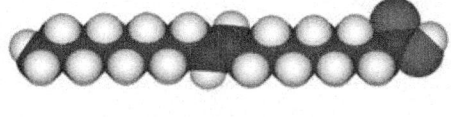

Fig. : Oleic acid. Fig. : Elaidic acid.

These differences can be very small, as in the case of the boiling point of straight-chain alkenes, such as *2-pentene*, which is 37°C in the *cis* isomer and 36°C in the *trans* isomer. The differences between *cis* and *trans* isomers can be larger if polar bonds are present, as in the *1,2-dichloroethenes*. The *cis* isomer in this case has a boiling point of 60.3°C, while the *trans* isomer has a boiling point of 47.5°C. In the *cis* isomer the two polar C-Cl *bond dipole moments* combine to give an overall molecular dipole, so that there are inter-molecular *dipole–dipole forces* (or Keesom forces), which add to the *London dispersion forces* and raise the boiling point. In the *trans* isomer on the other hand, this does not occur because the two C-Cl bond moments cancel and the molecule has a net zero dipole (it does however have a non-zero quadrupole).

The two isomers of butenedioic acid have such large differences in properties and reactivities that they were actually given completely different names. The *cis* isomer is called *maleic acid* and the *trans* isomer *fumaric acid*. Polarity is key in determining relative boiling point as it causes increased inter-molecular forces, thereby raising the boiling point. In the same manner, symmetry is key in determining relative melting point as it allows for better packing in the solid state, even if it does not alter the polarity of the molecule. One example of this is the relationship between *oleic acid* and *elaidic acid*; oleic acid, the *cis* isomer, has a melting point of 13.4°C, making it a liquid at room temperature, while the *trans* isomer, elaidic acid, has the much higher melting point of 43°C, due to the straighter *trans* isomer being able to pack more tightly, and is solid at room temperature.

Thus, *trans*-alkenes, which are less polar and more symmetrical, have lower boiling points and higher melting points, and *cis*-alkenes, which are generally more polar and less symmetrical, have higher boiling points and lower melting points.

In the case of geometric isomers that are a consequence of double bonds, and, in particular, when both substituents are the same, some general trends usually hold. These trends can be attributed to the fact that the dipoles of the substituents in a *cis* isomer will add up to give an overall molecular dipole. In a *trans* isomer, the dipoles of the substituents will cancel out due to their being on opposite site of the molecule. *Trans* isomers also tend to have lower densities than their *cis* counterparts.

As a general trend, *trans* alkenes tend to have higher *melting points* and lower *solubility* in inert solvents, as *trans* alkenes, in general, are more symmetrical than *cis* alkenes.

Vicinal coupling constants (J_{HH}), measured by *NMR spectroscopy*, are larger for *trans*-(range : 12–18 Hz; typical : 15 Hz) than for *cis*-(range : 0–12 Hz; typical : 8 Hz) isomers.

Stability

Usually, for acyclic systems *trans* isomers are more stable than *cis* isomers. This is typically due to the increased unfavourable steric interaction of the substituents in the *cis* isomer. Therefore, *trans* isomers have a less exothermic *heat of combustion*, indicating higher *thermo-chemical* stability. In the Benson *heat of formation group additivity* dataset, *cis* isomers suffer a 1.10 kcal/mol stability penalty. Exceptions to this rule exist, such as 1,2-difluoroethylene, 1,2-difluorodiazene, and several other halogen-and oxygen-substituted ethylenes. In these cases, the *cis* isomer is more stable than the *trans* isomer. This phenomenon is called the **cis effect**.

E/Z notation

Fig. : Bromine has a higher *CIP priority* than chlorine, so this alkene is the Z isomer.

The *cis/trans* system for naming alkene isomers should generally only be used when there are only two different substituents on the double bond, so there is no confusion about which substituents are being described relative to each other. For more complex cases, the *cis/trans* designation is generally based on the longest carbon chain as reflected in the root name of the molecule (*i.e.* an extension of standard organic nomenclature for the parent structure). The IUPAC standard designations *E/Z* are unambiguous in all cases, and therefore are especially useful for tri-and tetrasubstituted alkenes to avoid any confusion about which groups are being identified as *cis* or *trans* to each other. Z (from the German *zusammen*) means "together". E (from the German *entgegen*) means "opposite". That is, Z has the higher-priority groups *cis* to each other and E has the higher-priority groups *trans* to each other. Because the *cis/trans* and *E/Z* systems compare different groups on the alkene, it is not strictly true that Z corresponds to *cis* and E corresponds to *trans*. For example, *trans*-2-chlorobut-2-ene (the two methyl groups, C1 and C4, on the *2-butene* backbone are *trans* to each other) is (Z)-2-chlorobut-2-ene (the chlorine and C4 are together because C1 and C4 are opposite).

Whether a molecular configuration is designated E or Z is determined by the *Cahn-Ingold-Prelog priority rules*; higher atomic numbers are given higher priority. For each of the two atoms in the double bond, it is necessary to determine the priority of each substituent. If both the higher-priority substituents are on the same side, the arrangement is Z; if on opposite sides, the arrangement is E.

Inorganic Chemistry

Cis–trans isomerism can also occur in inorganic compounds, most notably in *diazenes* and *co-ordination compounds*.

Diazenes

Diazenes (and the related *diphosphenes*) can also exhibit *cis-trans* isomerism. As with organic compounds, the *cis* isomer is generally the more reactive of the two, being the only isomer that can reduce *alkenes* and *alkynes* to *alkanes*, but for a different reason : the *trans* isomer cannot line its hydrogens up suitably to reduce the alkene, but the *cis* isomer, being shaped differently, can.

Fig. : *trans*-diazene.   Fig. : *cis*-diazene.

Co-ordination Complexes

In inorganic *co-ordination complexes* with octahedral or square planar geometries, there are also *cis* isomers in which similar ligands are closer together and *trans* isomers in which they are further apart.

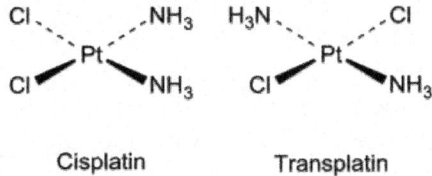

Cisplatin   Transplatin

Fig. : The two isomeric complexes, cisplatin and transplatin.

For example, there are two isomers of *square planar* $Pt(NH_3)_2Cl_2$, as explained by *Alfred Werner* in 1893. The *cis* isomer, whose full name is *cis*-diamminedichloroplatinum (II), was shown in 1969 by *Barnett Rosenberg* to have antitumor activity, and is now a chemotherapy drug known by the short name *cisplatin*. In contrast, the *trans* isomer (transplatin) has no useful anti-cancer activity. Each isomer can be synthesized using the *trans effect* to control which isomer is produced.

For *octahedral complexes* of formula MX_4Y_2, two isomers also exist. (Here M is a metal atom, and X and Y are two different types of *ligands*.) In the *cis* isomer, the two Y ligands are adjacent to each other at 90°, as is true for the two chlorine atoms shown in green in *cis*-$[Co(NH_3)_4Cl_2]$, at left. In the *trans* isomer shown at right, the two Cl atoms are on opposite sides of the central Co atom.

A related type of isomerism in octahedral MX_3Y_3 complexes is *facial-meridional* (or *fac/mer*) isomerism, in which different numbers of ligands are *cis* or *trans* to each other. Metal carbonyl compounds can be characterized as fac or mer using *infrared spectroscopy*.

CO-ORDINATION COMPLEX

In chemistry, a co-ordination complex or metal complex, consists of an atom or ion (usually metallic), and a surrounding array of bound molecules or anions, that are in turn known as ligands or complexing agents. Many metal-containing compounds (especially transition metals) consist of co-ordination complexes. Nomenclature and terminology.

Co-ordination complexes are so pervasive that the structure and reactions are described in many ways, sometimes confusingly. The atom within a ligand that is bonded to the central atom or ion is called the **donor atom**. A typical complex is bound to several donor atoms, which can be the same or different. *Polydentate* (multiple bonded) ligands consist of several donor atoms, several of which are bound to the central atom or ion. These complexes are called *chelate complexes*, the formation of such complexes is called chelation, complexation, and co-ordination.

The central atom or ion, together with all ligands comprise the *co-ordination sphere*. The central atoms or ion and the donor atoms comprise the first co-ordination sphere.

Co-ordination refers to the "co-ordinate covalent bonds" (*dipolar bonds*) between the ligands and the central atom. Originally, a complex implied a reversible association of *molecules, atoms,* or *ions* through such weak *chemical bonds*. As applied to co-ordination chemistry, this meaning has evolved. Some metal complexes are formed virtually irreversibly and many are bound together by bonds that are quite strong.

History

Co-ordination complexes were known–although not understood in any sense– since the beginning of chemistry, *e.g. Prussian blue* and *copper vitriol*. The key breakthrough occurred when *Alfred Werner* proposed in 1893 that Co (III) bears six *ligands* in an *octahedral geometry*. His theory allows one to understand the difference between coordinated and ionic in a compound, for example chloride in the cobalt *ammine* chlorides and to explain many of the previously inexplicable isomers.

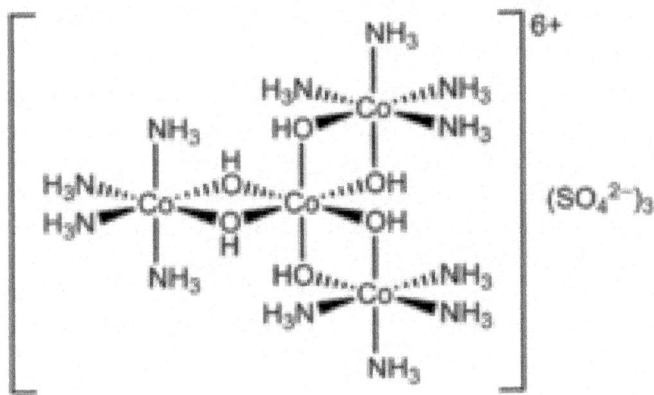

Fig. : Structure of hexol.

In 1914, Werner resolved the first co-ordination complex, called *hexol*, into optical isomers, overthrowing the theory that only carbon compounds could possess *chirality*.

Structures

The ions or molecules surrounding the central atom are called *ligands*. Ligands are generally bound to the central atom by a *co-ordinate covalent bond* (donating electrons from a *lone electron pair* into an empty metal orbital), and are said to be **coordinated** to the atom. There are also organic ligands such as *alkenes* whose *pi bonds* can co-ordinate to empty metal orbitals. An example is *ethene* in the complex known as *Zeise's salt*, $K[PtCl_3(C_2H_4)]$.

Geometry

In co-ordination chemistry, a structure is first described by its *co-ordination number*, the number of ligands attached to the metal (more specifically, the number of donor atoms). Usually one can count the ligands attached, but sometimes even the counting can become ambiguous. Co-ordination numbers are normally between two and nine, but large numbers of ligands are not uncommon for the lanthanides and actinides. The number of bonds depends on the size, charge, and *electron configuration* of the metal ion and the ligands. Metal ions may have more than one co-ordination number.

Typically the chemistry of transition metal complexes is dominated by interactions between s and p *molecular orbitals* of the ligands and the d orbitals of the metal ions. The s, p, and d orbitals of the metal can accommodate 18 electrons.

The maximum co-ordination number for a certain metal is thus related to the electronic configuration of the metal ion (to be more specific, the number of empty orbitals) and to the ratio of the size of the ligands and the metal ion. Large metals and small ligands lead to high co-ordination numbers, *e.g.* $[Mo(CN)_8]$. Small metals with large ligands lead to low co-ordination numbers, *e.g.* $Pt[P(CMe_3)]_2$. Due to their large size, *lanthanides*, *actinides*, and early transition metals tend to have high co-ordination numbers.

Different ligand structural arrangements result from the co-ordination number. Most structures follow the points-on-a-sphere pattern (or, as if the central atom were in the middle of a *polyhedron* where the corners of that shape are the locations of the ligands), where orbital overlap (between ligand and metal orbitals) and ligand-ligand repulsions tend to lead to certain regular geometries. The most observed geometries are listed below, but there are many cases that deviate from a regular geometry, *e.g.* due to the use of ligands of different types (which results in irregular bond lengths; the co-ordination atoms do not follow a points-on-a-sphere pattern), due to the size of ligands, or due to *electronic effects* :

- *Linear* for two-co-ordination
- *Trigonal planar* for three-co-ordination
- *Tetrahedral* or *square planar* for four-co-ordination
- *Trigonal bipyramidal* or *square pyramidal* for five-co-ordination
- *Octahedral* (orthogonal) or *trigonal prismatic* for six-co-ordination
- *Pentagonal bipyramidal* for seven-co-ordination
- *Square antiprismatic* for eight-co-ordination
- *Tri-capped trigonal prismatic* (Triaugmented triangular prism) for nine-co-ordination.

  Some exceptions and provisions should be noted :

- The idealized descriptions of 5-, 7-, 8-, and 9-co-ordination are often indistinct geometrically from alternative structures with slightly different L–M–L (ligand–metal–ligand) angles. The classic example of this is the difference between square pyramidal and trigonal bipyramidal structures.
- Due to special electronic effects such as (second-order) *Jahn–Teller* stabilization, certain geometries are stabilized relative to the other possibilities, *e.g.* for some compounds the trigonal prismatic geometry is stabilized relative to octahedral structures for six-co-ordination.

Isomerism

The arrangement of the ligands is fixed for a given complex, but in some cases it is mutable by a reaction that forms another stable *isomer*.

There exist many kinds of *isomerism* in co-ordination complexes, just as in many other compounds.

Stereoisomerism

Stereoisomerism occurs with the same bonds in different orientations relative to one another. Stereoisomerism can be further classified into :

Cis–trans Isomerism and Facial–meridional Isomerism

Cis–trans isomerism occurs in octahedral and *square planar* complexes (but not tetrahedral). When two ligands are mutually adjacent they are said to be **cis**, when opposite each other, **trans**. When three identical ligands occupy one face of an octahedron, the isomer is said to be facial, or **fac**. In a *fac* isomer, any two identical ligands are adjacent or *cis* to each other. If these three ligands and the metal ion are in one plane, the isomer is said to be meridional, or **mer**. A *mer* isomer can be considered as a combination of a *trans* and a *cis*, since it contains both trans and cis pairs of identical ligands.

Optical Isomerism

Optical isomerism occurs when a molecule is not superimposable with its mirror image. It is so called because the two isomers are each *optically active*, that is, they rotate the plane of *polarized light* in opposite directions. The symbol Λ (*lambda*) is used as a prefix to describe the left-handed propeller twist formed by three bidentate ligands, as shown. Like-wise, the symbol Δ (*delta*) is used as a prefix for the right-handed propeller twist.

Structural Isomerism

Structural isomerism occurs when the bonds are themselves different. There are four types of structural isomerism : ionisation isomerism, solvate or hydrate isomerism, linkage isomerism and co-ordination isomerism.

1. *Ionisation isomerism :* The isomers give different ions in solution although they have the same composition. This type of isomerism occurs when the counter ion of the complex is also a potential ligand. For example, pentaaminebromidocobalt (III) sulfate $[CoBr(NH_3)_5]SO_4$ is red violet and in solution gives a precipitate with barium chloride, confirming the presence of sulfate ion, while pentaaminesulfatecobalt (III) bromide $[CoSO_4(NH_3)_5]Br$ is red and tests negative for sulfate ion in solution, but instead gives a precipitate of AgBr with silver nitrate.

2. *Solvate or hydrate isomerism :* The isomers have the same composition but differ with respect to the number of solvent ligand molecules as well as the counter ion in the crystal lattice. For example $[Cr(H_2O)_6]Cl_3$ is violet coloured, $[CrCl(H_2O)_5]Cl_2 \cdot H_2O$ is blue-green, and $[CrCl_2(H_2O)_4]Cl \cdot 2H_2O$ is dark green

3. *Linkage isomerism* occurs with ambidentate ligands that can bind in more than one place. For example, NO_2 is an ambidentate ligand : It can bind to a metal at either the N atom or an O atom.

4. *Co-ordination isomerism :* This occurs when both positive and negative ions of a salt are complex ions and the two isomers differ in the distribution of

ligands between the cation and the anion. For example $[Co(NH_3)_6][Cr(CN)_6]$ and $[Cr(NH_3)_6][Co(CN)_6]$.

Electronic Properties

Many of the properties of transition metal complexes are dictated by their electronic structures. The electronic structure can be described by a relatively ionic model that ascribes formal charges to the metals and ligands. This approach is the essence of *crystal field theory* (CFT). Crystal field theory, introduced by *Hans Bethe* in 1929, gives a *quantum mechanically* based attempt at understanding complexes. But crystal field theory treats all interactions in a complex as ionic and assumes that the ligands can be approximated by negative point charges.

More sophisticated models embrace covalency, and this approach is described by *ligand field theory* (LFT) and *Molecular orbital theory* (MO). Ligand field theory, introduced in 1935 and built from molecular orbital theory, can handle a broader range of complexes and can explain complexes in which the interactions are *covalent*. The chemical applications of *group theory* can aid in the understanding of crystal or ligand field theory, by allowing simple, symmetry based solutions to the formal equations.

Chemists tend to employ the simplest model required to predict the properties of interest; for this reason, CFT has been a favourite for the discussions when possible. MO and LF theories are more complicated, but provide a more realistic perspective.

The electronic configuration of the complexes gives them some important properties :

Colour of Transition Metal Complexes

Transition metal complexes often have spectacular colours caused by electronic transitions by the absorption of light. For this reason they are often applied as *pigments*. Most transitions that are related to coloured metal complexes are either **d–d transitions** or **charge transfer bands**. In a d–d transition, an electron in a d orbital on the metal is excited by a photon to another d orbital of higher energy. A charge transfer band entails promotion of an electron from a metal-based orbital into an empty ligand-based orbital (*Metal-to-Ligand Charge Transfer* or MLCT). The converse also occurs : excitation of an electron in a ligand-based orbital into an empty metal-based orbital (*Ligand to Metal Charge Transfer* or LMCT). These phenomena can be observed with the aid of electronic spectroscopy; also known as *UV-Vis*. For simple compounds with high symmetry, the d–d transitions can be assigned using *Tanabe–Sugano diagrams*. These assignments are gaining increased support with *computational chemistry*.

Colours of Lanthanide Complexes

Superficially *lanthanide* complexes are similar to those of the transition metals in that some are coloured. However for the common Ln ions (Ln = lanthanide) the

Table : Colours of Various Example Co-ordination Complexes.

	Fe	Fe	Co	Cu	Al	Cr
Hydrated Ion	$[Fe(H_2O)_6]$ Pale green Soln	$[Fe(H_2O)_6]$ Yellow/brown Soln	$[Co(H_2O)_6]$ Pink Soln	$[Cu(H_2O)_6]$ Blue Soln	$[Al(H_2O)_6]$ Colourless Soln	$[Cr(H_2O)_6]$ Green Soln
OH, dilute	$[Fe(H_2O)_4(OH)_2]$ Dark green Ppt	$[Fe(H_2O)_3(OH)_3]$ Brown Ppt	$[Co(H_2O)_4(OH)_2]$ Blue/green Ppt	$[Cu(H_2O)_4(OH)_2]$ Blue Ppt	$[Al(H_2O)_3(OH)_3]$ White Ppt	$[Cr(H_2O)_3(OH)_3]$ Green Ppt
OH, concentrated	$[Fe(H_2O)_4(OH)_2]$ Dark green Ppt	$[Fe(H_2O)_3(OH)_3]$ Brown Ppt	$[Co(H_2O)_4(OH)_2]$ Blue/green Ppt	$[Cu(H_2O)_4(OH)_2]$ Blue Ppt	$[Al(OH)_4]$ Colourless Soln	$[Cr(OH)_6]$ Green Soln
NH$_3$, dilute	$[Fe(H_2O)_4(OH)_2]$ Dark green Ppt	$[Fe(H_2O)_3(OH)_3]$ Brown Ppt	$[Co(H_2O)_4(OH)_2]$ Blue/green Ppt	$[Cu(H_2O)_4(OH)_2]$ Blue Ppt	$[Al(H_2O)_3(OH)_3]$ White Ppt	$[Cr(H_2O)_3(OH)_3]$ Green Ppt
NH$_3$, concentrated	$[Fe(H_2O)_4(OH)_2]$ Dark green Ppt	$[Fe(H_2O)_3(OH)_3]$ Brown Ppt	$[Co(NH_3)_6]$ Straw coloured Soln	$[Cu(NH_3)_4(H_2O)_2]$ Deep blue Soln	$[Al(H_2O)_3(OH)_3]$ White Ppt	$[Cr(NH_3)_6]$ Green Soln
CO$_3$	$FeCO_3$ Dark green Ppt	$[Fe(H_2O)_3(OH)_3]$ Brown Ppt + bubbles	$CoCO_3$ Pink Ppt	$CuCO_3$ Blue/green Ppt		

colours are all pale, and hardly influenced by the nature of the ligand. The colours are due to 4f electron transitions. As the 4f orbitals in lanthanides are "buried" in the xenon core and shielded from the ligand by the 5s and 5p orbitals they are therefore not influenced by the ligands to any great extent leading to a much smaller *crystal field* splitting than in the transition metals. The absorption spectra of an Ln ion approximates to that of the free ion where the electronic states are described by spin-orbit coupling (also called L-S coupling or Russell-Saunders coupling). This contrasts to the transition metals where the ground state is split by the crystal field. Absorptions for Ln are weak as electric dipole transitions are parity forbidden (*Laporte Rule* forbidden) but can gain intensity due to the effect of a low-symmetry ligand field or mixing with higher electronic states (*e.g.* d orbitals). Also absorption bands are extremely sharp which contrasts with those observed for transition metals which generally have broad bands.

Magnetism

Metal complexes that have unpaired electrons are *magnetic*. Considering only monometallic complexes, unpaired electrons arise because the complex has an odd number of electrons or because electron pairing is destabilized. Thus, monomeric Ti (III) species have one "d-electron" and must be *(para) magnetic*, regardless of the geometry or the nature of the ligands. Ti (II), with two d-electrons, forms some complexes that have two unpaired electrons and others with none. This effect is illustrated by the compounds $TiX_2[(CH_3)_2PCH_2CH_2P(CH_3)_2]_2$: when X = Cl, the complex is paramagnetic (*high-spin* configuration), whereas when X = CH_3, it is diamagnetic (*low-spin* configuration). It is important to realize that ligands provide an important means of adjusting the *ground state* properties.

In bi-and polymetallic complexes, in which the individual centers have an odd number of electrons or that are high-spin, the situation is more complicated. If there is interaction (either direct or through ligand) between the two (or more) metal centers, the electrons may couple (*anti-ferromagnetic coupling*, resulting in a diamagnetic compound), or they may enhance each other (*ferromagnetic coupling*). When there is no interaction, the two (or more) individual metal centers behave as if in two separate molecules.

Reactivity

Complexes show a variety of possible reactivities :

- *Electron transfers :* A common reaction between co-ordination complexes involving ligands are *inner* and *outer sphere electron transfers*. They are two different mechanisms of *electron transfer redox* reactions, largely defined by the late *Henry Taube*. In an inner sphere reaction, a ligand with two *lone electron pairs* acts as a *bridging ligand,* a ligand to which both co-ordination centres can bond. Through this, electrons are transferred from one centre to another.

- *(Degenerate) ligand exchange :* One important indicator of reactivity is the rate of degenerate exchange of ligands. For example, the rate of interchange of co-ordinate water in $[M(H_2O)_6]$ complexes varies over 20 orders of magnitude. Complexes where the ligands are released and rebound rapidly are classified as labile. Such labile complexes can be quite stable thermodynamically. Typical labile metal complexes either have low-charge (Na), electrons in d-orbitals that are *anti-bonding* with respect to the ligands (Zn), or lack covalency (Ln, where Ln is any lanthanide). The lability of a metal complex also depends on the high-spin *vs.* low-spin configurations when such is possible. Thus, high-spin Fe (II) and Co (III) form labile complexes, whereas low-spin analogues are inert. Cr (III) can exist only in the low-spin state (quartet), which is inert because of its high formal oxidation state, absence of electrons in orbitals that are M–L antibonding, plus some "ligand field stabilization" associated with the d configuration.

- *Associative processes* : Complexes that have unfilled or half-filled orbitals often show the capability to react with substrates. Most substrates have a singlet ground-state; that is, they have lone electron pairs (*e.g.*, water, amines, ethers), so these substrates need an empty orbital to be able to react with a metal centre. Some substrates (*e.g.*, molecular oxygen) *have a triplet ground state*, which results that metals with half-filled orbitals have a tendency to react with such substrates (it must be said that the *dioxygen* molecule also has lone pairs, so it is also capable to react as a 'normal' Lewis base).

If the ligands around the metal are carefully chosen, the metal can aid in (*stoichiometric* or *catalytic*) transformations of molecules or be used as a sensor.

Classification

Metal complexes, also known as co-ordination compounds, include all metal compounds, aside from metal vapours, *plasmas*, and *alloys*. The study of "co-ordination chemistry" is the study of "inorganic chemistry" of all *alkali* and *alkaline earth metals, transition metals, lanthanides, actinides,* and *metalloids*. Thus, co-ordination chemistry is the chemistry of the majority of the periodic table. Metals and metal ions exist, in the condensed phases at least, only surrounded by ligands.

The areas of co-ordination chemistry can be classified according to the nature of the ligands, in broad terms :

- *Classical (or "Werner Complexes")* : Ligands in classical co-ordination chemistry bind to metals, almost exclusively, *via* their "*lone pairs*" of electrons residing on the main group atoms of the ligand. Typical ligands are H_2O, NH_3, Cl, CN, *en*

  Examples : $[Co(EDTA)]$, $[Co(NH_3)_6]Cl_3$, $[Fe(C_2O_4)_3]K_3$

- *Organometallic Chemistry* : Ligands are organic (alkenes, alkynes, alkyls) as well as "organic-like" ligands such as phosphines, hydride, and CO.

  Example : $(C_5H_5)Fe(CO)_2CH_3$

- *Bioinorganic Chemistry* : Ligands are those provided by nature, especially including the side chains of amino acids, and many *co-factors* such as *porphyrins*.

  Example : *hemoglobin* contains *heme*, a porphyrin complex of iron

Example : chlorophyll contains a porphyrin complex of magnesium

Many natural ligands are "classical" especially including water.

- *Cluster Chemistry :* Ligands are all of the above also include other metals as ligands.

  Example $Ru_3(CO)_{12}$

- In some cases there are combinations of different fields :

  Example : [Fe_4S_4(Scysteinyl)_4], in which a cluster is embedded in a biologically active species.

  Mineralogy, materials science, and *solid state chemistry* — as they apply to metal ions — are sub-sets of co-ordination chemistry in the sense that the metals are surrounded by ligands. In many cases these ligands are oxides or sulfides, but the metals are coordinated nonetheless, and the principles and guidelines discussed below apply. In *hydrates*, at least some of the ligands are water molecules. It is true that the focus of mineralogy, materials science, and solid state chemistry differs from the usual focus of co-ordination or inorganic chemistry. The former are concerned primarily with polymeric structures, properties arising from a collective effects of many highly interconnected metals. In contrast, co-ordination chemistry focuses on reactivity and properties of complexes containing individual metal atoms or small ensembles of metal atoms.

Older Classifications of Isomerism

Traditional classifications of the kinds of isomer have become archaic with the advent of modern structural chemistry. In the older literature, one encounters :

- *Ionisation isomerism* describes the possible isomers arising from the exchange between the outer sphere and inner sphere. This classification relies on an archaic classification of the inner and outer sphere. In this classification, the "outer sphere ligands," when ions in solution, may be switched with "inner sphere ligands" to produce an isomer.

- *Solvation isomerism* occurs when an inner sphere ligand is replaced by a *solvent* molecule. This classification is obsolete because it considers solvents as being distinct from other ligands. Some of the problems are discussed under *water of crystallization.*

Naming Complexes

The basic procedure for naming a complex :

1. When naming a complex ion, the ligands are named before the metal ion.
2. Write the names of the ligands in alphabetical order. (Numerical prefixes do not affect the order.)

   - Multiple occurring monodentate ligands receive a prefix according to the number of occurrences : *di-, tri-, tetra-, penta-,* or *hexa*. Polydentate ligands (*e.g.,* ethylenediamine, oxalate) receive *bis-, tris-, tetrakis-,* etc.

- Anions end in *ido*. This replaces the final 'e' when the anion ends with '-ate', *e.g. sulfate* becomes *sulfato*. It replaces 'ide' : *cyanide* becomes *cyanido*.
- Neutral ligands are given their usual name, with some exceptions : NH_3 becomes *ammine*; H_2O becomes *aqua* or *aquo*; CO becomes *carbonyl*; NO becomes *nitrosyl*.

3. Write the name of the central atom/ion. If the complex is an anion, the central atom's name will end in-*ate*, and its Latin name will be used if available (except for mercury).

4. If the central atom's oxidation state needs to be specified (when it is one of several possible, or zero), write it as a Roman numeral (or 0) in parentheses.

5. Name cation then anion as separate words (if applicable, as in last example)

   Examples :

   $[NiCl_4]$ → tetrachloridonickelate (II) ion

   $[CuCl_5NH_3]$ → amminepentachloridocuprate (II) ion

   $[Cd(CN)_2(en)_2]$ → dicyanidobis (ethylenediamine) cadmium (II)

   $[CoCl(NH_3)_5]SO_4$ → pentaamminechloridocobalt (III) sulfate.

The co-ordination number of ligands attached to more than one metal (bridging ligands) is indicated by a subscript to the Greek symbol μ placed before the ligand name. Thus the *dimer* of *aluminium trichloride* is described by Al_2Cl_{42}.

Application of Co-ordination Compounds

1. They are used in photography, *i.e.*, AgBr forms a soluble complex with *sodium thiosulfate* in photography.

2. $K[Ag(CN)_2]$ is used for *electroplating* of silver, and $K[Au(CN)_2]$ is used for *gold plating*.

3. Some ligands *oxidise* Co to Co ion.

4. *Ethylenediaminetetraacetic acid* (EDTA) is used for estimation of Ca and Mg in *hard water*.

5. Silver and gold are extracted by treating zinc with their cyanide complexes.

SUPER-CONDUCTIVITY

Super-conductivity is a phenomenon of exactly zero *electrical resistance* and expulsion of *magnetic fields* occurring in certain materials when *cooled* below a characteristic *critical temperature*. It was discovered by Dutch physicist *Heike Kamerlingh Onnes* on April 8, 1911 in *Leiden*. Like *ferromagnetism* and *atomic spectral lines*, super-conductivity is a *quantum mechanical* phenomenon. It is characterized by the *Meissner effect*, the complete ejection of *magnetic field lines* from the interior of the super-conductor as it transitions into the super-conducting state. The occurrence of the Meissner effect indicates that super-conductivity cannot be understood simply as the idealization of *perfect conductivity* in *classical physics*.

Explanation

The electrical resistivity of a metallic *conductor* decreases gradually as temperature is lowered and at the same time its conductivity becomes infinite. Thus a current in a super-conductor flows without any change in magnitude. In ordinary *conductors*, such as *copper* or *silver*, this decrease is limited by impurities and other defects. Even near *absolute zero*, a real sample of a normal conductor shows some resistance. In a super-conductor, the resistance drops abruptly to zero when the material is cooled below its critical temperature. An *electric current* flowing through a loop of *super-conducting wire* can persist indefinitely with no power source.

In 1986, it was discovered that some *cuprate-perovskite ceramic* materials have a critical temperature above 90 K (−183°C). Such a high transition temperature is theoretically impossible for a *conventional super-conductor*, leading the materials to be termed *high-temperature super-conductors*. *Liquid nitrogen* boils at 77 K, and super-conduction at higher temperatures than this facilitates many experiments and applications that are less practical at lower temperatures. In conventional super-conductors, electrons are held together in *Cooper pairs* by an attraction mediated by lattice *phonons*. The best available model of high-temperature super-conductivity is still somewhat crude. There are currently two main hypotheses–the *resonating-valence-bond theory*, and spin fluctuation which has the most support in the research community. The second hypothesis proposed that electron pairing in high-temperature super-conductors is mediated by short-range spin waves known as paramagnons.

Classification

There are many criteria by which super-conductors are classified. The most common are :

- *Response to a magnetic field* : A super-conductor can be *Type I*, meaning it has a single critical field, above which all super-conductivity is lost; or *Type II*, meaning it has two critical fields, between which it allows partial penetration of the magnetic field.

- *By theory of operation* : It is *conventional* if it can be explained by the *BCS theory* or its derivatives, or *unconventional*, otherwise.

- *By critical temperature* : A super-conductor is generally considered *high temperature* if it reaches a super-conducting state when cooled using *liquid nitrogen*–that is, at only $T_c > 77$ K)–or *low temperature* if more aggressive cooling techniques are required to reach its critical temperature.

- *By material* : Super-conductor material classes include *chemical elements* (e.g. *mercury* or *lead*), alloys (such as *niobium-titanium*, *germanium-niobium*, and *niobium nitride*), ceramics (*YBCO* and *magnesium diboride*), or *organic super-conductors* (*fullerenes* and *carbon nanotubes*; though perhaps these examples should be included among the chemical elements, as they are composed entirely of *carbon*).

Elementary Properties of Super-conductors

Most of the physical properties of super-conductors vary from material to material, such as the *heat capacity* and the critical temperature, critical field, and critical current density at which super-conductivity is destroyed.

On the other hand, there is a class of properties that are independent of the underlying material. For instance, all super-conductors have *exactly* zero resistivity to low applied currents when there is no magnetic field present or if the applied field does not exceed a critical value. The existence of these "universal" properties implies that super-conductivity is a *thermodynamic phase*, and thus possesses certain distinguishing properties which are largely independent of microscopic details.

Zero Electrical DC Resistance

The simplest method to measure the *electrical resistance* of a sample of some material is to place it in an *electrical circuit* in series with a *current source I* and measure the resulting *voltage V* across the sample. The resistance of the sample is given by *Ohm's law* as $R = V/I$. If the voltage is zero, this means that the resistance is zero.

Super-conductors are also able to maintain a current with no applied voltage whatsoever, a property exploited in *super-conducting electromagnets* such as those found in *MRI* machines. Experiments have demonstrated that currents in super-conducting coils can persist for years without any measurable degradation. Experimental evidence points to a current lifetime of at least 100,000 years. Theoretical estimates for the lifetime of a persistent current can exceed the estimated lifetime of the *universe*, depending on the wire geometry and the temperature.

In a normal conductor, an electric current may be visualized as a fluid of *electrons* moving across a heavy *ionic* lattice. The electrons are constantly colliding with the ions in the lattice, and during each collision some of the *energy* carried by the current is absorbed by the lattice and converted into *heat*, which is essentially the vibrational *kinetic energy* of the lattice ions. As a result, the energy carried by the current is constantly being dissipated. This is the phenomenon of electrical resistance.

The situation is different in a super-conductor. In a conventional super-conductor, the electronic fluid cannot be resolved into individual electrons. Instead, it consists of bound *pairs* of electrons known as *Cooper pairs*. This pairing is caused by an attractive force between electrons from the exchange of *phonons*. Due to *quantum mechanics*, the *energy spectrum* of this Cooper pair fluid possesses an *energy gap*, meaning there is a minimum amount of energy ΔE that must be supplied in order to excite the fluid. Therefore, if ΔE is larger than the *thermal energy* of the lattice, given by kT, where k is *Boltzmann's constant* and T is the *temperature*, the fluid will not be scattered by the lattice. The Cooper pair fluid is thus a *super-fluid*, meaning it can flow without energy dissipation.

In a class of super-conductors known as *type II super-conductors*, including all known *high-temperature super-conductors*, an extremely small amount of resistivity appears at temperatures not too far below the nominal super-conducting transition when an electric current is applied in conjunction with a strong magnetic field, which may be caused by the electric current. This is due to the motion of *magnetic vortices* in the electronic super-fluid, which dissipates some of the energy carried by the current. If the current is sufficiently small, the vortices are stationary, and the resistivity vanishes. The resistance due to this effect is tiny compared with that of non-super-conducting materials, but must be taken into account in sensitive experiments. However, as the temperature decreases far enough below the nominal super-conducting transition, these vortices can become frozen into a disordered but stationary phase known as a "vortex glass". Below this vortex glass transition temperature, the resistance of the material becomes truly zero.

Super-conducting Phase Transition

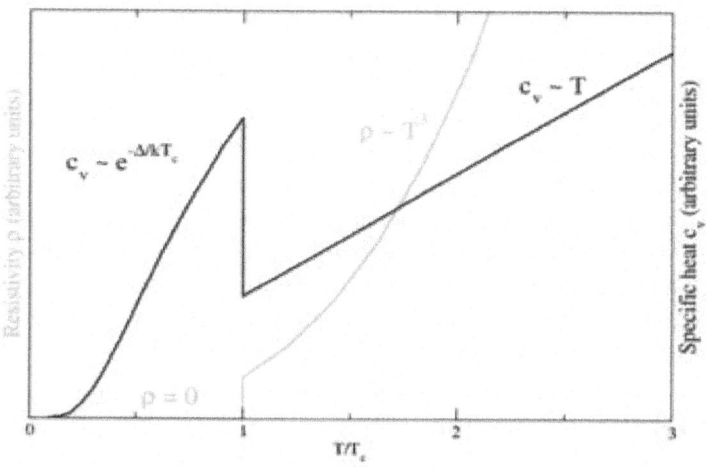

Fig. : Behaviour of heat capacity (c_v, blue) and resistivity (ρ, green) at the super-conducting phase transition.

In super-conducting materials, the characteristics of super-conductivity appear when the *temperature T* is lowered below a **critical temperature** T_c. The value of this critical temperature varies from material to material. Conventional super-conductors usually have critical temperatures ranging from around 20 K to less than 1 K. Solid *mercury*, for example, has a critical temperature of 4.2 K. As of 2009, the highest critical temperature found for a conventional super-conductor is 39 K for *magnesium diboride* (MgB_2), although this material displays enough exotic properties that there is some doubt about classifying it as a "conventional" super-conductor. *Cuprate* super-conductors can have much higher critical temperatures : $YBa_2Cu_3O_7$, one of the first cuprate super-conductors to be discovered, has a critical temperature of 92 K, and mercury-based cuprates have been found with critical temperatures in excess of 130 K. The explanation for these high critical temperatures remains unknown. Electron pairing due to *phonon* exchanges explains super-conductivity in conventional super-conductors, but it does not explain super-conductivity in the newer super-conductors that have a very high critical temperature.

Similarly, at a fixed temperature below the critical temperature, super-conducting materials cease to super-conduct when an external *magnetic field* is applied which is greater than the *critical magnetic field*. This is because the *Gibbs free energy* of the super-conducting phase increases quadratically with the magnetic field while the free energy of the normal phase is roughly independent of the magnetic field. If the material super-conducts in the absence of a field, then the super-conducting phase free energy is lower than that of the normal phase and so for some finite value of the magnetic field (proportional to the square root of the difference of the free energies at zero magnetic field) the two free energies will be equal and a phase transition to the normal phase will occur. More generally, a higher temperature and a stronger magnetic field lead to a smaller fraction of the electrons in the super-conducting band and consequently a longer *London penetration depth* of external magnetic fields and currents. The penetration depth becomes infinite at the phase transition.

The onset of super-conductivity is accompanied by abrupt changes in various physical properties, which is the hallmark of a *phase transition*. For example, the electronic *heat capacity* is proportional to the temperature in the normal (non-super-conducting) regime. At the super-conducting transition, it suffers a discontinuous jump and thereafter ceases to be linear. At low temperatures, it varies instead as e for some constant, α. This exponential behaviour is one of the pieces of evidence for the existence of the *energy gap*.

The *order* of the super-conducting phase transition was long a matter of debate. Experiments indicate that the transition is second-order, meaning there is no *latent heat*. However in the presence of an external magnetic field there is latent heat, because the super-conducting phase has a lower entropy below the critical temperature than the normal phase. It has been experimentally demonstrated that, as a consequence, when the magnetic field is increased beyond the critical field, the resulting phase transition leads to a decrease in the temperature of the super-conducting material.

Calculations in the 1970s suggested that it may actually be weakly first-order due to the effect of long-range fluctuations in the electromagnetic field. In the 1980s it was shown theoretically with the help of a *disorder field theory*, in which the *vortex lines* of the super-conductor play a major role, that the transition is of second order within the *type II* regime and of first order (*i.e.*, *latent heat*) within the *type I* regime, and that the two regions are separated by a *tricritical point*. The results were strongly supported by Monte Carlo computer simulations.

History of super-conductivity

Super-conductivity was discovered on April 8, 1911 by *Heike Kamerlingh Onnes*, who was studying the resistance of solid *mercury* at *cryogenic* temperatures using the recently produced *liquid helium* as a *refrigerant*. At the temperature of 4.2 K, he observed that the resistance abruptly disappeared. In the same experiment, he also observed the *super-fluid* transition of helium at 2.2 K, without recognizing its significance. The precise date and circumstances of the discovery were only reconstructed a century later, when Onnes's notebook was found. In subsequent decades, super-conductivity was observed in several other materials. In 1913, *lead* was found to super-conduct at 7 K, and in 1941 *niobium nitride* was found to super-conduct at 16 K.

Great efforts have been devoted to finding out how and why super-conductivity works; the important step occurred in 1933, when *Meissner* and *Ochsenfeld* discovered that super-conductors expelled applied magnetic fields, a phenomenon which has come to be known as the *Meissner effect*. In 1935, *Fritz* and *Heinz London* showed that the Meissner effect was a consequence of the minimization of the electromagnetic *free energy* carried by super-conducting current.

London Theory

The first phenomenological theory of super-conductivity was *London theory*. It was put forward by the brothers Fritz and Heinz London in 1935, shortly after the discovery that magnetic fields are expelled from super-conductors. A major triumph of the equations of this theory is their ability to explain the *Meissner effect*, wherein a material exponentially expels all internal magnetic fields as it crosses the super-conducting threshold. By using the London equation, one can obtain the dependence of the magnetic field inside the super-conductor on the distance to the surface.

There are two London equations :

$$\frac{\partial j_s}{\partial t} = \frac{n_s e^2}{m} E, \qquad \nabla \times j_s = -\frac{n_s e^2}{m} B.$$

The first equation follows from *Newton's second law* for super-conducting electrons.

Conventional Theories

During the 1950s, theoretical *condensed matter* physicists arrived at a solid understanding of "conventional" super-conductivity, through a pair of remarkable and important theories : the phenomenological *Ginzburg-Landau theory* and the microscopic *BCS theory*.

In 1950, the *phenomenological Ginzburg-Landau theory* of super-conductivity was devised by *Landau* and *Ginzburg*. This theory, which combined Landau's theory of second-order *phase transitions* with a *Schrödinger*-like wave equation, had great success in explaining the macroscopic properties of super-conductors. In particular, *Abrikosov* showed that Ginzburg-Landau theory predicts the division of super-conductors into the two categories now referred to as Type I and Type II. Abrikosov and Ginzburg were awarded the 2003 Nobel Prize for their work (Landau had received the 1962 Nobel Prize for other work, and died in 1968).

The four-dimensional extension of the Ginzburg-Landau theory, the *Coleman-Weinberg model*, is important in *quantum field theory* and *cosmology*.

Also in 1950, Maxwell and Reynolds *et. al.* found that the critical temperature of a super-conductor depends on the *isotopic mass* of the constituent *element*. This important discovery pointed to the *electron-phonon* interaction as the microscopic mechanism responsible for super-conductivity.

The complete microscopic theory of super-conductivity was finally proposed in 1957 by *Bardeen, Cooper* and *Schrieffer*. This BCS theory explained the super-conducting current as a *super-fluid* of *Cooper pairs*, pairs of electrons interacting through the exchange of phonons. For this work, the authors were awarded the Nobel Prize in 1972.

The BCS theory was set on a firmer footing in 1958, when *N. N. Bogolyubov* showed that the BCS wavefunction, which had originally been derived from a variational argument, could be obtained using a canonical transformation of the electronic *Hamiltonian*. In 1959, *Lev Gor'kov* showed that the BCS theory reduced to the Ginzburg-Landau theory close to the critical temperature.

Generalizations of BCS theory for conventional super-conductors form the basis for understanding of the phenomenon of *super-fluidity*, because they fall into the *Lambda transition* universality class. The extent to which such generalizations can be applied to *unconventional super-conductors* is still controversial.

Further History

The first practical application of super-conductivity was developed in 1954 with *Dudley Allen Buck*'s invention of the *cryotron*. Two super-conductors with greatly different values of critical magnetic field are combined to produce a fast, simple, switch for computer elements.

In 1962, the first commercial super-conducting wire, a *niobium-titanium* alloy, was developed by researchers at *Westinghouse*, allowing the construction of the first practical *super-conducting magnets*. In the same year, *Josephson* made the important theoretical prediction that a super-current can flow between two pieces of super-conductor separated by a thin layer of insulator. This phenomenon, now called the *Josephson effect*, is exploited by super-conducting devices such as *SQUIDs*. It is used in the most accurate available measurements of the *magnetic flux quantum* $\Phi_0 = h/(2e)$, where h is the *Planck constant*. Coupled with the *quantum Hall resistivity*, this leads to a precise measurement of the Planck constant. Josephson was awarded the Nobel Prize for this work in 1973.

In 2008, it was proposed that the same mechanism that produces super-conductivity could produce a *super-insulator* state in some materials, with almost infinite *electrical resistance*.

Meissner Effect

When a super-conductor is placed in a weak external *magnetic field* **H**, and cooled below its transition temperature, the magnetic field is ejected. The Meissner effect does not cause the field to be completely ejected but instead the field penetrates the super-conductor but only to a very small distance, characterized by a parameter λ, called the *London penetration depth*, decaying exponentially to zero within the bulk of the material. The *Meissner effect* is a defining characteristic of super-conductivity. For most super-conductors, the London penetration depth is on the order of 100 nm.

The Meissner effect is sometimes confused with the kind of *diamagnetism* one would expect in a perfect electrical conductor : according to *Lenz's law*, when a *changing* magnetic field is applied to a conductor, it will induce an electric current in the conductor that creates an opposing magnetic field. In a perfect conductor, an arbitrarily large current can be induced, and the resulting magnetic field exactly cancels the applied field.

The Meissner effect is distinct from this—it is the spontaneous expulsion which occurs during transition to super-conductivity. Suppose we have a material in its normal state, containing a constant internal magnetic field. When the material is cooled below the critical temperature, we would observe the abrupt expulsion of the internal magnetic field, which we would not expect based on Lenz's law.

The Meissner effect was given a phenomenological explanation by the brothers *Fritz* and *Heinz London*, who showed that the electromagnetic *free energy* in a super-conductor is minimized provided

$$\nabla^2 H = \lambda^{-2} H$$

where **H** is the magnetic field and λ is the London penetration depth.

This equation, which is known as the *London equation*, predicts that the magnetic field in a super-conductor *decays exponentially* from whatever value it possesses at the surface.

A super-conductor with little or no magnetic field within it is said to be in the Meissner state. The Meissner state breaks down when the applied magnetic field is too large. Super-conductors can be divided into two classes according to how this breakdown occurs. In *Type I super-conductors*, super-conductivity is abruptly destroyed when the strength of the applied field rises above a critical value H_c. Depending on the geometry of the sample, one may obtain an intermediate state consisting of a baroque pattern of regions of normal material carrying a magnetic field mixed with regions of super-conducting material containing no field. In *Type II super-conductors*, raising the applied field past a critical value H_{c1} leads to a mixed state (also known as the vortex state) in which an increasing amount of *magnetic flux* penetrates the material, but there remains no resistance to the flow of electric current as long as the current is not too large. At a second critical field strength H_{c2}, super-conductivity is destroyed. The mixed state is actually caused by vortices in the electronic super-fluid, sometimes called *fluxons* because the flux carried by these vortices is *quantized*. Most pure *elemental* super-conductors, except *niobium* and *carbon nanotubes*, are Type I, while almost all impure and compound super-conductors are Type II.

London Moment

Conversely, a spinning super-conductor generates a magnetic field, precisely aligned with the spin axis. The effect, the *London moment*, was put to good use in *Gravity Probe B*. This experiment measured the magnetic fields of four super-conducting gyroscopes to determine their spin axes. This was critical to the experiment since it is one of the few ways to accurately determine the spin axis of an otherwise featureless sphere.

HIGH-TEMPERATURE SUPER-CONDUCTIVITY

High-temperature super-conductors (abbreviated **high-T_c** or **HTS**) are materials that behave as *super-conductors* at unusually high temperatures. The first high-T_c super-conductor was discovered in 1986 by IBM researchers *Georg Bednorz* and *K. Alex Müller*, who were awarded the 1987 *Nobel Prize in Physics* "for their important break-through in the discovery of super-conductivity in ceramic materials".

For an explanation about T_c (the *critical temperature* for super-condictivity).

Whereas "ordinary" or metallic super-conductors usually have transition temperatures (temperatures below which they super-conduct) below 30 K (−243.2°C), HTS have been observed with transition temperatures as high as 138 K (−135°C). Until 2008, only certain compounds of copper and oxygen (so-called "*cuprates*") were believed to have HTS properties, and the term high-temperature super-conductor was used interchangeably with cuprate super-conductor for compounds such as bismuth strontium calcium copper oxide (*BSCCO*) and yttrium barium copper oxide (*YBCO*). However, several iron-based compounds (the iron *pnictides*) are now known to be super-conducting at high temperatures.

History

The phenomenon of super-conductivity was discovered by *Kamerlingh Onnes* in 1911, in metallic mercury below 4 K (−269.15°C). For seventy-five years after that, researchers attempted to observe super-conductivity at higher and higher temperatures. In the late 1970s, super-conductivity was observed in certain metal oxides at temperatures as high as 13 K (−260.1°C), which were much higher than those for elemental metals. In 1986, J. Georg Bednorz and K Alex Müller, working at the *IBM* research lab near *Zurich, Switzerland* were exploring a new class of *ceramics* for super-conductivity. Bednorz encountered a *barium*-doped compound of *lanthanum* and copper oxide whose resistance dropped down to zero at a temperature around 35 K (−238.2°C). Their results were soon confirmed by many groups, notably *Paul Chu* at the *University of Houston* and Shoji Tanaka at the *University of Tokyo*.

Shortly after, *P. W. Anderson*, at *Princeton University* came up with the first theoretical description of these materials, using the *resonating valence bond theory*, but a full understanding of these materials is still developing today. These super-conductors are now known to possess a d-wave pair symmetry. The first proposal that high-temperature cuprate super-conductivity involves d-wave pairing was made in 1987 by Bickers, Scalapino and Scalettar, followed by three subsequent theories in 1988 by Inui, Doniach, Hirschfeld and Ruckenstein, using spin-fluctuation theory, and by Gros, Poilblanc, Rice and Zhang, and by Kotliar and Liu identifying d-wave pairing as a natural consequence of the RVB theory. The confirmation of the d-wave nature of the cuprate super-conductors was made by a variety of experiments, including the direct observation of the d-wave nodes in the excitation spectrum through Angle Resolved Photoemission Spectroscopy, the observation of a half-integer flux in tunneling experiments, and indirectly from the temperature dependence of the penetration depth, specific heat and thermal conductivity.

The super-conductor with the highest transition temperature that has been confirmed by multiple independent research groups (a prerequisite to be called a discovery, verified by *peer review*) is mercury barium calcium copper oxide ($HgBa_2Ca_2Cu_3O_8$) at around 133 K.

After more than twenty years of intensive research the origin of high-tempera-ture super-conductivity is still not clear, but it seems that instead of *electron-phonon* attraction mechanisms, as in conventional super-conductivity, one is dealing with genuine *electronic* mechanisms (*e.g.* by anti-ferromagnetic correlations), and instead of *s-wave* pairing, d-waves are substantial. One goal of all this research is *room-temperature super-conductivity*.

Crystal Structures of High-temperature Ceramic Super-conductors

The structure of high-T_c copper oxide or cuprate super-conductors are often closely related to *perovskite* structure, and the structure of these compounds has been de-scribed as a distorted, oxygen deficient multi-layered perovskite structure. One of the properties of the crystal structure of oxide super-conductors is an alternating multi-layer of CuO_2 planes with super-conductivity taking place between these lay-ers. The more layers of CuO_2 the higher T_c. This structure causes a large anisotropy in normal conducting and super-conducting properties, since electrical currents are carried by holes induced in the oxygen sites of the CuO_2 sheets. The electrical conduction is highly anisotropic, with a much higher conductivity parallel to the CuO_2 plane than in the perpendicular direction. Generally, Critical temperatures depend on the chemical compositions, cations substitutions and oxygen content. They can be classified as *super-stripes*; *i.e.*, particular realizations of super-lattices at atomic limit made of super-conducting atomic layers, wires, dots separated by spacer layers, that gives multiband and multigap super-conductivity.

YBaCuO Super-conductors

The first super-conductor found with $T_c > 77$ K (*liquid nitrogen* boiling point) is yttrium barium copper oxide ($YBa_2Cu_3O_{7-x}$); the proportions of the 3 different metals in the $YBa_2Cu_3O_7$ super-conductor are in the mole ratio of 1 to 2 to 3 for yttrium to barium to copper, respectively. Thus, this particular super-conductor is often referred to as the 123 super-conductor.

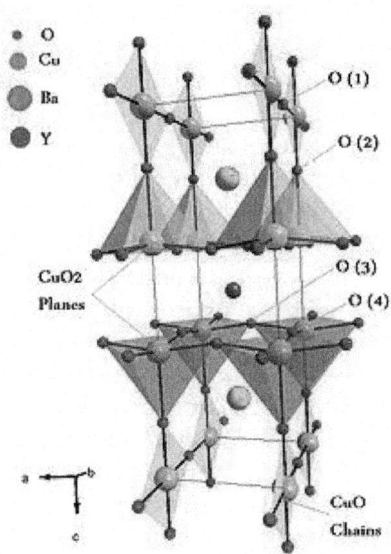

Fig. : YBCO unit cell.

The unit cell of $YBa_2Cu_3O_7$ consists of three pseudocubic elementary perovskite unit cells. Each perovskite unit cell contains a Y or Ba atom at the center : Ba in the bottom unit cell, Y in the middle one, and Ba in the top unit cell. Thus, Y and Ba are stacked in the sequence [Ba–Y–Ba] along the c-axis. All corner sites of the unit cell are occupied by Cu, which has two different co-ordinations, Cu(1) and Cu(2), with respect to oxygen. There are four possible crystallographic sites for oxygen : O(1), O(2), O(3) and O(4). The co-ordination polyhedra of Y and Ba with respect to oxygen are different. The tripling of the perovskite unit cell leads to nine oxygen atoms, whereas $YBa_2Cu_3O_7$ has seven oxygen atoms and, therefore, is referred to as an oxygen-deficient perovskite structure. The structure has a stacking of different layers : $(CuO)(BaO)(CuO_2)(Y)(CuO_2)(BaO)(CuO)$. One of the key feature of the unit cell of $YBa_2Cu_3O_{7-x}$ (YBCO) is the presence of two layers of CuO_2. The role of the Y plane is to serve as a spacer between two CuO_2 planes. In YBCO, the Cu–O chains are known to play an important role for super-conductivity. T_c is maximal near 92 K when $x \approx 0.15$ and the structure is orthorhombic. Super-conductivity disappears at $x \approx 0.6$, where the structural transformation of YBCO occurs from orthorhombic to tetragonal.

Bi-, Tl-and Hg-based High-Tc Super-conductors

The crystal structure of Bi-, Tl-and Hg-based high-T_c super-conductors are very similar. Like YBCO, the perovskite-type feature and the presence of CuO_2 layers also exist in these super-conductors. However, unlike YBCO, Cu–O chains are not present in these super-conductors. The YBCO super-conductor has an orthorhombic structure, whereas the other high-T_c super-conductors have a tetragonal structure.

The **Bi–Sr–Ca–Cu–O** system has three super-conducting phases forming a homologous series as $Bi_2Sr_2Ca_{n-1}Cu_nO_{4+2n+x}$ (n = 1, 2 and 3). These three phases are Bi-2201, Bi-2212 and Bi-2223, having transition temperatures of 20, 85 and 110 K, respectively, where the numbering system represent number of atoms for Bi, Sr, Ca and Cu respectively. The two phases have a tetragonal structure which consists of two sheared crystallographic unit cells. The unit cell of these phases has double Bi–O planes which are stacked in a way that the Bi atom of one plane sits below the oxygen atom of the next consecutive plane. The Ca atom forms a layer within the interior of the CuO_2 layers in both Bi-2212 and Bi-2223; there is no Ca layer in the Bi-2201 phase. The three phases differ with each other in the number of CuO_2 planes; Bi-2201, Bi-2212 and Bi-2223 phases have one, two and three CuO_2 planes, respectively. The c axis of these phases increases with the number of CuO_2 planes. The co-ordination of the Cu atom is different in the three phases. The Cu atom forms an octahedral co-ordination with respect to oxygen atoms in the 2201 phase, whereas in 2212, the Cu atom is surrounded by five oxygen atoms in a pyramidal arrangement. In the 2223 structure, Cu has two co-ordinations with respect to oxygen : one Cu atom is bonded with four oxygen atoms in square planar configuration and another Cu atom is coordinated with five oxygen atoms in a pyramidal arrangement.

Tl–Ba–Ca–Cu–O super-conductor : The first series of the Tl-based super-conductor containing one Tl–O layer has the general formula $TlBa_2Ca_{n-1}Cu_nO_{2n+3'}$ whereas the second series containing two Tl–O layers has a formula of $Tl_2Ba_2Ca_{n-1}Cu_nO_{2n+4}$ with n = 1, 2 and 3. In the structure of $Tl_2Ba_2CuO_6$ (Tl-2201), there is one CuO_2 layer with the stacking sequence (Tl–O) (Tl–O) (Ba–O) (Cu–O) (Ba–O) (Tl–O) (Tl–O). In $Tl_2Ba_2CaCu_2O_8$ (Tl-2212), there are two Cu–O layers with a Ca layer in between. Similar to the $Tl_2Ba_2CuO_6$ structure, Tl–O layers are present outside the Ba–O layers. In $Tl_2Ba_2Ca_2Cu_3O_{10}$ (Tl-2223), there are three CuO2 layers enclosing Ca layers between each of these. In Tl-based super-conductors, T_c is found to increase with the increase in CuO_2 layers. However, the value of T_c decreases after four CuO_2 layers in $TlBa_2Ca_{n-1}Cu_nO_{2n+3'}$ and in the $Tl_2Ba_2Ca_{n-1}Cu_nO_{2n+4}$ compound, it decreases after three CuO_2 layers.

Hg–Ba–Ca–Cu–O super-conductor : The crystal structure of $HgBa_2CuO_4$ (Hg-1201), $HgBa_2CaCu_2O_6$ (Hg-1212) and $HgBa_2Ca_2Cu_3O_8$ (Hg-1223) is similar to that of Tl-1201, Tl-1212 and Tl-1223, with Hg in place of Tl. It is noteworthy that the T_c of the Hg compound (Hg-1201) containing one CuO_2 layer is much larger as compared to the one-CuO_2-layer compound of thallium (Tl-1201). In the Hg-based super-conductor, T_c is also found to increase as the CuO_2 layer increases. For Hg-1201, Hg-1212 and Hg-1223, the values of T_c are 94, 128 and the record value at ambient pressure 134 K, respectively, as shown in table below. The observation that the T_c of Hg-1223 increases to 153 K under high pressure indicates that the T_c of this compound is very sensitive to the structure of the compound.

Table : Critical temperature (T_c), crystal structure and lattice constants of some high-T_c super-conductors.

Formula	Notation	T_c (K)	No. of Cu-O planes in unit cell	Crystal structure
$YBa_2Cu_3O_7$	123	92	2	Orthorhombic
$Bi_2Sr_2CuO_6$	Bi-2201	20	1	Tetragonal
$Bi_2Sr_2CaCu_2O_8$	Bi-2212	85	2	Tetragonal
$Bi_2Sr_2Ca_2Cu_3O_6$	Bi-2223	110	3	Tetragonal
$Tl_2Ba_2CuO_6$	Tl-2201	80	1	Tetragonal
$Tl_2Ba_2CaCu_2O_8$	Tl-2212	108	2	Tetragonal
$Tl_2Ba_2Ca_2Cu_3O_{10}$	Tl-2223	125	3	Tetragonal
$TlBa_2Ca_3Cu_4O_{11}$	Tl-1234	122	4	Tetragonal
$HgBa_2CuO_4$	Hg-1201	94	1	Tetragonal
$HgBa_2CaCu_2O_6$	Hg-1212	128	2	Tetragonal
$HgBa_2Ca_2Cu_3O_8$	Hg-1223	134	3	Tetragonal

Preparation of High-Tc Super-conductors

The simplest method for preparing high-T_c super-conductors is a solid-state thermo-chemical reaction involving mixing, *calcination* and *sintering*. The appropriate amounts of precursor powders, usually oxides and carbonates, are mixed thoroughly using a ball mill. Solution chemistry processes such as *coprecipitation*, *freeze-drying* and *sol-gel* methods are alternative ways for preparing a homogeneous mixture. These powders are calcined in the temperature range from 800°C to 950°C for several hours. The powders are cooled, reground and calcined again. This process is repeated several times to get homogeneous material. The powders are subsequently compacted to pellets and sintered. The sintering environment such as temperature, annealing time, atmosphere and cooling rate play a very important role in getting good high-T_c super-conducting materials. The $YBa_2Cu_3O_{7-x}$ compound is prepared by calcination and sintering of a homogeneous mixture of Y_2O_3, $BaCO_3$ and CuO in the appropriate atomic ratio. Calcination is done at 900–950°C, whereas sintering is done at 950°C in an oxygen atmosphere. The oxygen stoichiometry in this material is very crucial for obtaining a super-conducting $YBa_2Cu_3O_{7-x}$ compound. At the time of sintering, the semiconducting tetragonal $YBa_2Cu_3O_6$ compound is formed, which, on slow cooling in oxygen atmosphere, turns into super-conducting $YBa_2Cu_3O_{7-x}$. The uptake and loss of oxygen are reversible in $YBa_2Cu_3O_{7-x}$. A fully oxidized orthorhombic $YBa_2Cu_3O_{7-x}$ sample can be transformed into tetragonal $YBa_2Cu_3O_6$ by heating in a vacuum at temperature above 700°C.

The preparation of Bi-, Tl-and Hg-based high-T_c super-conductors is difficult compared to YBCO. Problems in these super-conductors arise because of the existence of three or more phases having a similar layered structure. Thus, syntactic intergrowth and defects such as stacking faults occur during synthesis and it becomes difficult to isolate a single super-conducting phase. For Bi–Sr–Ca–Cu–O, it is relatively simple to prepare the Bi-2212 ($T_c \approx 85$ K) phase, whereas it is very difficult to prepare a single phase of Bi-2223 ($T_c \approx 110$ K). The Bi-2212 phase appears only after few hours of sintering at 860–870°C, but the larger fraction of the Bi-2223 phase is formed after a long reaction time of more than a week at 870°C. Although the substitution of Pb in the Bi–Sr–Ca–Cu–O compound has been found to promote the growth of the high-T_c phase, a long sintering time is still required.

Properties

"High-temperature" has two common definitions in the context of super-conductivity :

1. Above the temperature of 30 K that had historically been taken as the upper limit allowed by *BCS theory*. This is also above the 1973 record of 23 K that had lasted until copper-oxide materials were discovered in 1986.

2. Having a transition temperature that is a larger fraction of the *Fermi temperature* than for conventional super-conductors such as elemental *mercury* or *lead*. This definition encompasses a wider variety of *unconventional super-conductors* and is used in the context of theoretical models.

The label high-Tc may be reserved by some authors for s materials with critical temperature greater than the boiling point of *liquid nitrogen* (77 K or −196°C). However, a number of materials–including the original discovery and recently discovered pnictide super-conductors–had critical temperatures below 77 K but are commonly referred to in publication as being in the high-Tc class.

Technological applications could benefit from both the higher critical temperature being above the boiling point of liquid nitrogen and also the higher critical magnetic field (and critical current density) at which super-conductivity is destroyed. In magnet applications the high critical magnetic field may prove more valuable than the high T_c itself. Some cuprates have an upper critical field of about 100 tesla. However, cuprate materials are brittle ceramics which are expensive to manufacture and not easily turned into wires or other useful shapes.

After two decades of intense experimental and theoretical research, with over 100,000 published papers on the subject, several common features in the properties of high-temperature super-conductors have been identified. as of 2011, no widely accepted theory explain their properties. Relative to *conventional superconductors*, such as elemental mercury or lead that are adequately explained by the BCS theory, cuprate super-conductors (and other *unconventional super-conductors*) remain distinctive. There also has been much debate as to high-temperature super-conductivity coexisting with *magnetic ordering* in YBCO, *iron-based super-conductors*, several ruthenocuprates and other exotic super-conductors, and the search continues for other families of materials. HTS are *Type-II super-conductors*, which allow *magnetic fields* to penetrate their interior in *quantized* units of flux, meaning that much higher magnetic fields are required to suppress super-conductivity. The layered structure also gives a directional dependence to the magnetic field response.

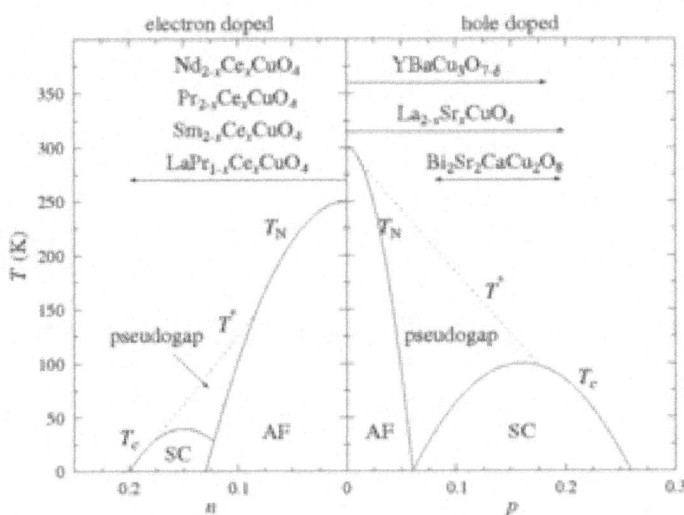

Cuprates

Simplified doping dependent phase diagram of cuprate super-conductors for both *electron* (n) and *hole* (p) doping. The phases shown are the *anti-ferromagnetic* (AF) phase close to zero doping, the *super-conducting* phase around optimal doping, and the *pseudogap* phase. Doping ranges possible for some common compounds are also shown. After.

Cuprate super-conductors are generally considered to be quasi-two-dimensional materials with their super-conducting properties determined by electrons moving within weakly coupled copper-oxide (CuO_2) layers. Neighbouring layers containing ions such as *lanthanum, barium, strontium,* or other atoms act to stabilize the structure and dope electrons or holes onto the copper-oxide layers. The undoped 'parent' or 'mother' compounds are *Mott insulators* with long-range anti-ferromagnetic order at low enough temperature. Single *band* models are generally considered to be sufficient to describe the electronic properties.

The cuprate super-conductors adopt a perovskite structure. The copper-oxide planes are *checkerboard lattices* with squares of O ions with a Cu ion at the centre of each square. The *unit cell* is rotated by 45° from these squares. Chemical formulae of super-conducting materials generally contain fractional numbers to describe the doping required for super-conductivity. There are several families of cuprate super-conductors and they can be categorized by the elements they contain and the number of adjacent copper-oxide layers in each super-conducting block. For example, YBCO and BSCCO can alternatively be referred to as Y123 and Bi2201/Bi2212/Bi2223 depending on the number of layers in each super-conducting block (*n*). The super-conducting transition temperature has been found to peak at an optimal doping value (*p* =0.16) and an optimal number of layers in each super-conducting block, typically *n* = 3.

Possible mechanisms for super-conductivity in the cuprates are still the subject of considerable debate and further research. Certain aspects common to all materials have been identified. Similarities between the *anti-ferromagnetic* low-temperature state of the undoped materials and the super-conducting state that emerges upon doping, primarily the d_{x-y} orbital state of the Cu ions, suggest that electron-electron interactions are more significant than electron-phonon interactions in cuprates–making the super-conductivity unconventional. Recent work on the *Fermi surface* has shown that nesting occurs at four points in the anti-ferromagnetic *Brillouin zone* where spin waves exist and that the super-conducting energy gap is larger at these points. The weak isotope effects observed for most cuprates contrast with conventional super-conductors that are well described by BCS theory.

Similarities and differences in the properties of hole-doped and electron doped cuprates :

- Presence of a pseudogap phase up to at least optimal doping.
- Different trends in the Uemura plot relating transition temperature to the super-fluid density. The inverse square of the *London penetration depth* appears to be proportional to the critical temperature for a large number of underdoped cuprate super-conductors, but the constant of proportionality is different for hole-and electron-doped cuprates. The linear trend implies that the physics of these materials is strongly two-dimensional.
- Universal hourglass-shaped feature in the spin excitations of cuprates measured using inelastic neutron diffraction.
- *Nernst effect* evident in both the super-conducting and pseudogap phases.

Iron-based Super-conductors

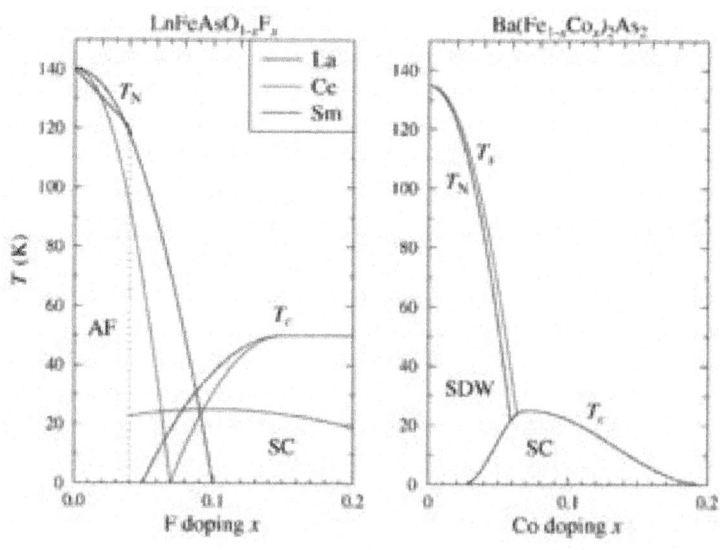

Simplified doping dependent phase diagrams of iron-based super-conductors for both Ln-1111 and Ba-122 materials. The phases shown are the anti-ferromagnetic/*spin density wave* (AF/SDW) phase close to zero doping and the super-conducting phase around optimal doping. The Ln-1111 phase diagrams for La and Sm were determined using *muon spin spectroscopy*, the phase diagram for Ce was determined using *neutron diffraction*. The Ba-122 phase diagram is based on.

Iron-based super-conductors contain layers of *iron* and a *pnictogen* — such as *arsenic* or *phosphorus* — or a *chalcogen*. This is currently the family with the second highest critical temperature, behind the cuprates. Interest in their super-conducting properties began in 2006 with the discovery of super-conductivity in LaFePO at 4 K and gained much greater attention in 2008 after the analogous material LaFeAs(O,F) was found to super-conduct at up to 43 K under pressure.

Since the original discoveries several families of iron-based super-conductors have emerged :

- LnFeAs(O,F) or $LnFeAsO_{1-x}$ with T_c up to 56 K, referred to as 1111 materials. A *fluoride* variant of these materials was subsequently found with similar T_c values.

- $(Ba,K)Fe_2As_2$ and related materials with pairs of iron-arsenide layers, referred to as 122 compounds. T_c values range up to 38 K. These materials also super-conduct when iron is replaced with *cobalt*

- LiFeAs and NaFeAs with T_c up to around 20 K. These materials super-conduct close to stoichiometric composition and are referred to as 111 compounds.

- FeSe with small off-*stoichiometry* or *tellurium* doping.

Most undoped iron-based super-conductors show a tetragonal-orthorhombic structural phase transition followed at lower temperature by magnetic ordering, similar to the cuprate super-conductors. However, they are poor metals rather than Mott insulators and have five *bands* at the *Fermi surface* rather than one. The phase diagram emerging as the iron-arsenide layers are doped is remarkably similar, with the super-conducting phase close to or overlapping the magnetic phase. Strong evidence that the T_c value varies with the As-Fe-As bond angles has already emerged and shows that the optimal T_c value is obtained with undistorted $FeAs_4$ tetrahedra. The symmetry of the pairing wavefunction is still widely debated, but an extended s-wave scenario is currently favoured.

Other Materials Sometimes Referred to as High-temperature Super-conductors

Magnesium diboride is occasionally referred to as a high-temperature super-conductor because its T_c value of 39 K is above that historically expected for *BCS* super-conductors. However, it is more generally regarded as the highest T_c conventional super-conductor, the increased T_c resulting from two separate bands being present at the *Fermi level*.

Fulleride super-conductors where alkali-metal atoms are intercalated into C_{60} molecules show super-conductivity at temperatures of up to 38 K for Cs_3C_{60}.

Some *organic super-conductors* and *heavy fermion* compounds are considered to be high-temperature super-conductors because of their high T_c values relative to their Fermi energy, despite the T_c values being lower than for many conventional super-conductors. This description may relate better to common aspects of the super-conducting mechanism than the super-conducting properties.

Theoretical work by *Neil Ashcroft* in 1968 predicted that solid *metallic hydrogen* at extremely high pressure should become super-conducting at approximately room-temperature because of its extremely high *speed of sound* and expected strong *coupling* between the conduction electrons and the lattice *vibrations*. This prediction is yet to be experimentally verified.

All known high-T_c super-conductors are Type-II super-conductors. In contrast to *Type-I super-conductors*, which expel all magnetic fields due to the *Meissner effect*, Type-II super-conductors allow magnetic fields to penetrate their interior in quantized units of flux, creating "holes" or "tubes" of *normal metallic* regions in the super-conducting bulk called *vortices*. Consequently, high-T_c super-conductors can sustain much higher magnetic fields.

Ongoing Research

The question of how super-conductivity arises in high-temperature super-conductors is one of the major unsolved problems of theoretical *condensed matter physics*. The mechanism that causes the electrons in these crystals to form pairs is not known. Despite intensive research and many promising leads, an explanation has so far eluded scientists. One reason for this is that the materials in question are generally very complex, multi-layered crystals (for example, *BSCCO*), making theoretical modelling difficult.

Improving the quality and variety of samples also gives rise to considerable research, both with the aim of improved characterisation of the physical properties of existing compounds, and synthesizing new materials, often with the hope of increasing T_c. Technological research focuses on making HTS materials in sufficient quantities to make their use economically viable and optimizing their properties in relation to *applications*.

Possible Mechanism

There have been two representative theories for HTS. Firstly, it has been suggested that the HTS emerges from anti-ferromagnetic spin fluctuations in a doped system. According to this theory, the pairing wave function of the cuprate HTS should have a d_{x-y} symmetry. Thus, determining whether the pairing wave function has *d*-wave symmetry is essential to test the spin fluctuation mechanism. That is, if the HTS order parameter (pairing wave function) does not have *d*-wave symmetry, then a pairing mechanism related to spin fluctuations can be ruled out. (Similar arguments can be made for iron-based super-conductors but the different material properties allow a different pairing symmetry.) Secondly, there was the **inter-layer coupling model**, according to which a layered structure consisting of BCS-type (s-wave symmetry) super-conductors can enhance the super-conductivity by it-self. By introducing an additional tunnelling interaction between each layer, this model successfully explained the anisotropic symmetry of the order parameter as well as the emergence of the HTS. Thus, in order to solve this unsettled problem, there have been numerous experiments such as *photoemission spectroscopy, NMR, specific heat* measurements, etc. But, unfortunately, the results were ambiguous, some reports supported the d symmetry for the HTS whereas others supported the s symmetry. This muddy situation possibly originated from the indirect nature of the experimental evidence, as well as experimental issues such as sample quality, impurity scattering, twinning, etc.

Techniques related to the *holographic principle* in *string theory* have been applied to study the *strange metal* behaviour of high-temperature cuprates. These suggest a *universal* behaviour of the super-conducting phase transition, by describing phase transitions as *field theories* on the boundary of a higher dimensional *gravitational theory*. Calculations using holographic methods have been able to reproduce the *conductivity* function for quasi 2-dimensional materials such as cuprate super-conductors.

Junction Experiment Supporting the d Symmetry

There was a clever experimental design to overcome the muddy situation. An experiment based on flux quantization of a three-grain ring of $YBa_2Cu_3O_7$ (YBCO) was proposed to test the symmetry of the order parameter in the HTS. The symmetry of the order parameter could best be probed at the junction interface as the Cooper pairs tunnel across a Josephson junction or weak link. It was expected that a half-integer flux, that is, a spontaneous magnetization could only occur for a junction of d symmetry super-conductors. But, even if the junction experiment is the strongest method to determine the symmetry of the HTS order parameter, the results have been ambiguous. J. R. Kirtley and C. C. Tsuei thought that the ambiguous results came from the defects inside the HTS, so that they designed an experiment where both clean limit (no defects) and dirty limit (maximal defects) were considered simultaneously. In the experiment, the spontaneous magnetiza-tion was clearly observed in YBCO, which supported the d symmetry of the order parameter in YBCO. But, since YBCO is orthorhombic, it might inherently have an admixture of s symmetry. So, by tuning their technique further, they found that there was an admixture of s symmetry in YBCO within about 3%. Also, they found that there was a pure d_{x-y} order parameter symmetry in the tetragonal $Tl_2Ba_2CuO_6$.

Qualitative Explanation of the Spin-fluctuation Mechanism

Despite all these years, the mechanism of high-T_c super-conductivity is still highly controversial, mostly due to the lack of exact theoretical computations on such strongly interacting electron systems. However, most rigorous theoretical calcula-tions, including phenomenological and diagrammatic approaches, converge on magnetic fluctuations as the pairing mechanism for these systems. The qualitative explanation is as follows :

In a super-conductor, the flow of electrons cannot be resolved into individual electrons, but instead consists of many pairs of bound electrons, called Cooper pairs. In conventional super-conductors, these pairs are formed when an electron moving through the material distorts the surrounding crystal lattice, which in turn attracts another electron and forms a bound pair. This is sometimes called the "water bed" effect. Each Cooper pair requires a certain minimum energy to be displaced, and if the thermal fluctuations in the crystal lattice are smaller than this energy the pair can flow without dissipating energy. This ability of the electrons to flow without resistance leads to super-conductivity.

In a high-T_c super-conductor, the mechanism is extremely similar to a conventional super-conductor. Except, in this case, phonons virtually play no role and their role is replaced by spin-density waves. As all conventional super-conductors are strong phonon systems, all high-T_c super-conductors are strong spin-density wave systems, within close vicinity of a magnetic transition to, for example, an anti-ferromagnet. When an electron moves in a high-T_c super-conductor, its spin creates a spin-density wave around it. This spin-density wave in turn causes a nearby electron to fall into the spin depression created by the first electron (water-bed effect again). Hence, again, a Cooper pair is formed. When the system temperature is lowered, more spin density waves and Cooper pairs are created, eventually leading to super-conductivity. Note that in high-T_c systems, as these systems are magnetic systems due to the Coulomb interaction, there is a strong Coulomb repulsion between electrons. This Coulomb repulsion prevents pairing of the Cooper pairs on the same lattice site. The pairing of the electrons occur at near-neighbor lattice sites as a result. This is the so-called d-wave pairing, where the pairing state has a node (zero) at the origin.

Examples

Examples of high-T_c cuprate super-conductors include $La_{1.85}Ba_{0.15}CuO_4$, and YBCO (*Yttrium-Barium-Copper-Oxide*), which is famous as the first material to achieve super-conductivity above the boiling point of liquid nitrogen.

Table : Transition temperatures of well-known super-conductors (Boiling point of liquid nitrogen for comparison)

Transition temperature (in kelvin)	Transition temperature (in celsius)	Material	Class
195	–78	*Sublimation* point of *dry ice*	
133	–140	$HgBa_2Ca_2Cu_3O_x$	Copper–oxide super-conductors
110	–163	$Bi_2Sr_2Ca_2Cu_3O_{10}$	
90	–183	$YBa_2Cu_3O_7$ (*YBCO*) Also, boiling point of *liquid oxygen*	
77	–196	Boiling point of *liquid nitrogen*	

(Contd...)

Transition temperature (in kelvin)	Transition temperature (in celsius)	Material	Class
55	–218	SmFeAs(O,F)	Iron–based super-conductors
41	–232	CeFeAs(O,F)	
26	–247	LaFeAs(O,F)	
20	–253	Boiling point of *liquid hydrogen*	
18	–255	*Nb$_3$Sn*	Metallic low–temperature super-conductors
10	–263	*NbTi*	
9.2	–263.8	*Nb*	
4.2	–268.8	Boiling point of *liquid helium*	
4.2	–268.8	Hg (*mercury*)	Metallic low–temperature super-conductors

TECHNOLOGICAL APPLICATIONS OF SUPER-CONDUCTIVITY

Some of the **technological applications of** *super-conductivity* include :

- The production of sensitive *magnetometers* based on *SQUIDs*
- Fast *digital circuits* (including those based on *Josephson junctions* and *rapid single flux quantum* technology),
- Powerful *super-conducting electromagnets* used in *maglev trains*, *Magnetic Resonance Imaging* (MRI) and *Nuclear magnetic resonance* (NMR) machines, magnetic confinement *fusion* reactors (*e.g. tokamaks*), and the beam-steering and focusing magnets used in *particle accelerators*
- Low-loss power cables
- *RF and microwave filters* (*e.g.*, for *mobile phone* base stations, as well as military ultra-sensitive/selective receivers)
- Fast *fault current limiters*
- High sensitivity *particle detectors*, including the *transition edge sensor*, the super-conducting *bolometer*, the *super-conducting tunnel junction* detector, the *kinetic inductance detector*, and the *super-conducting nanowire single-photon detector*
- *Railgun* and *coilgun* magnets
- *Electric motors* and *generators*.

Magnetic Resonance Imaging (MRI) and Nuclear Magnetic Resonance (NMR)

The biggest application for super-conductivity is in producing the large volume, stable, and high magnetic fields required for MRI and NMR. This represents a multi-billion US$ market for companies such as *Oxford Instruments* and *Siemens*. The magnets typically use *low temperature super-conductors* (LTS) because *high-temperature super-conductors* are not yet cheap enough to cost-effectively deliver the high, stable and large volume fields required, notwithstanding the need to cool LTS instruments to *liquid helium* temperatures. Super-conductors are also used in high field scientific magnets.

High-temperature super-conductivity (HTS)

The commercial applications so far for *high temperature super-conductors* (HTS) have been limited.

HTS can super-conduct at temperatures above the boiling point of *liquid nitrogen*, which makes them cheaper to cool than low temperature super-conductors (LTS). However, the problem with HTS technology is that the currently known high temperature super-conductors are brittle ceramics which are expensive to manufacture and not easily formed into wires or other useful shapes. Therefore, the applications for HTS have been where it has some other intrinsic advantage, *e.g.* in

- Low thermal loss current leads for LTS devices (low thermal conductivity),
- RF and microwave filters (low resistance to RF), and
- Increasingly in specialist scientific magnets, particularly where size and electricity consumption are critical (while HTS wire is much more expensive than LTS in these applications, this can be offset by the relative cost and convenience of cooling); the ability to ramp field is desired (the higher and wider range of HTS's *operating temperature* means faster changes in field can be managed); or cryogen free operation is desired (LTS generally requires *liquid helium* that is becoming more scarce and expensive).

HTS-based Systems

HTS has application in scientific and industrial magnets, including use in NMR and MRI systems. Commercial systems are now available in each category.

Also one intrinsic attribute of HTS is that it can withstand much higher magnetic fields than LTS, so HTS at liquid helium temperatures are being explored for very high-field inserts inside LTS magnets.

Promising future industrial and commercial HTS applications include *Induction heaters, transformers, fault current limiters, power storage, motors* and *generators, fusion* reactors devices.

Early applications will be where the benefit of smaller size, lower weight or the ability to rapidly switch current (fault current limiters) outweighs the added cost. Longer-term as conductor price falls HTS systems should be competitive in a much wider range of applications on energy efficiency grounds alone. (For a relatively technical and US-centric view of state of play of HTS technology in power systems and the development status of Generation 2 conductor.

Holbrook Super-conductor Project

The *Holbrook Super-conductor Project* is a project to design and build the world's first production *super-conducting transmission* power cable. The cable was commissioned in late June 2008. The suburban *Long Island* electrical sub-station is fed by about 600-meter-long underground cable system consists of about 99 miles of *high-temperature super-conductor wire* manufactured by *American Super-conductor*, installed underground and chilled with *liquid nitrogen* greatly reducing the costly right-of-way required to deliver additional power.

Tres Amigas Project

American Super-conductor was chosen for The *Tres Amigas Project*, the United States' first renewable energy market hub. The Tres Amigas renewable energy market hub will be a multi-mile, triangular electricity pathway of super-conductor electricity pipelines capable of transferring and balancing many gigawatts of power between three U.S. power grids (the Eastern Inter-connection, the Western Inter-connection and the Texas Inter-connection). Unlike traditional powerlines, it will transfer power as DC instead of AC current. It will be located in Clovis, New Mexico.

Magnesium Diboride

Magnesium diboride is a much cheaper super-conductor than either *BSCCO* or *YBCO* in terms of cost per current-carrying capacity per length (cost/, in the same ballpark as LTS, and on this basis many manufactured wires are already cheaper than copper. Furthermore, MgB_2 super-conducts at temperatures higher than LTS (its critical temperature is 39 K, compared with less than 10 K for NbTi and 18.3 K for Nb_3Sn), introducing the possibility of using it at 10-20 K in cryogen-free magnets or perhaps eventually in liquid hydrogen. However MgB_2 is limited in the magnetic field it can tolerate at these higher temperatures, so further research is required to demonstrate its competitiveness in higher field applications.

SUPER-FLUIDITY

Super-fluidity is a *state of matter* in which the *matter* behaves like a fluid with zero *viscosity*; where it appears to exhibit the ability to self-propel and travel in a way that defies the forces of *gravity* and *surface tension*. While this characteristic was originally discovered in *liquid helium*, it is also found in astrophysics, high-energy physics, and theories of *quantum gravity*. The phenomenon is related to the *Bose–Einstein condensation*, but it is not identical : not all Bose-Einstein condensates can be regarded as super-fluids, and not all super-fluids are Bose–Einstein condensates.

Fig. : Helium II will "creep" along surfaces in order to find its own level – after a short while, the levels in the two containers will equalize. The *Rollin film* also covers the interior of the larger container; if it were not sealed, the helium II would creep out and escape.

Fig. : The liquid helium is in the super-fluid phase. As long as it remains super-fluid, it creeps up the wall of the cup as a thin film. It comes down on the outside, forming a drop which will fall into the liquid below. Another drop will form — and so on — until the cup is empty.

Super-fluidity of Liquid Helium

In liquid helium, the super-fluidity effect was discovered by *Pyotr Kapitsa* and *John F. Allen*. It has since been described through *phenomenological* and microscopic theories. In liquid helium-4, the super-fluidity occurs at far higher temperatures than it does in helium-3. Each atom of helium-4 is a *boson* particle, by virtue of its integer spin. A helium-3 atom, however, is a *fermion* particle; it can form bosons only by pairing with itself at much lower temperatures. This process is similar to the electron pairing in *super-conductivity*.

Ultra-cold Atomic Gases

Super-fluidity in an ultra-cold fermionic gas was experimentally proved by *Wolf-gang Ketterle* and his team who observed *quantum vortices* in Li at a temperature of 50 nK at *MIT* in April 2005. Such vortices had previously been observed in an ultra-cold bosonic gas using Rb in 2000, and more recently in two-dimensional gases. As early as 1999 *Lene Hau* created such a condensate using sodium atoms for the purpose of slowing light, and later stopping it completely. Her team then subsequently used this system of compressed light to generate the super-fluid analogue of shock waves and tornadoes : "These dramatic excitations result in the formation of solitons that in turn decay into quantized vortices — created far out of equilibrium, in pairs of opposite circulation — revealing directly the process of super-fluid breakdown in Bose-Einstein condensates. With a double light-roadblock setup, we can generate controlled collisions between shock waves resulting in completely unexpected, non-linear excitations. We have observed hybrid structures consisting of vortex rings embedded in dark solitonic shells. The vortex rings act as 'phantom propellers' leading to very rich excitation dynamics."

Super-fluid in Astrophysics

The idea that super-fluidity exists inside *neutron stars* was first proposed by *Arkady Migdal*. By analogy with electrons inside *super-conductors* forming *Cooper pairs* due to electron-lattice interaction, it is expected that *nucleons* in a neutron star at sufficiently high density and low temperature can also form Cooper pairs due to the long-range attractive nuclear force and lead to super-fluidity and super-conductivity.

Super-fluidity in High-energy Physics and Quantum Gravity

Super-fluid vacuum theory is an approach in *theoretical physics* and *quantum mechanics* where the physical *vacuum* is viewed as super-fluid.

The ultimate goal of the approach is to develop scientific models that unify *quantum mechanics* (describing three of the four known fundamental interactions) with *gravity*. This makes *SVT* a candidate for the theory of *quantum gravity* and an extension of the *Standard Model*. It is hoped that development of such theory would unify into a single consistent model of all fundamental interactions, and to describe all known interactions and elementary particles as different manifestations of the same entity, super-fluid vacuum.

FLUX PINNING

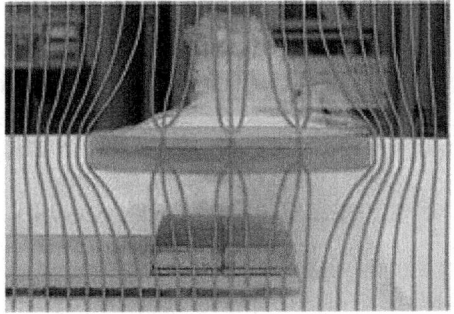

Fig. : Flux Pinning : Flux Tube diagram.

Flux pinning is the phenomenon where a super-conductor is pinned in space above a magnet. The super-conductor must be a *type-II super-conductor* due to the fact that *type-I super-conductors* cannot be penetrated by magnetic fields. The act of magnetic penetration is what makes flux pinning possible. At higher temperatures the super-conductor allows magnetic flux to enter in quantized packets surrounded by a super-conducting current vortex. These sites of penetration are known as *flux tubes*. The number of flux tubes per unit area is proportional to the magnetic field with a constant of proportionality equal to the *magnetic flux quantum*. On a simple 3-inch diameter, 1-micrometer thick disk, next to a magnetic field of 350 Oe, there are approximately 100 billion flux tubes that hold 70,000 times the super-conductor's weight. At low temperatures the flux tubes are pinned in place and cannot move. This pinning is what holds the super-conductor in place thereby allowing it to levitate. This phenomenon is closely related to the *Meissner effect*, though with one crucial difference — the Meissner effect shields the super-conductor from all magnetic fields causing repulsion, unlike the pinned state of the super-conductor disk which pins flux, and the super-conductor in place.

Importance of Flux Pinning

Flux pinning is desirable in *high-temperature ceramic* super-conductors to prevent "flux creep", which can create a pseudo-*resistance* and depress both critical *current density* and critical field.

Degradation of a high-temperature super-conductor's properties due to flux creep is a limiting factor in the use of these super-conductors. *SQUID* magnetometers suffer reduced precision in a certain range of applied field due to flux creep in the super-conducting magnet used to bias the sample, and the maximum field strength of high-temperature super-conducting magnets is drastically reduced by the depression in critical field.

Flux Pinning in the Future

The worth of flux pinning is seen through many implementations such as lifts, frictionless joints, and transportation. The thinner the super-conducting layer, the stronger the pinning that occurs when exposed to magnetic fields. Since the super-conductor is pinned above the magnet away from any surfaces, there is the potential for a frictionless joint. Transportation is another area flux pinning technology could revolutionize and reform. *MagSurf* was developed by a French university utilizing flux pinning to create a hovercraft-like effect that could support a human, demonstrating the usefulness of the technology.

TYPE-I SUPER-CONDUCTOR

The interior of a bulk *super-conductor* cannot be penetrated by a weak *magnetic field*, a phenomenon known as the *Meissner effect*. When the applied magnetic field becomes too large, super-conductivity breaks down. Super-conductors can be divided into two types according to how this breakdown occurs. In **type-I super-conductors**, super-conductivity is abruptly destroyed *via* a *first order phase transition* when the strength of the applied field rises above a critical value H_c. This type of super-conductivity is normally exhibited by pure metals, *e.g.* aluminium, lead, and mercury. Depending on the demagnetization factor, one may obtain an intermediate state. This state, first described by *Lev Landau*, is a phase separation into macroscopic non-super-conducting and super-conducting domains.

This behaviour is different from *type-II super-conductors* which exhibit two critical magnetic fields. The first, lower critical field occurs when *magnetic flux vortices* penetrate the material but the material remains super-conducting outside of these microscopic vortices. When the vortex density becomes too large, the entire material becomes non-super-conducting; this corresponds to the second, higher critical field.

The ratio of the *London penetration depth* λ to the *super-conducting coherence length* ξ determines whether a super-conductor is type-I or type-II. Type-I super-conductors are those with $0 < \lambda/\xi < 1/\sqrt{2}$, and type-II super-conductors are those with $\lambda/\xi > 1/\sqrt{2}$.

TYPE-II SUPER-CONDUCTOR

In *super-conductivity*, a **type-II super-conductor** is characterized by the formation of *magnetic vortices* in an applied *magnetic field*. This occurs above a certain critical field strength H_{c1}. The vortex density increases with increasing field strength. At a higher critical field H_{c2}, super-conductivity is completely destroyed.

History

The idea of two types of *super-conductors* was proposed by *Lev Landau* and *Vitaly Ginzburg* in their paper on *Ginzburg-Landau theory*. In their argument, a *type-I super-conductor* had positive *free energy* of the super-conductor-normal metal boundary. At that time, all known super-conductors were type-I, and initially type-II behaviour was considered unphysical. Type-II super-conducting behaviour was first observed in experiments by *Lev Shubnikov*, who investigated super-conducting alloys in a magnetic field, and later by Zavaritskii. The theory for the behaviour of the type-II super-conducting state in magnetic field was developed by *Alexei Alexeyevich Abrikosov*, who was elaborating on the ideas by *Lars Onsager* and *Richard Feynman* of quantum vortices in *super-fluids* and *Fritz London*'s idea of *magnetic flux* quantization in super-conductors. The *Nobel Prize in Physics* was awarded for the theory of type-II super-conductivity in 2003.

Vortex State

Ginzburg–Landau theory defines two parameters : The *super-conducting coherence length* and the *London magnetic field penetration depth*. In a type-II super-conductor, the *coherence length* is smaller than the penetration depth. This leads to negative energy of the interface between super-conducting and normal phases. The existence of the negative interface energy was known since the mid-1930s from the early works by the London brothers. A negative *interface energy* suggests that the system should be unstable against maximizing the number of such interfaces, which was not observed in first experiments on super-conductors, before the experiments of Shubnikov in 1936 where two critical fields were found. As was later discussed by A.A. Abrikosov these interfaces manifest as lines of magnetic flux passing through the material, turning a region of the super-conductor normal. This normal region is separated from the rest of the super-conductor by a circulating super-current. In analogy with fluid dynamics, the swirling super-current creates what is known as a *vortex*, or an *Abrikosov vortex*, after *Alexei Alexeyevich Abrikosov*. He found that the vortices arrange themselves into a regular array known as a *vortex lattice*.

In the extreme type-II limit, the problem of type-II super-conductor in magnetic field is exactly equivalent to that of vortex state in rotating *super-fluid helium*, which was discussed earlier by *Richard Feynman* in 1955.

Materials

Type-II super-conductors are usually made of metal *alloys* or complex oxide *ceramics*. All *high temperature super-conductors* are type-II super-conductors. While most elemental super-conductors are type-I, *niobium, vanadium,* and *technetium* are elemental type-II super-conductors. *Boron*-doped *diamond* and *silicon* are also type-II super-conductors. Metal alloy super-conductors also exhibit type-II behaviour (*e.g. niobium-titanium* and *niobium-tin*).

Other type-II examples are the *cuprate-perovskite* ceramic materials which have achieved the highest super-conducting critical temperatures. These include $La_{1.85}Ba_{0.15}CuO_4$, *BSCCO,* and *YBCO* (*Yttrium-Barium-Copper-Oxide*), which is famous as the first material to achieve super-conductivity above the boiling point of *liquid nitrogen* (77 K). Due to strong *vortex pinning*, the cuprates are close to *ideally hard super-conductors*.

Important Uses

Strong super-conducting electromagnets (used in *MRI* scanners, *NMR* machines, and *particle accelerators*) often use *niobium-titanium* or, for higher fields, *niobium-tin*.

JOULE HEATING

Joule heating, also known as **ohmic heating** and **resistive heating,** is the process by which the passage of an *electric current* through a *conductor* releases *heat*. The amount of heat released is proportional to the square of the current such that

$Q \propto I^2 \cdot R$

This relationship is known as **Joule's first law.** The *SI unit* of *energy* was subsequently named the *joule* and given the symbol *J*. The commonly known unit of power, the *watt*, is equivalent to one joule per second. Joule heating is independent of the direction of current, unlike heating due to the *Peltier effect*.

Background

History

Resistive heating was first studied by *James Prescott Joule* in 1841. Joule immersed a length of wire in a fixed *mass* of *water* and measured the *temperature* rise due to a known current flowing through the wire for a 30 *minute* period. By varying the current and the length of the wire he deduced that the heat produced was *proportional* to the *square* of the current multiplied by the *electrical resistance* of the wire.

Microscopic Description

Joule heating is caused by interactions between the moving *particles* that form the current (usually, but not always, *electrons*) and the *atomic ions* that make up the body of the conductor. *Charged* particles in an *electric circuit* are *accelerated* by an *electric field* but give up some of their *kinetic energy* each time they collide with an ion. The increase in the kinetic or *vibrational* energy of the ions manifests itself as heat and a rise in the temperature of the conductor. Hence energy is transferred from the electrical *power supply* to the conductor and any materials with which it is in *thermal contact*.

Power Loss and Noise

Joule heating is referred to as *ohmic heating* or *resistive heating* because of its relationship to *Ohm's Law*. It forms the basis for the large number of practical applications involving *electric heating*. However, in applications where heating is an unwanted *by-product* of current use (*e.g., load losses* in *electrical transformers*) the diversion of energy is often referred to as *resistive loss*. The use of *high voltages* in *electric power transmission* systems is specifically designed to reduce such losses in cabling by operating with commensurately lower currents. The *ring circuits*, or ring mains, used in UK homes are another example, where power is delivered to outlets at lower currents, thus reducing Joule heating in the wires. Joule heating does not occur in *super-conducting* materials, as these materials have zero electrical resistance in the super-conducting state.

Resistors create electrical noise, called *Johnson–Nyquist noise*. There is an intimate relationship between Johnson–Nyquist noise and Joule heating, explained by the *fluctuation-dissipation theorem*.

Formulas

Direct Current

The most general and fundamental formula for Joule heating is :

$P = VI$

where :

- P is the *power* (energy per unit time) converted from electrical energy to thermal energy,
- I is the current travelling through the resistor or other element,
- V is the *voltage drop* across the element.

   The explanation of this formula ($P=VI$) is :

   (*Energy dissipated per unit time*) = (*Energy dissipated per charge passing through resistor*) × (*Charge passing through resistor per unit time*)

   When *Ohm's law* is also applicable, the formula can be written in other equivalent forms :

$$P = IV = I^2 R = V^2 / R$$

where R is the *resistance*.

Alternating Current

When current varies, as it does in AC circuits,

$$P(t) = I(t)V(t)$$

where t is time and P is the instantaneous power being converted from electrical energy to heat. Far more often, the *average* power is of more interest than the instantaneous power :

$$P_{avg} = I_{rms}V_{rms} = I_{rms}^2 R = V_{rms}^2 / R$$

where "avg" denotes *average (mean)* over one or more cycles, and "rms" denotes *root mean square*.

These formulas are valid for an ideal resistor, with zero *reactance*. If the reactance is nonzero, the formulas are modified :

$$P_{avg} = I_{rms}V_{rms}\cos\phi = I_{rms}^2 = \mathrm{Re}(Z) = V_{rms}^2\,\mathrm{Re}(Y^*)$$

where ϕ is the phase difference between current and voltage, Re means *real part*, Z is the *complex impedance*, and Y^* is the *complex conjugate* of the *admittance* (equal to $1/Z^*$).

Differential Form

In plasma physics, the Joule heating often needs to be calculated at a particular location in space. The differential form of the Joule heating equation gives the power per unit volume.

$$P = J \cdot E$$

Here, J is the current density, and E is the electric field.

Reason for High-voltage Transmission of Electricity

In electric power transmission, high voltage is used to reduce Joule heating of the *overhead power lines*. The valuable electric energy is intended to be used by consumers, not for heating the power lines. Therefore, this Joule heating is referred to as a type of *transmission loss*.

A given quantity of electric power can be transmitted through a transmission line either at low voltage and high current, or with a higher voltage and lower current. *Transformers* can convert a high transmission voltage to a lower voltage for use by customer loads. Since the power lost in the wires is proportional to the conductor resistance and the square of the current, using low current at high voltage reduces the loss in the conductors due to Joule heating (or alternatively allows smaller conductors to be used for the same relative loss).

Applications

There are many practical uses of Joule heating. Some of the commonest are as follows :

- An *incandescent light bulb* glows when the filament is heated by Joule heating, so hot that it glows white with *thermal radiation* (also called *blackbody radiation*).
- *Electric stoves* and other *electric heaters* usually work by Joule heating.
- *Soldering irons* and *cartridge heaters* are very often heated by Joule heating.
- *Electric fuses* rely on the fact that if enough current flows, enough heat will be generated to melt the fuse wire.
- *Electronic cigarettes* usually work by Joule heating, vapourizing propylene glycol and vegetable glycerine.
- *Thermistors* and *resistance thermometers* are resistors whose resistance changes when the temperature changes. These are sometimes used in conjunction with Joule heating (also called self-heating in this context) : If a large current is running through the non-linear resistor, the resistor's temperature rises and therefore, its resistance changes. Therefore, these components can be used in a circuit-protection role similar to *fuses*, or for *feedback* in circuits, or for many other purposes. In general, self-heating can turn a resistor into a non-linear and hysteretic circuit element.

Less-common Applications

- *Food processing* equipment may make use of Joule heating in food production. In this case, the food material serves as an electrical resistor, and heat is released internally.

Heating Efficiency

As a heating technology, Joule heating has a *coefficient of performance* of 1.0, meaning that every 1 watt of electrical power is converted to 1 joule of heat. By comparison, a *heat pump* can have a coefficient of more than 1.0 since it also absorbs additional heating energy from the environment, moving this thermal energy to where it is needed.

The definition of the efficiency of a heating process requires defining the boundaries of the system to be considered. When heating a building, the overall efficiency is different when considering heating effect per unit of electric energy delivered on the customer's side of the meter, compared to the overall efficiency when also considering the losses in the power plant and transmission of power.

Hydraulic Equivalent

In the *energy balance of groundwater flow* a hydraulic equivalent of Joule's law is used :

$$\frac{dE}{dx} = \frac{v_x^2}{K}$$

where :

dE/dx = loss of hydraulic energy (E) due to friction of flow in x-direction per unit of time–comparable to Q/t

v_x = flow velocity in x-direction–comparable to I

K = *hydraulic conductivity* of the soil–the hydraulic conductivity is inversely proportional to the hydraulic resistance which compares to R

LIST OF SUPER-CONDUCTORS

A table showing major parameters of major super-conductors of simple structure (numerous metallic alloys are not shown). X : Y means material X doped with element Y, T_C is the highest reported transition temperature in *kelvins* and H_C is a critical magnetic field in *teslas*. "Non-metals" here refers to materials which are normally not considered as metals, but become super-conducting upon heavy doping. "BCS" means whether or not the super-conductivity is explained within the *BCS theory*.

Formula	$T_c(K)$	$H_c(T)$	Type	BCS	References
Metals					
Al	1.20	0.01	I	yes	
Cd	0.52	0.0028	I	yes	
Ga	1.083	0.0058	I	yes	
Hf	0.165		I	yes	
a-Hg	4.15	0.04	I	yes	
β-Hg	3.95	0.04	I	yes	
Ga	1.1	0.005	I	yes	
In	3.4	0.03	I	yes	
Ir	0.14	0.0016	I	yes	
a-La	4.9		I	yes	
β-La	6.3		I	yes	
Mo	0.92	0.0096	I	yes	
Nb	9.26	0.82	II	yes	
Os	0.65	0.007	I	yes	
Pa	1.4		I	yes	
Pb	7.19	0.08	I	yes	
Re	2.4	0.03	I	yes	
Ru	0.49	0.005	I	yes	
Sn	3.72	0.03	I	yes	
Ta	4.48	0.09	I	yes	
Tc	7.46–11.2	0.04	II	yes	

(Contd...)

(*Contd...*)

Formula	$T_c(K)$	$H_c(T)$	Type	BCS	References
a-Th	1.37	0.013	I	yes	
Ti	0.39	0.01	I	yes	
Tl	2.39	0.02	I	yes	
a-U	0.68		I	yes	
β-U	1.8		I	yes	
V	5.03	1	II	yes	
a-W	0.015	0.00012	I	yes	
β-W	1–4				
Zn	0.855	0.005	I	yes	
Zr	0.55	0.014	I	yes	
Compounds					
Ba_8Si_{46}	8.07	0.008	II	yes	
C_6Ca	11.5	0.95	II		
$C_6Li_3Ca_2$	11.15		II		
C_8K	0.14		II		
C_8KHg	1.4		II		
C_6K	1.5		II		
C_3K	3.0		II		
C_3Li	<0.35		II		
C_2Li	1.9		II		
C_3Na	2.3–3.8		II		
C_2Na	5.0		II		
C_8Rb	0.025		II		
C_6Sr	1.65		II		
C_6Yb	6.5		II		
$C_{60}Cs_2Rb$	33		II	yes	
$C_{60}K_3$	19.8	0.013	II	yes	
$C_{60}Rb_X$	28		II	yes	
Diamond : B	11.4	4	II	yes	
FeB4	2.9		I		
InN	3		II	yes	
In_2O_3	3.3	~3	II	yes	
Si : B	0.4	0.4	II	yes	
SiC : B	1.4	0.008	I	yes	
SiC : Al	1.5	0.04	II	yes	

(*Contd...*)

(Contd...)

Formula	$T_c(K)$	$H_c(T)$	Type	BCS	References
Binary alloys					
LaB_6	0.45			yes	
MgB_2	39	74	II	yes	
Nb_3Al	18		II	yes	
Nb_3Ge	23.2	37	II	yes	
NbO	1.38		II	yes	
NbN	16		II	yes	
Nb_3Sn	18.3	30	II	yes	
$NbTi$	10	15	II	yes	
YB_6	8.4		II	yes	
TiN	5.6			yes	
ZrN	10			yes	
ZrB_{12}	6.0		I	yes	

MOLECULAR WIRE

Molecular wires (or sometimes called molecular nanowires) are molecular-scale objects which conduct electrical current. They are the fundamental building blocks for molecular electronic devices. Their typical diameters are less than three nanometers, while their bulk lengths may be macroscopic, extending to centimeters or more.

Materials

The most common types of molecular wires are based on organic molecules. Higher conductivities originate from highly conjugated systems, while alkane chains are important in understanding basic charge transport and tunneling. A molecular wire occurring in nature is *DNA*, but it is believed not to conduct in natural form, and while doping studies have so far not produced convincing long-range conductivity in DNA wires, research into producing conducting DNA has been quite intense. Prominent inorganic examples include polymeric materials such as $Li_2Mo_6Se_6$ and $Mo_6S_{9-x}I_x$, and single-molecule *extended metal atom chains* which comprise strings of *transition metal* atoms directly bonded to each other. Molecular wires containing paramagnetic inorganic moieties are interesting, in particular, because they can lead to observations of *Kondo peaks*.

Structure

Unlike the more usual nanowires (which are very thin crystals), molecular nanowires are composed of repeating molecular units, which may be organic (*e.g.* DNA) or inorganic (*e.g.* $Mo_6S_{9-x}I_x$). In the case of *DNA*, the repeat units are the nucleotides with a backbone made of sugars and phosphate groups joined by ester bonds. Attached to each sugar is one of four types of bases. In case of $Mo_6S_{9-x}I_x$, the repeat units are $Mo_6S_{9-x}I_x$ clusters, which are joined together by flexible sulfur or iodine bridges. Molecular nanowires can be manipulated and investigated as single molecules, but often aggregate in solution into swatches or bundles. In the case of the Mo chalcogenide-halides, they grow in the form of ordered strands, in which the individual strands are linked by very weak Van der Waals forces. Individual molecules can be manipulated, ordered and their length can be controlled with atomic force microsope tips. *EMAC* molecular wires consist of distinct molecules whose length can be controlled on the atomic scale.

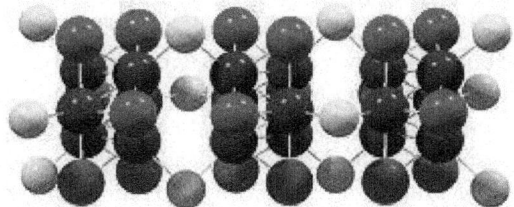

Fig. : The structure of a $Mo_6S_{9-x}I_x$ molecular wire. Mo atoms are blue, iodine atoms are red and sulphur atoms are yellow.

Conduction of Electrons

Molecular wires conduct electricity. They typically have non-linear current-voltage characteristics, and do not behave as simple ohmic conductors. The conductance follows typical power law behaviour as a function of temperature or electric field, whichever is the greater, arising from their strong one-dimensional character. Numerous theoretical ideas have been used in an attempt to understand the conductivity of one-dimensional systems, where strong interactions between electrons lead to departures from normal metallic (*Fermi liquid*) behaviour. Important concepts are those introduced by *Tomonaga, Luttinger* and *Wigner*. Effects caused by classical Coulomb repulsion (called *Coulomb blockade*), interactions with vibrational degrees of freedom (called *phonons*) and *Quantum Decoherence* have also been found to be important in determining the properties of molecular wires.

Use of Nanowires in Molecular Electronics

To be of use for connecting molecules together, MWs need to display some very important characteristics. The connectors between elements need to be able to self-assemble following well-defined routes and form reliable electrical contacts between them. To reproducibly self-assemble a complex circuit based on single molecules, it is essential that the connectors which join them have recognitive ability. They should be able to connect to diverse materials, such as gold metal surfaces (for connections to outside world), biomolecules (for nanosensors, nanoelectrodes, molecular switches) and most importantly, they must allow branching. The connectors should also be available of pre-determined diameter and length. They should also have covalent bonding to ensure reproducible transport and contact properties. DNA-like molecules have specific molecular-scale recognition and can be used very effectively in molecular scaffold fabrication. Very complex shapes have recently been demonstrated, but unfortunately metal coated DNA which is electrically conducting is much too thick to connect to individual molecules. Thinner coated DNA lacks electronic connectivity, and are not suitable for connecting molecular electronics elements. Some varieties of *carbon nanotubes* (CNTs) are conducting, and connectivity at their ends can be achieved by attachment of connecting groups. Unfortunately manufacturing CNTs with pre-determined properties is impossible at present, and the functionalized ends are typically not conducting, limiting their usefulness as molecular connectors. Individual CNTs can be soldered in an electron microscope, but the contact is not covalent and cannot be self-assembled. Recently possible routes for the construction of larger functional circuits using $Mo_6S_{9-x}I_x$ MWs have been demonstrated, either *via* gold nanoparticles as linkers, or by direct connection to thiolated molecules. The two approaches may lead to different possible applications. The use of GNPs offers the possibility of branching and construction of larger circuits.

Other Uses

Molecular wires can be incorporated into *polymers*, enhancing their mechanical and/or conducting properties. The enhancement of these properties relies on uniform dispersion of the wires into the host polymer. Recent advances in the use of MoSI wires have been made in such composites, relying on their superior solubility within the polymer host compared to other nanowires or nanotubes. Bundles of wires can be used to enhance tribological properties of polymers, with applications in actuators and potentiometers.

Chapter 17

BA-DOPED IRON OXIDE AS A NEW MATERIAL FOR NO$_2$ DETECTION

Christian Lopez[1], Chiara Baroni[2] and Jean-Marc Tulliani[2,*]

[1] Laboratory of Electrochemistry and Physical-chemistry of Materials and Interfaces, UMR 5279, CNRS-Grenoble INP-Université de Savoie-Université Joseph Fourier, BP75, 38402 Saint Martind d'Hères, France; E-Mail: christian.lopez@lepmi.grenoble-inp.fr

[2] Department of Applied Science and Technology, Politecnico di Torino, INSTM Reference Laboratory for Ceramics Engineering, Corso Duca degli Abruzzi 24, 10129 Torino, Italy; E-Mail: chiara.baroni@polito.it

* Author to whom correspondence should be addressed; E-Mail: jeanmarc.tulliani@polito.it; Tel.: +39-11-090-4700; Fax: +39-11-090-4624.

ABSTRACT

Various compositions of barium-doped hematite between pure hematite (α-Fe$_2$O$_3$) and pure barium hexaferrite (BaFe$_{12}$O$_{19}$) were synthesized by solid state reaction. The XRD analyses confirmed the progressive evolution of the two crystalline phases. Tests as humidity sensors show that the electrical resistance of samples containing high proportions of hexaferrite phase is strongly influenced. Electrochemical impedance spectroscopy (EIS) analyses under air or argon revealed an intrinsic semiconducting behavior for hematite and samples doped with 3 and 4 wt % equivalent BaO. The samples containing higher proportions of barium exhibited an extrinsic semiconducting behavior characterized by a variation of the conductivity with the oxygen partial pressure. This study allowed us to define the percolation threshold of the barium hexaferrite crystalline phase in the hematite matrix. The value was estimated to hematite doped with 5 wt % BaO, *i.e.*, 36 wt % of barium hexaferrite phase. EIS analyses under various NO$_2$ partial pressures confirmed the sensitivity of these materials. The linearity of the response was particularly evident for the 5, 10 and 14 wt % samples.

Keywords

Hematite; barium hexaferrite; NO_2 detection; impedance measurements.

1. INTRODUCTION

Nitrogen oxides (NO and NO_2: NO_x), released from combustion facilities and automobiles, are a main cause of air pollution. They are responsible for acid rains, photochemical smog and are also potentially eutrophying agents, *i.e.*, can cause an oversupply of nutrient in soils and water bodies. Therefore, they are known to be harmful to the environment, to people, and also to historical monuments and buildings. The current directive 2008/50/EC of the European Union and the future Decision 2011/850/EU (from 1 January 2014) on ambient air quality has set at 40 µg/m³ the annual limit value, and at 200 µg/m³ the hourly limit value, not to be exceeded more than 18 times in a calendar year, for the protection of human health against the effects of gaseous NO_2 [1,2]. Therefore, reliable, simple, effective and low-cost methods to monitor them have been highly demanded for atmospheric environmental measurements and controls.

Many of the systems usually employed for the monitoring of air pollutants are based on traditional photometric techniques like chemiluminescence [1], even if, more recently, electroanalysis techniques (amperometric approach), have been proposed and tested [3]. The main drawbacks associated to these techniques are represented by the use of expensive bench scale laboratory equipment including calibrating facilities.

Semiconducting metal oxide (SMO) sensors are one of the most widely studied groups of chemiresistive gas sensors due to their unique advantages such as low cost, small size, measurement simplicity, durability, ease of fabrication, and low detection limits (<ppm levels) [3]. Moreover, most SMO based sensors tend to be long-lived and somewhat resistant to poisoning [3]. The SMO undergoes reduction or oxidation while reacting with the target gas and this process causes an exchange of electrons at a certain characteristic rate, thereby affecting the sensor's resistance and yielding a certain signal [3].

Concerning the sensing materials for NO_2 detection, tungsten oxide based materials have received a great attention in the last two decades. For example, pure and doped with various metal oxides WO_3 sensors have been used as potential NO_x sensors [3]. WO_3-based sensors produced by screen-printing can be highly responsive to NO_2 down to 1 ppm, when operated at 250 °C [4]. WO_3 thick films fired at 700 °C in Ar/O_2 flow have also proved to be operated at 100°C with excellent properties, such as fast response times, saturated stable sensitivity and rapid recovery characteristics to NO_2 gas in air [5]. WO_3-based nanocrystalline (3.0–9.0 nm size) thick films sensors with TiO_2, can be used successfully for detecting and monitoring of NO_2 in exhaust gases in parts per million level at 350 °C [6]. Zinc oxide thin and thick films have been also extensively studied for more than two decades. Several methods have been used to fabricate ZnO films and also their physical properties depend greatly on the method and condition of deposition [7]. Screen-printed ZnO, SnO_2 and Sb_2O_3 thick-films sensors sintered at 800, 1000 and 1200 °C showed high sensitivity and excellent selectivity for ppm levels of NO_2 gas [8]. Very recently, hybrid ZnO tetrapods + titanyl phthalocyanine exhibited a high sensor response ($\frac{\Delta R}{R_{air}} \approx 56$) under 100 ppb of NO_2 at room temperature, but with slow response and recovery times (several tens of minutes) [9]. In contrast, monocrystalline SnO_2 nanowires were sensitive in the range 18.9–1000 ppm of NO_2 at 250 °C with fast response and recovery times of 7 s and 8 s, respectively [10]. Gas sensors based on indium oxide nanowires, In_2O_3 or $In_xO_yN_z$ films grown by the metal organic CVD technique also showed good selectivity to NO_2 with little interference from other gases [11,12]. In addition, Indium Tin Oxide (ITO) thin films were found to exhibit high sensitivity toward NO_2 and NO associated to a good selectivity with respect to CO and CH_4 [12]. TeO_2 thin films proved to be effective in NO_2 detection in the range of 1–120 ppm too [13]. The results showed the best sensitivity to NO_2 at room temperature, but with a response time of about 6 min for 1 ppm to about 1.2 min for 120 ppm NO_2 concentration and longer recovery times.

Carbon nanotubes [14], YSZ [15], SnO_2, Nb or In_2O_3 doped hematite [16,17] have been also proposed for NO_2 detection. Alkaline or earthy-alkaline-doped hematite (α-Fe_2O_3) materials have been investigated recently in the literature, as NO_x sensors [18] and 5 wt % BaO addition to hematite seemed to lead to a promising sensing material. Therefore, the aim of this work is to study in more detail barium-doped hematite as an electrochemical sensor for NO_2 detection.

2. EXPERIMENTAL SECTION

α-Fe_2O_3 powder (Aldrich > 99%, particle size distribution below 2 m) was mixed in ethanol with barium nitrate used as precursor of 3, 4, 5, 10 and 14 wt % equivalent of barium oxide respect to hematite (Fluka > 99%), until stoichiometrical composition of $BaFe_{12}O_{19}$, in a planetary mill for 1 h. After drying overnight, the mixtures were uniaxially pressed at 370 MPa and calcined at 900 °C for 1 h. These samples were then planetary milled for 6 h in ethanol with polyethylene glycol (PEG 4000, Sigma-Aldrich, Milan, Italy) to increase the densities of the different samples. After drying overnight, the powders were ready to use. The grain size of the produced powders was then determined by means of a laser granulometre (Fritsch analysette 22). The different mixtures of Ba-doped α-Fe_2O_3 were pressed ad 370 MPa again and sintered at 1300 °C for 1 h.

Geometrical density evolution in function of percentage of added barium oxide was studied and the samples were characterized by X-ray diffraction (PW1710, Philips Eindhoven, The Netherland), in the 5°–70° 2 theta range, after calcination at 900 °C and sintering at 1300 °C. The pellets were also observed by means of a scanning electron microscope (SEM, S2300, Hitachi Tokyo, Japan).

Interdigitated gold electrodes (ESL 520A) were screen-printed onto the surface of the pellets of hematite and doped hematite fired at 1300 °C and the sensors humidity response was studied in the range 0%–100% relative humidity (RH) because it is known in the literature that water molecules can interfere in gas detection, both with respect to adsorption of other species and to surface catalysis [19,20]. In a chamber, compressed air was separated into two fluxes: one was dehydrated over a chromatography alumina bed, while the second one was directed through two water bubblers, generating, respectively, a dry and a humid flow. Two precision microvalves allowed to recombine the two fluxes into one by means of a mixer and to adjust the RH content while keeping constant the testing conditions, in particular a flow rate of 0.05 L/s. The relative humidity was not increased in a continuous mode but was varied by steps every 3 min. The measurements were performed at room temperature. A commercial humidity and temperature probe was used as a reference for temperature and RH values (Delta Ohm HD2101.1), accuracy: ±0.1% in the 0%–100% RH range and -50–250 °C temperature range. Each tested sensor was alimented by an external alternating voltage ($V = 3.6$ V at the rate of 1 kHz) and then constituted a variable resistance of this electrical circuit. A multimeter (Keithley 2000) was used to measure the tension V_{DC} at the output of the circuit. The sensor resistance was determined by substituting them, in the circuit, by known resistances and then plotting a calibrating curve $R = f(V_{DC})$ [21,22].

The electrical behavior of the materials was studied by AC impedance spectroscopy (4192A LF Impedance Analyzer, Hewlett Packard, Palo Alto, CA, USA) after painting platinum electrodes (ESL 5545, Electro-Science Laboratories, King of Prussia, PA, USA) onto the two faces of the pellets (1 cm in diameter and 1 mm in thickness). The measurements were performed in a furnace under dry synthetic air (Messer air 80/20) and argon (Messer argon 5.0), between 100 and 700 °C, in the 5 Hz-13 MHz frequency range. Excitation voltage was fixed at 100 mV. The previous samples were also studied under NO_2. AC impedance measurements were performed in a mixed flux of helium (Messer 4.6) and nitrogen dioxide (Messer 1000 ppm NO_2 1.8 in N_2 5.0) under a constant flow rate of 40 mL/min and a 0-500 ppm range NO_2 concentration. Excitation voltage was fixed at 100 mV in the 5 Hz-13 MHz frequency range.

Sensor response *(SR)* was calculated from impedance measurements considering the resistance at 1000 Hz. *SR* is a function of NO_2 concentration and is determined as follows [23] (Equation (1)):

$$SR = \frac{R_{(Pgas)} - R_{(Pgas \to 0)}}{R_{(Pgas \to 0)}} \qquad (1)$$

$R_{(Pgas)}$ represents the value of resistance in presence of the gas studied (NO_2 in the present case) and $R_{(Pgas \to 0)}$ the resistance without this gas.

3. RESULTS AND DISCUSSION

3.1. Microstructure

After the first thermal treatment at 900 °C and the 6 h planetary milling, the powders showed a mean diameter of 3.40 μm for 3, 4, 5 and 10 wt % equivalent of Ba-doped hematite samples, and of 9.00 μm for 14 wt % of Ba-doped hematite. These values are comparable to the mean diameter of the starting α-Fe_2O_3 powder

(2 um). X-ray diffraction patterns of the Ba-doped hematite after the first thermal treatment at 900 °C for 1 h and sintering at 1300 °C, confirmed the presence of two crystalline phases in all the samples: hematite (JCPDS card n.33-0664) and barium hexaferrite, $BaFe_{12}O_{19}$ (JCPDS card n.39-1433). As expected, we observe (Figure 1) the intensity of the barium hexaferrite peaks increasing with BaO additions from the pure hematite sample (H) to the pure barium hexaferrite sample (14% BA *i.e.,* α-Fe_2O_3 + 14 wt % BaO).

Table 1 summarizes the six compositions studied in the present work. Theoretical densities were calculated considering that in all the mixtures the barium oxide was completely transformed into barium hexaferrite crystalline phase by reaction with hematite. The densities increase with increasing the sintering temperature (Figure 2a) but also decrease with barium oxide content (Figure 2b).

Figure 3 illustrates the microstructures of pure hematite, 3% BA, 4% BA, 5% BA, 10% BA and 14% BA samples: the rather high densities of the materials fired at 1300 °C are confirmed. In the Ba-doped hematite samples, two different microstructures are present: a first one showing hexagonal grains, characteristic of hematite, and a second one characterized by lamellar grains, corresponding to the $BaFe_{12}O_{19}$ phase.

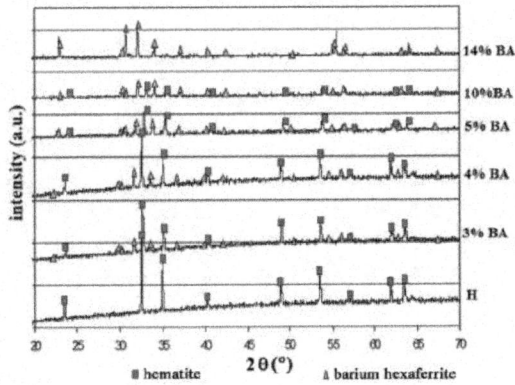

Figure 1. XRD spectra of various barium-doped hematite samples.

Table 1. Density of the barium-doped hematite pellets with polyethylene glycol (PEG) addition.

Label	Sample	Expected BaFe₁₂O₁₉ (wt %)	Relative density (%)
H	α-Fe₂O₃	0	96.9
3% BA	α-Fe₂O₃ + 3 wt % BaO fired at 1300 °C	18.12	95.2
4% BA	α-Fe₂O₃ + 4 wt % BaO fired at 1300 °C	21.75	94.8
5% BA	α-Fe₂O₃ + 5 wt % BaO fired at 1300 °C	36.24	94.8
10% BA	α-Fe₂O₃ + 10 wt % BaO fired at 1300 °C	72.48	93.1
14% BA	α-Fe₂O₃ + 14 wt % BaO fired at 1300 °C	100.00	92.0

The grain sizes are rather different between the various samples (Figure 3): the grain size decrease when increasing barium hexaferrite content until 10% BA.

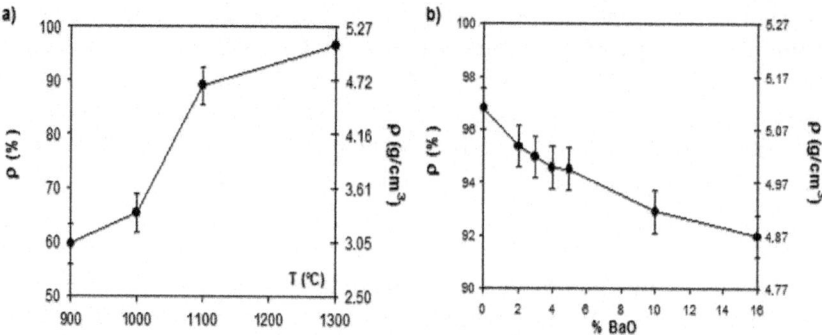

Figure 2. Geometrical density evolutions: (a) with temperature for 5% BA; (b) with barium oxide content for samples fired at 1300 °C.

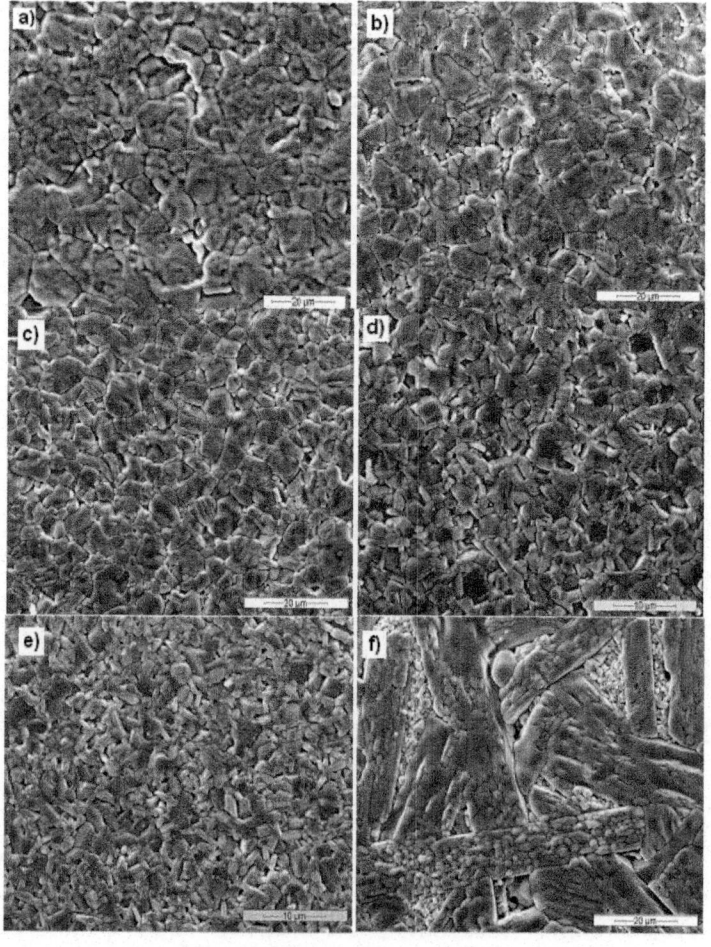

Figure 3. SEM micrographs of sintered samples: (a) H (1000×); (b) 3% BA (1000×); (c) 4% BA (1000×); (d) 5% BA (2000×); (e) 10% BA (2000×); (f) 14% BA (1000×).

3.2. Humidity Sensitivity

The correlation between humidity sensor response and the material porosity is known in the literature, and it is possible to evaluate the pore radius at which capillary condensation occurs at different temperatures (T) by means of the Kelvin equation [19] (Equation (2)):

$$r_k = \frac{2\gamma M}{\rho RT \ln\left(\dfrac{P}{P_s}\right)} \qquad (2)$$

where r_k is pore radius; M are respectively the water surface tension, density and molecular weight, while P and Ps are the water vapor pressures in the surrounding atmosphere and at saturation, respectively.

Because the porous structure of ceramics with open pores tends to favor water and gases adsorption and condensation, and though, on semiconducting materials these features are less critical [24], the efforts were oriented on the reduction of the sample porosity, as in general, dense ceramics show negligible humidity-sensitivity [25].

The retained solution was to increase the green density of the pellets by adding polyethylene glycol (PEG 4000) to the doped powder, during the 6 h planetary milling step and prior to uniaxial pressing. The samples were then sintered at 1300 °C and characterized like the previous ones. A significant increase of the density was then observed after PEG addition and sintering on the 5% BA composition (Table 2).

Interdigitated gold electrodes were screen-printed on the surface of the different samples and the sensors humidity responses were studied in the range 0%–100% relative humidity (RH) at 20 °C. Humidity measurements were realized every three minutes.

As an illustration of the influence of density regarding the humidity sensitivity the electrical response of the 5% BA samples with and without PEG 4000 under water vapor (Figure 4) shows that the pellet with PEG ($R = 1400\ \Omega$) is quite insensitive to humidity. This result validates our choice to increase the density of the samples if we consider humidity as an interfering gas regarding nitrogen dioxide sensitivity.

Table 2. Density of the 5 wt % barium-doped hematite pellets with and without PEG addition.

Sample	Geometrical density (g/cm³)	Relative density (%)
5% BA	4.32 ± 0.05	83.0
5% BA + PEG 4000	5.08 ± 0.05	95.0

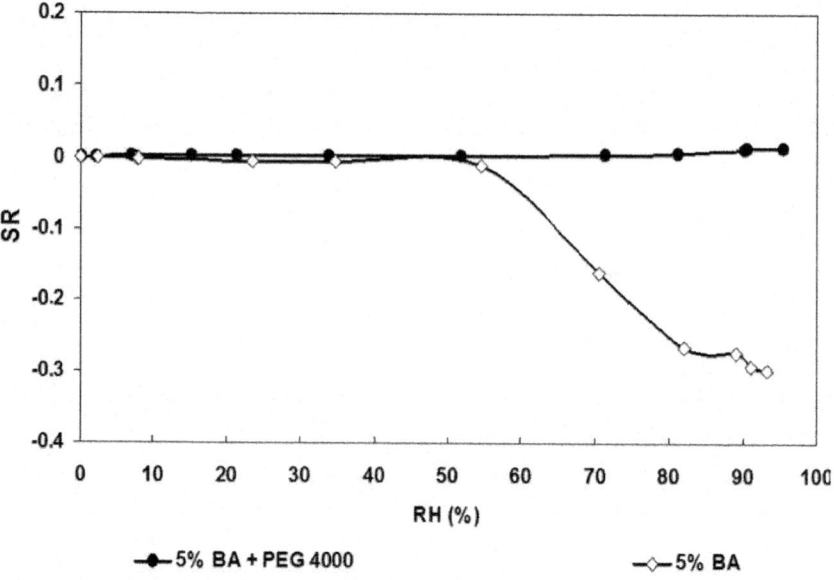

Figure 4. Humidity sensors (5% BA) response after 3 min of valve switch in function of RH, at 20 °C, for the samples fired at 1300 °C, with and without PEG 4000 addition.

Subsequently, all the samples responses were characterized under humidity (Figure 5) at room temperature. They show no sensitivity to water vapor, with the exception of the 10% BA sample which exhibits a slightly increase of the SR above 70 RH % and the 14% BA sample for which an increase of the sensor response (negative values) is evidenced above 35 RH %.

3.3. Electrical Study

Impedance measurements were performed in dry synthetic air (20% O_2) and under argon (<2 ppm O_2), between 100 and 700 °C, in the 5 Hz–13 MHz frequency range.

The conductivities evolutions in Arrhenius representations of pure hematite and the 5 compositions of barium-doped hematite samples showed that the samples behaved as semiconductors. Figure 6 shows two types of electrical behaviors. Pure hematite as well as 3% BA and 4% BA samples present a linear dependence of the conductivity with superimposition of the measurements under air and argon.

The n-type semiconductor behavior of hematite is clearly reported in literature [26]. Oxygen vacancies ($V_O^{\bullet\bullet}$) can be produced by heating this material following the disorder Equation (3) [27]:

$$O_O^x = \frac{1}{2}O_{2(g)} + V_O^{\bullet\bullet} + 2e'$$

$$(3)$$

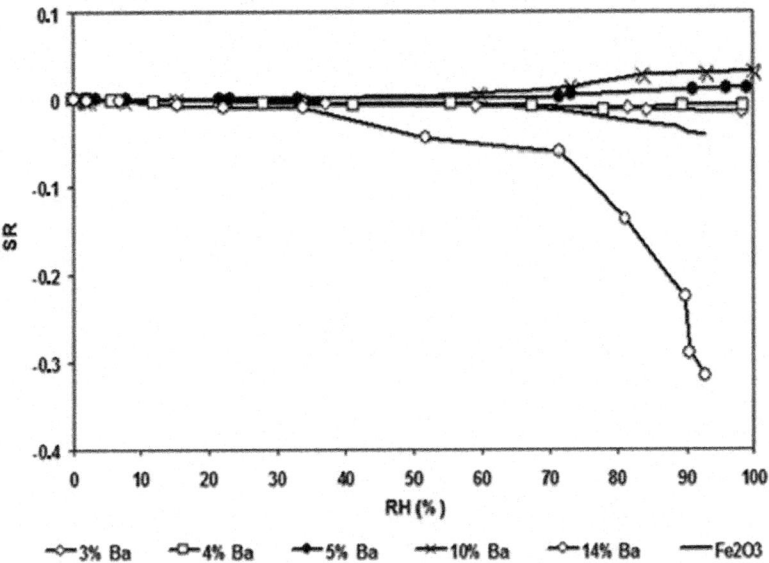

Figure 5. Humidity sensors response of various barium-doped hematite samples, after 3 min of valve switch, in function of relative humidity (RH), at 20 °C.

Gardner *et al.* [28] measured the oxygen deficit resulting from this equilibrium shift. Considering this structural disorder and the independence of the hematite conductivity regarding the oxygen partial pressure shown in the Figure 6a, we can conclude that this material is an intrinsic semiconductor [27,29]. The large value of activation energy measured in this work (Table 3) is the consequence of our experimental procedure where samples were sintered at 1300 °C and cooled under air laboratory. Under those conditions, the oxygen vacancies created during the heat treatment disappear by re-oxidation [28] and hematite becomes a slight n-type semiconductor [30]. The value of 0.71 eV obtained in the present work is close to the one reported by Gardner [28].

Table 3. Activation energies calculated from the linear regions of the Figure 6.

	Activation Energy (eV)			
Sample	Below 500 °C		Above 500 °C	
	Argon	Air	Argon	Air
H		0.71		
3% BA		0.61		
4% BA		0.54		
5% BA	0.36	0.51	0.27	
10% BA	0.35	0.40	0.26	
14% BA	0.43	0.52	0.34	

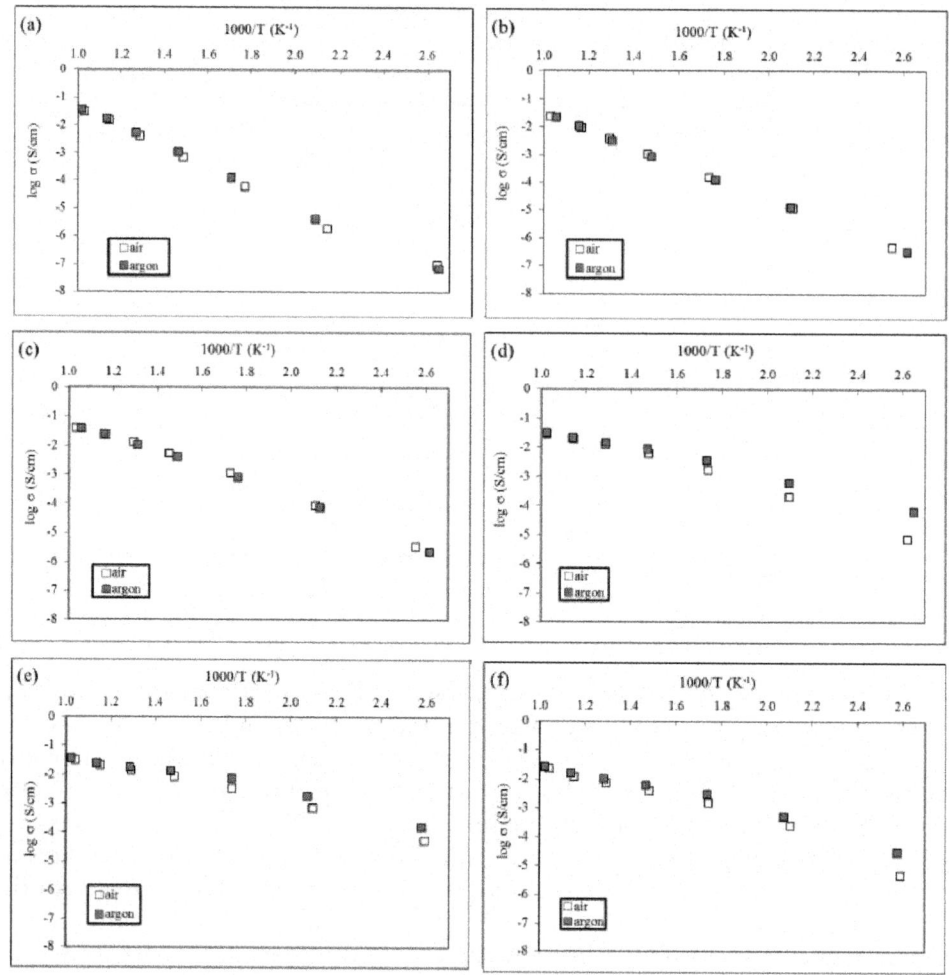

Figure 6. Arrhenius plots in air and argon of samples: **(a)** H; **(b)** 3% BA; **(c)** 4% BA; **(d)** 5% BA; **(e)** 10% BA; **(f)** 14% BA.

In the case of 5% BA, 10% BA and 14% BA samples, the Figure 6 exhibits two linear regions, below and above 500 °C ($1000/T = 1.3$ K^{-1}): no influence of the oxygen partial pressure above 500 °C is observed, while a strong influence of the oxygen partial pressure below 500 °C is evidenced. The conductivities increase as the oxygen partial pressure decreases (the conductivities under argon are greater than the conductivities under air). This corresponds to the behavior of an n-type extrinsic semiconductor.

Kim *et al.* [27] suggested a doping process with reduction of a part of Fe(III) to Fe(II), oxygen vacancies formation and concentration increase of negative charge carriers (electrons). These authors measured the electrical conductivity of pure and CdO-α-Fe$_2$O$_3$ system with various mol % of CdO in the 300–1300 °C temperature range and 10^{-9}–10^{-1} atm oxygen partial pressure range. They also suggested that the semiconductivity became intrinsic for temperatures above 500 °C. We observe the same phenomenon and the activation energies deduced from the Figure 6d–f are close to the values obtained by Gardner *et al.* [28] on pure hematite in the same temperature range (0.1 to 0.3 eV).

The increase of barium percentages in hematite is correlated with the increase of the conductivity for the different materials. It is particularly evident since we compare the Figure 6a,f. The XRD analyses presented earlier underlined the presence of two crystalline phases: α-Fe_2O_3 (hematite) and $BaFe_{12}O_{19}$ (barium hexaferrite). If we consider that the Figure 6a represents the pure hematite electrical behavior and the Figure 6f the pure barium hexaferrite one, we can conclude that this latter phase is a better electronic conductor than the hematite phase. Considering now the intermediate compositions presented in the Figure 6b–e, it is noticeable that the transition of electrical behavior observed between the Figures 6c and 6d corresponds, respectively, to a composition transition between 21 to 36 wt % of the most conductive phase (barium hexaferrite) in the material. Therefore, for a composition below 36 wt % of barium hexaferrite, hematite is the main crystalline phase and the materials exhibit the intrinsic semiconducting behavior observed in the Figure 6a–c. The Figure 6c,d define a percolation threshold of the barium hexaferrite crystalline phase in the hematite matrix. Thus, the materials which composition exceeds 36 wt % of this latter phase, exhibit the n type extrinsic semiconducting behavior observed on the Figure 6d–f.

3.4. NO₂ Sensitivity

The response to NO₂ was investigated in a mixed flux of helium and N₂/NO₂ under a flow rate of 40 mL/min (0–500 ppm of NO₂), between 60 and 350 °C, in the 5 Hz–13 MHz frequency range. As an example, Figure 7 represents the evolution of impedance spectra regarding the NO₂ partial pressure.

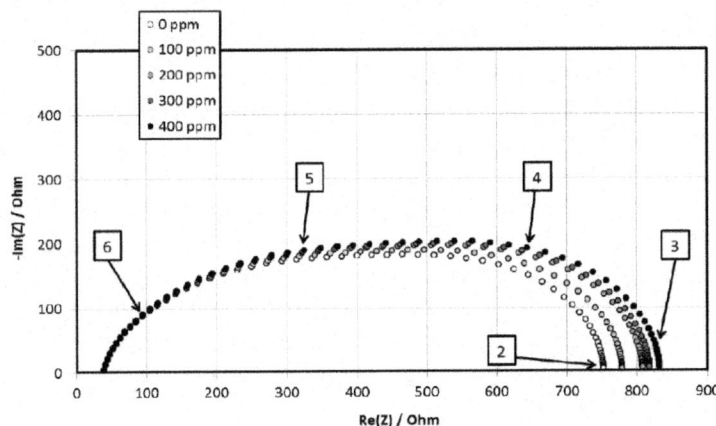

Figure 7. Impedance spectra evolution of 5% BA sample at 200 °C for various nitrogen oxide partial pressures.

The gas composition clearly influences the low frequency impedance. Consequently the values of SR were calculated from resistance measured at 1000 Hz. Figure 8 represents the evolution of the SR (P_{NO2}) of pure and doped hematite at 200 °C. We observed that for temperatures above 300 °C most of the materials exhibited a poor NO_2 sensitivity while for temperatures below 200 °C the impedance of most of them were too high (and also the uncertainties) to permit a quantitative determination of the resistance. Therefore, considering the highest sensitivity S (slope of the representation SR (P_{NO2}) gathered in the Table 4) and the lowest uncertainties, the optimal working temperature is estimated to 200 °C [16,31,32].

Table 4. Sensitivity (S) of hematite and various barium-doped hematite pellets measured at 200 °C.

Sample	S (ppm^{-1})	$R(P_{NO2\rightarrow 0})$ (Ohm)
H	$0.95 \times 10^-$	115,200
3% BA	$1.19 \times 10^-$	11,485
4% BA	$1.00 \times 10^-$	3943
5% BA	$2.86 \times 10^-$	747
10% BA	3.67×10^{-5}	695
14% BA	$6.06 \times 10^-$	380

In the continuously regenerating diesel particulate filter (CRDPF) technology, NO_2 is used to combust the soot collected in a particulate filter because it is a stronger oxidant than O_2, promoting low temperature oxidation of soot in the range 200–500 °C [33]. This oxidation of carbon by NO_2 is then achievable under normal driving conditions, particularly in heavy duty engine applications [34]. However, typically, NO_2 is 5% to 15% of the total NO_x in the diesel exhaust, but oxidation catalysts like Pt could oxidize NO to NO_2 increasing NO_2 concentrations to 50% of the total NO_x, in the temperature range of 300–350 °C [33]. Then, emissions of NO_x can be higher than 300 ppm [33,34], in function of the driving conditions, therefore, the materials studied in this work could be proposed as potential NO_x sensors for diesel exhaust gases.

All the materials present a resistance increase with the NO_2 partial pressure. This result has already been observed with hematite and various ferrites [16,32,35]. The 3% BA and 4% BA samples (Figure 8b,c) exhibit a SR evolution close to the pure hematite one (Figure 8a). Such a similarity was observed in the electrical study reported in the present paper. Otherwise, the Figure 8d–f show that beyond the percolation threshold described previously, the sensitivity to NO_2 increases with the barium hexaferrite content in the sample. Nevertheless, considering the results of humidity sensitivity presented in the present paper the 5% BA sample seems to be the best compromise regarding the highest NO_2 sensitivity with the lowest humidity interference.

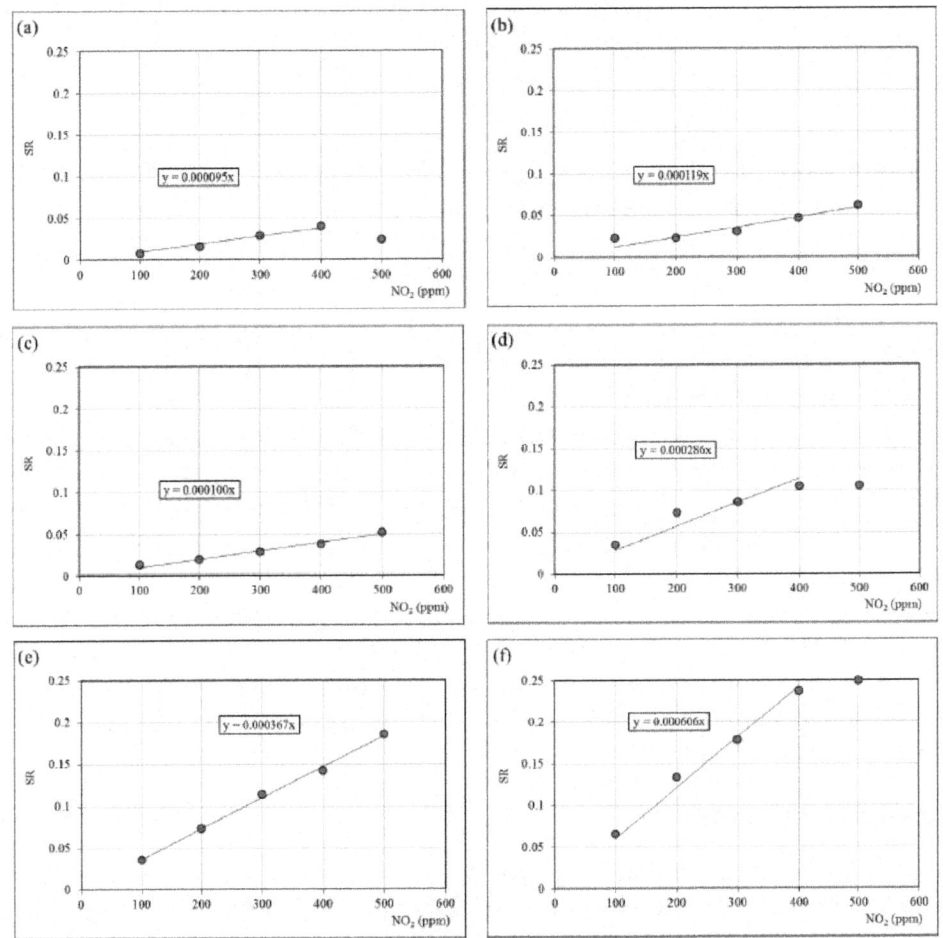

Figure 8. Sensor response (SR) evolution at 200 °C on the samples (a) H; (b) 3% BA; (c) 4% BA; (d) 5% BA; (e) 10% BA; (f) 14% BA in 0–500 ppm NO₂. Resistance measured at 1000 Hz.

In an n-type semiconductor, NO_2 does not react with pre-adsorbed oxygen and resistance changes occur by a direct chemisorption process [36]. A series of reactions (Equations (4)–(8)) giving nitrates and nitrites is proposed in reference [37]:

$$NO_2 + e^- \text{ (c.b.)} \leftrightarrow NO_2^- \tag{4}$$

$$NO_2 + V_O^+ \leftrightarrow NO_2^- + V_O^{2+} \tag{5}$$

$$2NO_2 + O_2^- + e^- \text{ (c.b.)} \leftrightarrow 2NO_3^- \tag{6}$$

$$2NO_2 + O_2^- + V_O^+ \text{ (c.b.)} \leftrightarrow 2NO_3^- + V_O^+ \tag{7}$$

$$NO_2 + O^- \leftrightarrow NO_3^- \tag{8}$$

In surface reactions (4) to (7), electrons from the conduction band (c.b.) are trapped when surface species are formed, but not in reaction (8). If nitrates are formed by the last mechanism, a further equilibrium has to be considered, that is to say NO^{3-} dissociation (Equations (9)–(10)):

$$NO_3^- + e^- \text{ (c.b.)} \leftrightarrow NO + 2O^- \tag{9}$$

$$NO_3^- + V_O^+ \leftrightarrow NO + 2O^- + V_O^{2+} \tag{10}$$

when O^- pressure increases, the resistance of the material also increases in an n-type semiconductor, as experimentally observed.

4. CONCLUSIONS

Ba-doped hematite were investigated as NO_2 sensing materials. Electrical characterizations were performed by ac impedance spectroscopy under various temperatures and gas atmospheres.

Pure hematite, 3 wt % and 4 wt % barium-doped hematite exhibit an intrinsic semiconducting behavior. They also show a poor NO_2 sensitivity. 5, 10 and 14 wt % barium-doped hematite exhibit an n-type extrinsic semiconducting behavior.

They are also good candidates for NO_2 detection. The nitrogen dioxide sensitivity increases with the barium hexaferrite content in the material. Finally, the best compromise regarding the highest NO_2 sensitivity with the lowest humidity interference is the 5 wt % barium-doped hematite.

The electrical study also underlines a percolation threshold of the barium hexaferrite crystalline phase in the hematite matrix. The value was estimated to 36 wt % barium hexaferrite.

Acknowledgments

C. Baroni is grateful to Piedmont Region for partial financial support of her Ph.D grant, to INSTM (Inter-University National Consortium on Material Science and Technology), Italy and to the Italian Ministry of Foreign Affairs in collaboration with French Government.

Conflicts of Interest

The authors declare no conflict of interest.

REFERENCES

1. The European Parliament and the Council of the European Union. Directive 2008/50/EC of the European Parliament and of the Council of 21 May 2008 on ambient air quality and cleaner air for Europe. Available online: http://eur-lex.europa.eu/LexUriServ/LexUriServ.do?uri=OJ:L: 2008:152:0001:0044:EN:PDF (accessed on 22 September 2013).

2. The European Commission. Laying down rules for Directives 2004/107/EC and 2008/50/EC of the European Parliament and of the Council as regards the reciprocal exchange of information and reporting on ambient air quality. Available online: http://eur-lex.europa.eu/LexUriServ/ LexUriServ.do?uri=OJ:L:2011:335:0086:0106:EN:PDF (accessed on 22 September 2013).

3. Kanan, S.M.; El-Kadri, O.M.; Abu-Yousef, I.A.; Kanan, M.C. Semiconducting metal oxide based sensors for selective gas pollutant detection. *Sensors* **2009**, *9*, 8158–8196.

4. Ivanov, P.; Hubalek, J.; Malysz, K.; Prášek, J.; Vilanova, X.; Llobet, E.; Correig, X. A route toward more selective and less humidity sensitive screen-printed SnO_2 and WO_3 gas sensitive layers. *Sens. Actuators Chem.* **2004**, *100*, 221–227.

5. Chung, Y.K.; Kim, M.H.; Um, W.S.; Lee, H.S.; Song, J.K.; Choi, S.C.; Yi, K.M.; Lee, M.J.; Chung, K.W. Gas sensing properties of WO_3 thick film for NO_2 gas dependent on process condition. *Sens. Actuators Chem.* **1999**, *60*, 49–56.

6. Lee, D.S.; Han, S.D.; Huh, J.S.; Lee, D.D. Nitrogen oxides-sensing characteristics of WO_3-based nanocrystalline thick film gas sensor. *Sens. Actuators Chem.* **1999**, *60*, 57–63.

7. Ismail, B.; Abaab, M.; Rezig, B. Structural and electrical properties of ZnO films prepared by screen printing technique. *Thin Solid Film.* **2001**, *383*, 92–94.

8. Yasushi, Y.; Katsuji, Y.; Yumi, M.; Masami, O. A Thick-film NO_2 sensor fabricated using Zn–Sn–Sb–O composite material. *Jpn. J. Appl. Phys.* **2003**, *42*, 7594–7598.

9. Coppedè, N.; Villani, M.; Mosca, R.; Iannotta, S.; Zappettini, A.; Calestani, D. Low temperature sensing properties of a nano hybrid material based on ZnO nanotetrapods and titanyl phthalocyanine. *Sensors* **2013**, *13*, 3445–3453.

10. Tonezzer, M.; Hieu, N.V. Size-dependent response of single-nanowire gas sensors. *Sens. Actuators Chem.* **2012**, *163*, 146–152.

11. Ando, M.; Steffes, H.; Chabicovsky, R.; Haruta, M.; Stangl, G. Optical and electrical H_2- and NO_2-sensing properties of $Au/InxOyNz$ films. *IEEE Sens. J.* **2004**, *4*, 232–236.

12. Sberveglieri, G.; Benussi, P.; Coccoli, G.; Groppelli, S.; Nelli, P. Reactively sputtered indium tin oxide polycrystalline thin films as NO and NO_2 gas sensors. *Thin Solid Film.* **1990**, *186*, 349–360.

13. Siciliano, T.; Giulio, M.D.; Tepore, M.; Filippo, E.; Micocci, G.; Tepore, A. Room temperature NO_2 sensing properties of reactivity sputtered TeO_2 thin films. *Sens. Actuators Chem.* **2009**, *137*, 644–648.

14. Suehiro, J.; Zhou, G.; Imakiire, H.; Ding, W.; Hara, M. Controlled fabrication of carbon nanotube NO_2 gas sensor using dielectrophoretic impedance measurement. *Sens. Actuators Chem.* **2005**, *108*, 398–403.

15. Zhuiykov, S.; Ono, T.; Yamazoe, N.; Miura, N. High-temperature NO_x sensors using zirconia solid electrolyte and zinc-family oxide sensing electrode. *Solid State Ion.* **2002**, *152–153*, 801–807.

16. Sun, H.T.; Cantalini, C.; Faccio, M.; Pelino, M. NO_2 gas sensitivity of sol-gel-derived α-Fe_2O_3 thin films. *Thin Solid Film.* **1995**, *269*, 97–101.

17. Cantalini, C.; Sun, H.T.; Faccio, M.; Ferri, G.; Pelino, M. Niobium-doped α-Fe_2O_3 semiconductor ceramic sensors for the measurement of nitric oxide gases. *Sens. Actuators Chem.* **1995**, *25*, 673–677.

18. Tulliani, J.-M.; Baroni, C.; Lopez, C.; Dessemond, L. New NOx sensors based on hematite doped with alkaline and alkaline-earth elements. *J. Eur. Ceram. Soc.* **2011**, *31*, 2357–2364.

19. Korotcenkov, G. Metal oxides for solid-state gas sensors: What determines our choice? *Mater. Sci. Eng.* **2007**, *139*, 1–23.

20. Barsan, N.; Weimar, U. Understanding the fundamental principles of metal oxide based gas sensors; the example of CO sensing with SnO_2 sensors in the presence of humidity. *J. Phys. Condens. Matter* **2003**, *15*, R813–R839.

21. Tulliani, J.M.; Bonville, P. Influence of the dopants on the electrical resistance of hematite-based humidity sensors. *Ceram. Int.* **2005**, *31*, 507–514.

22. Esteban-Cubillo, A.; Tulliani, J.M.; Pecharromán, C.; Moya, J.S. Iron-oxide nanoparticles supported on sepiolite as a novel humidity sensor. *J. Eur. Ceram. Soc.* **2007**, *27*, 1983–1989.

23. Nenov, T.G.; Yordanov, S.P. *Ceramic Sensor: Technology and Applications*; CRC PressTechnomic Publishing Company, Inc.: Lancaster, PA, USA, 1996.

24. Traversa, E.; Ceramic sensors for humidity detection: the state-of-the-art and future developments. *Sens. Actuators Chem.* **1995**, *23*, 135–156.

25. Traversa, E.; Bearzotti, A. Humidity sensitive electrical properties of dense ZnO with non-ohmic electrode. *J. Ceram. Soc. Jpn.* **1995**, *103*, 11–15.

26. Santilli, C.V.; Bonnet, J.P.; Dordor, P.; Onillion, M.; Hagenmuller, P. Influence of structural defects on the electrical properties of α-Fe$_2$O$_3$ ceramics. *Ceram. Int.* **1990**, *16*, 25–32.

27. Kim, K.H.; Lee, S.H.; Choi, J.S. Electrical conductivity of pure and doped α-ferric oxides. *J. Phys. Chem. Solids* **1985**, *46*, 331–338.

28. Gardner, R.F.G.; Sweett, F.; Tanner, D.W. The electrical properties of alpha ferric oxide-II. Ferric oxide of high purity. *J. Phys. Chem. Solids* **1963**, *24*, 1183–1196.

29. Chang, R.H.; Wagner, J.B. Direct-current conductivity and iron tracer diffusion in hematite at high temperatures. *J. Am. Ceram. Soc.* **1972**, *55*, 211–213.

30. Stone, H.E.N. Electrical conductivity and sintering in iron oxides at high temperatures. *J. Mater. Sci.* **1968**, *3*, 321–325.

31. Han, J.S.; Davey, D.E.; Mulcahy, D.E.; Yu, A.B. Effect of pH value of the precipitation solution on the CO sensitivity of α-Fe$_2$O$_3$. *Sens. Actuators Chem.* **1999**, *61*, 83–91.

32. Ivanovskaya, M.; Gurlo, A.; Bogdanov, P. Mechanism of O$_3$ and NO$_2$ detection and selectivity of the In$_2$O$_3$ sensors. *Sens. Actuators Chem.* **2001**, *77*, 264–267.

33. Liu, Z.; Ge, Y.; Tan, J.; He, C.; Shah, A.N.; Ding, Y.; Yu, L.; Zhao, W. Impacts of continuously regenerating trap and particle oxidation catalyst on the NO$_2$ and particulate matter emissions emitted from diesel engine. *J. Environ. Sci.* **2012**, *24*, 624–631.

34. Liu, Z.; Shah, A.N.; Ge, D.Y.; Tan, J.; Jiang, L.; Yu, L.; Zhao, W.; Wang, C.; Zeng, T. Effects of continuously regenerating diesel particulate filters on regulated emissions and number-size distribution of particles emitted from a diesel engine, *J. Environ. Sci.* **2011**, *23*, 798–807.

35. Aono, H.; Hirazawa, H.; Naohara, T.; Maehara, T. Surface study of fine MgFe$_2$O$_4$ ferrite powder prepared by chemical methods. *Appl. Surf. Sci.* **2008**, *254*, 2319–2324.

36. Berger, O.; Hoffmann, T.; Fischer, W.-J.; Melev, V. Tungsten-oxide thin films as a novel materials with high sensitivity to NO$_x$, O$_y$, and H$_2$S. Part II: Application as gas sensors. *J. Mater. Sci. Mater. Electron.* **2004**, *15*, 483–493.

37. Chiorino, A.; Ghiotti, G.; Prinetto, F.; Carotta, M.C.; Gnani, D.; Martinelli, G. Preparation and characterization of SnO$_2$ and MoO$_x$-SnO nanosized powders for thick film gas sensors. *Sens. Actuators Chem.* **1999**, *58*, 338–349.

INDEX

This page left intentionally blank.